Ecology of
Aquatic Systems

Ecology of
Aquatic Systems

Second Edition

Michael Dobson
Director of the Freshwater Biological Association
Ambleside, Cumbria
UK

Chris Frid
Professor of Marine Biology
School of Biological Sciences
University of Liverpool
Liverpool
UK

OXFORD
UNIVERSITY PRESS

OXFORD
UNIVERSITY PRESS

Great Clarendon Street, Oxford OX2 6DP

Oxford University Press is a department of the University of Oxford.
It furthers the University's objective of excellence in research, scholarship,
and education by publishing worldwide in

Oxford New York

Auckland Cape Town Dar es Salaam Hong Kong Karachi
Kuala Lumpur Madrid Melbourne Mexico City Nairobi
New Delhi Shanghai Taipei Toronto

With offices in

Argentina Austria Brazil Chile Czech Republic France Greece
Guatemala Hungary Italy Japan Poland Portugal Singapore
South Korea Switzerland Thailand Turkey Ukraine Vietnam

Oxford is a registered trade mark of Oxford University Press
in the UK and in certain other countries

Published in the United States
by Oxford University Press Inc., New York

British Library Cataloguing in Publication Data

Data available

Library of Congress Cataloging in Publication Data

Dobson, Michael

Ecology of aquatic systems / Michael Dobson and Chris Frid. — 2nd ed.
p. cm.
ISBN 978-0-19-929754-2
1. Fishery resources. 2. Aquatic resources. I. Frid, Chris. II. Frid, Chris.
Ecology of aquatic systems. III. Title.
SH327.5.F74 2008
577.6—dc22
2008036480

Typeset by Graphicraft Limited, Hong Kong
Printed in Italy
by L.E.G.O S.p.A

ISBN 978–0–19–929754–2

3 5 7 9 10 8 6 4 2

Preface

Aquatic ecology has traditionally been split into separate marine and freshwater components, each of which progressed more or less independently of the other for many years. This fundamental division is reflected in the way that the discipline is often taught, the two being treated as completely separate subjects, but it masks two very important features of aquatic systems: first, that they are highly diverse and second, that the biology and hence ecology of the organisms are linked to the physical and chemical properties of water. The diversity of aquatic systems is apparent not just in the range of organisms in each environment but also in the richness of physical habitat types. A major research initiative in aquatic ecology today, therefore, is emphasis upon the common properties underlying their diversity, an approach which recognizes that, from an ecological perspective, there is no fundamental difference between marine and freshwater systems.

In this book, we attempt to give a flavour of these features. While emphasizing the diversity and unique features of the major aquatic habitats by treating them under separate chapter headings, we also show that the same underlying themes crop up again and again, in every field of aquatic ecology. There is, of course, no way that a book of this size can do justice to the subject encompassed by its title, but it is our hope that it introduces enough of the excitement of aquatic ecology to stimulate further, more detailed study. To this end, we have tried to keep our explanations simple and, wherever appropriate, to define the terms and ideas which the reader may come across elsewhere.

Although written primarily as an undergraduate textbook, we hope that the book will be of wider use. Therefore, we expect from the reader some basic ecological knowledge, but no prior experience of aquatic ecology, in either of its disciplines. We hope that some will read the book from cover to cover, but accept that many will not, preferring to dip into it as a source of specific information, and we have designed its layout accordingly. We appreciate that some may find this 'reference book' format annoying, but trust that, for the majority of our target audience, it will be a positive feature.

With the second edition, we have been able to take on board the comments and criticisms received over the decade since the first edition appeared. Of course, all specialists in the field have their own preferences and all have different opinions about what really is important and what is less so. We are as guilty of this as anybody else, so our own interests may come through more strongly than those of others; we have, however, tried to be fair in our coverage. Although it could be argued that there have been relatively few fundamental changes in our understanding of aquatic systems over the past ten years, what have changed are methods and preferences in the communication of information. The first edition was liberally supplied with references to primary and secondary literature; the advice we have been receiving more recently is that a textbook of this type should have few literature citations in the text, and therefore we have minimized these. We have, however, retained citations to literature if they refer to seminal works, classic examples or comprehensive reviews. We have also preferred to include citations for recent ideas and examples that will not otherwise be widely cited in the secondary literature.

As with any work of this type, we are indebted to many people for their assistance. Most important are those whose published ideas and examples we have called upon. We are also grateful to Simon Rundle, Jan Vermaat and the anonymous reviewers who worked their way through the text over many months; we did not always agree with their comments or suggestions, but each one forced us to pause and consider alternative approaches, so all were valuable. For the practicalities, we thank Jonathan Crowe and Suzy Armitage at Oxford University Press, and Sarah Johnson and Julie McNicol at the Freshwater Biological Association. Martin Attrill, Stan Dobson, Margaret Gill and Andrew Thomas each allowed us to use unpublished data or photographs. Finally, special personal thanks to our partners, Deirdre and Susan, for support and understanding and, from MD, to Hannah and Lawrence for their patience.

Contents

Figure Acknowledgements

Figure	Acknowledgement
2.4	Reproduced from P S Corbert. 1964. *Canadian Entomologist*. Figure 4, page 269. By permission of Entomological Society of Canada.
2.5a	Reproduced with the kind permission of Steve Trewhella.
2.7	Reproduced from Robert Guralnick. 2004. *Life-History Patterns in the Brooding Freshwater Bivalve Pisidium (Sphaeriidae). Journal of Molluscan Studies, 70. pp.341–351*. By permission of Malacological Society of London.
3.1b	Reproduced from Arrington & Winemuller. 2006. *Journal of the North American Benthological Society. Figure 1 page 128*. By permission of Journal of the North American Benthological Society.
3.4	Reproduced from Calow & Petts. 1992. The Rivers Handbook, Vol 1. *Figure 20.6, page 437*. By permission of Blackwell Publishing Ltd.
3.5	Reproduced from Hall et al. 1980. *Ecology, Vol 61 (Part 4)*. Figure 5, page 983. By permission of ESA Publications.
3.6	Reproduced from Giller et al. 1991. *Verhandlung der Internationale Vereinigung für Theoretische und Angewandte Limnologie, Vol 24*. Figure 1, page 1725. By permission of E. Schweizerbart'sche Verlagsbuchhandlung. http://www.schweizerbart.de.
3.7	Reproduced from J. L. Pretty et al. 2005. *Freshwater Biology, Vol 50*. Figure 1, page 582. By permission of Blackwell Publishing Ltd.
3.9	Reproduced from J. L. Pretty & M. Dobson. 2004. *Leaf transport and retention in a high gradient stream, Hydol. Earth Syst. Sci., 8*. Figure 1a. By permission of Copernicus Publications.
3.11	Reproduce from Jonsson & Malmqvist. 2000. *Oikos, 89*. Figure 2, page 521. By Permission of Blackwell Publishing Ltd.
3.12	Reproduced from March et al. 2002. *Freshwater Biology*, Vol 47. Figure 3 and 4, page 385. By permission of Blackwell Publishing Ltd.
3.16	Reproduced from U. Humpesch. 1992. *Freshwater Forum Vol 2 Part 1*. Figure 2, page 35. By permission of the Freshwater Biological Association.
3.17a	Reproduced from Healey & Richardson. 1996. Archiv für Hydrobiologie, Spplement 113 (Large Rivers 10). Figure 6, page 287. By permission of E. Schweizerbart'sche Verlagsbuchhandlung. http://www.schweizerbart.de.
3.17b	Reproduced from Healey & Richardson, *Archiv für Hydrobiologie, Spplement 113 (Large Rivers 10)* (1996) Figure 4 upper panel, page 285. By permission of E. Schweizerbart'sche Verlagsbuchhandlung http://www.schweizerbart.de.
Box 3.3 Fig 1	Reproduced from D. C. Sigee. 2005. *Freshwater Microbiology. By permission of* John Wiley & Sons, Ltd.
Box 3.4 Fig 1	Reproduced from Dobson et al. 2004. *Hydrobiologia, Vol 519*. Fig 1d, page 209. With kind permission from Springer Science and Business Media.
Box 3.5 Fig 1	Reproduced from Minshall et al. 1985. *Canadian Journal of Fisheries and Aquatic Science Vol. 42*. Fig 3, Page 1051. By permission of NRC Research Press.
4.7	Reproduced from A. Nelson-Smith. 1977. *The Coastline. By permission of* John Wiley & Sons, Ltd.
4.13	Reproduced from D. S. McLuskly. 1989. *The Estuarine Ecosystem, 2nd Edition. By permission of* Blackie and Son Ltd.

Figure	Acknowledgement
5.11	Reproduced from J. H. Connell 1961. *Ecology Vol 42*. Figure 5, page 722. By permission of ESA Publications.
5.13	Reproduced from Lalli & Parsons. 1993. *Biological Oceanography: An Introduction. By permission of* Butterworth-Heinemann.
5.14	Reproduced from A. H. Taylor. 1995. *ICES J. mar, Sci., 200*. Figure 1, page 712, Figure 5, page 715. By permission of Elsevier (Academic Press Ltd).
5.16	Reproduced from Levinton, J.S. 1995. *Marine Biology: Function, Biodiversity, Ecology.* By permission of Oxford University Press, Inc.
6.3	Reprinted from *Biological Oceanographic Processes, 3rd Edition*, Parsons et al, Copyright Elsevier (1984).
6.5	Reproduced from J. E. G. Raymont. 1983. *Plankton and productivity in oceans Vol 2 Zooplankton*. By permission of Pergamon Press Ltd.
6.6	Reproduced from P. H. Wiebe. 1970. *Limnology and Oceanography, 18*. By permission of American Society of Limnology and Oceanography Inc.
Box 6.2 Fig 1	This article was published in *Biological Oceanographic Processes*, 3rd Edition, Parsons et al, Copyright Elsevier (1984).
7.5	Reproduced from Kosarev & Yablonskaya. 1994. *The Caspian Sea*. By permission of Backhuys Publishers.
7.6	Reproduced from Bondarenko et al. 1996. *Freshwater Biology, Vol 3 (part 3)*. Figure 2, page 519. By permission of Blackwell Publishing Ltd.
7.9	Reproduced from Bronmark & Hansson. 2005. *Biology of Lakes and Ponds*. By permission of Oxford University Press.
7.10	Reproduced from Bronmark & Hansson. 2005. *Biology of Lakes and Ponds*. By permission of Oxford University Press.
7.11(c) & (d)	Reproduced from Adapted from S Dodson, *Bioscience, 39 (1989)* Copyright, American Institute of Biological Sciences.
7.13	Reproduced from Winder et al. 2003. *Freshwater Biology, Vol 48*. Fig 3, page 387. By permission of Blackwell Publishing Ltd.
7.16	Reproduced from Meijer et al. 1990. *Hydrobiologia, 200/201*. Fig 6, page 310. By permission of The Atrium.
7.17b	Reproduced from R Lowe-McConnell. 1994. *Freshwater Forum, Vol 4, Part 2*. Fig 3, page 84. By permission of Blackwell Publishing Ltd.
7.18	Reproduced from Chapman et al. 2003. *Conservation Biology, 17*. By permission of Blackwell Publishing Ltd.
Box 7.2 Fig 1	Reproduced from J. M. Melack. 2002. *Verh. Internat. Verein. Limnol, 28*. Fig 1, page 32. Verhandlung der Internationale Vereinigung für Theoretische und Angewandte Limnologie, Vol 24. Figure 1, page 1725. By permission of E.Schweizerbart'sche Verlagsbuchhandlung. http://www.schweizerbart.de.
8.6	Reproduced from P Denny. 1995. *Wetlands of the World I:Inventory, Ecology and management. Kluwer, Doredrecht*. Fig 9, page 16. With kind permission from Springer Science and Business Media.
8.10	Reproduced from Jenny L. Chapman. 1992. Michael Reiss *Ecology: Principles and Applications*. By permission of Cambridge University Press.
8.12	Reproduced from Aselmann & Crutzen. 1989. Journal of Atmospheric Chemistry, 8. Fig 1a. With kind permission from Springer Science and Business Media.
Box 8.3 Fig 1	Reproduced from B. M. Svensson. 2000. *Oikos, 74*. By permission of Blackwell Publishing Ltd.
Box 8.3 Fig 2	Reproduced from R Rydin. 2003. *Advances in Bryology, 5*. Fig 2, page 159. By permission of E.Schweizerbart'sche Verlagsbuchhandlung. http://www.schweizerbart.de.
Box 8.5 Fig 1	Reproduced from Kratz & DeWitt. 1986. *Ecology, 67*. By permission of ESA Publications.

The Aquatic Environment

1.1 Introduction

Aquatic systems are those in which the primary medium inhabited by organisms is water, whether as an open water body or dominating the spaces between particles in a substratum. This broad definition covers a wide variety of habitat types, although generally they are divided into two separate groups: marine and freshwater systems. Research into aquatic systems has generally reflected this division, limnology—the study of inland waters, and oceanography—the study of marine systems, proceeding more or less independently for many years.

This division masks two important features of aquatic systems. First, they are extremely diverse, with a wide range of aquatic habitat types distinguishable, each with its own unique components and processes. Some systems are transitional between marine and freshwater environments, yet, at the same time, show unique, distinctive features. Of these, estuaries and salt marshes are generally classed as marine environments, while saline lakes are considered in relation to freshwater environments (or even ignored). Secondly,

there is much overlap among marine and freshwater systems: ecologically, a turbulent river may have more in common with a rocky marine coastline than with a garden pond, while there are many broad similarities between large fresh water lakes and the open sea. Despite these common features, the separation of the two disciplines has meant that, even now, limnologists and oceanographers use different terminology for the same processes, and even the same terminology for different processes!

The aim of this book is to explore these two features. We wish to emphasize the great diversity of aquatic systems, while demonstrating that all operate within the same functional principles, deriving from the physical properties of water. We also wish to show that the fundamental division between marine and freshwater systems is untenable from an ecological perspective. Taxonomically, they are very different, but ecologically, similarities across the freshwater–marine divide outweigh the differences. Furthermore, the fundamental and dominant determinant of

1

aquatic community structure, across all aquatic habitats, is the physical environment generated by the movement of water. Biotic processes, such as competition and predation, can be important locally or over the short term, but the physical environment dominates overall. Other abiotic features of water, such as salinity, acidity and pollution load, have major effects upon the taxonomic components of water bodies, but their effect upon ecological processes is more muted. In contrast, similar ecological processes can be identified in aquatic habitats whose water chemistry is very different but whose patterns of water movement are similar.

Throughout the diversity of aquatic habitats, therefore, a common and recurring feature is movement of water, relative both to the substra-tum that marks its edges and to water masses within other parts of the same water body. It is this movement which will be returned to throughout the book, as it is the physical similarities and differences among water bodies which determine the ecological comparisons and contrasts which can be made between different aquatic systems. Some of these ecological similarities are introduced in Chapter 2, while others are to be found in an appropriate habitat chapter, although similarities between habitats are emphasized by cross-referencing where appropriate. Before these ecological themes are referred to, however, it is important to appreciate the diversity of aquatic environments and of the life which inhabits them, as well as the chemical components of that environment which impinge upon living organisms.

1.2 Distribution of the world's aquatic habitats

1.2.1 The hydrological cycle

The Earth's surface and atmosphere contains about 1.4×10^9 km^3 of water, the overwhelming majority of which is in the oceans, which cover around 71% of the world's surface and average 3.7 km in depth (**Figure 1.1**). If the ice sheets and glaciers of polar and mountainous regions and the water within the ground are added, this accounts for approximately 100% of the water (**Table 1.1**). This simplistic appraisal neglects, however, the importance of water movement on a global scale. The world's water is in a very dynamic state, with all of its aquatic features connected in an active hydrological cycle (**Figure 1.2**), generated by processes which lead to the evaporation of surface water, its transport by atmospheric circulation and its subsequent precipitation. Between this atmosphere–surface link is lateral and vertical movement within aquatic systems themselves,

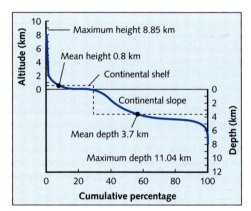

Figure 1.1 A hypsographic curve illustrating the cumulative percentage of the Earth below a given elevation. Features to note are that about 71% of the Earth's surface is below sea level and a rise in sea level of only 100 m would flood more than 5% of the current land area.

dominated by terrestrial run-off and oceanic circulation. An important feature of the hydrological cycle, therefore, is that water moves both between its various components, as evaporation and precipitation, and within each component. The oceans and atmosphere are dominated by circulatory currents, and even apparently immobile ice sheets are in a state of flux, albeit more slowly than liquid water. The

Table 1.1 Distribution of the world's water		
Volume	(km³ × 10³)	%
Oceans	1,338,000	96.5
Glacial ice	24,400	1.8
Groundwater	23,400	1.7
Freshwater lakes	105	0.008
Saline lakes	85	0.006
Atmosphere	13	0.001
Rivers	2	0.0001

The approximate volume of water in various aquatic systems is given, along with the percentage of the total water available.

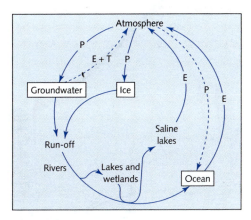

Figure 1.2 The hydrological cycle. Although movement is mainly one way (anticlockwise in this case), there are mechanisms which act in the reverse direction, of which the most important are shown here as dashed lines. Major storage compartments are boxed. P, precipitation (rain and snow); E, evaporation; T, transpiration.

mobile nature of water is nowhere more apparent than in places where it is moving relative to the land: rivers as linear, one-way channels, and wave and tidal action against the seashore.

Within this cycle, biological habitats are rather unevenly distributed. The oceans are inhabited by living organisms which recognize, and have created, a wide variety of different habitats in what is effectively a single body of water. Habitat diversity of inland waters is augmented by their fragmented nature, but the vast majority of non-marine aquatic organisms live in the rivers, lakes and wetlands which, in terms of the volume of water they contain, are very minor components of the hydrological cycle.

Ice is a virtually sterile environment, whose inhabitants are limited to a few specialist forms living on the surface, and groundwater, equally, supports a very low diversity and biomass of organisms, except where it occurs close to surface water features.

1.2.2 Oceans

The oceans are clearly the most widely distributed of the Earth's aquatic habitats, containing almost all of its water (Table 1.1). By far the largest component of the oceanic system is the water column itself, the pelagic zone. Geographically separate parts of the pelagic zone interact through the processes of ocean circulation, although different mechanisms, operating in shallow and deep waters respectively, ensure that ocean basins contain two relatively discrete environments, separated by depth.

Surface ocean circulation

Surface ocean currents are generated by wind friction on the sea surface, and the global pattern of currents (Figure 1.3) therefore resembles that of atmospheric movement (Box 1.1). Coriolis force influences the movement of surface waters—the surface layer moves at 45° to the wind direction (deflected to the right in the northern hemisphere and the left in the southern hemisphere) and this deflection increases with depth, as succeeding layers of water are deflected relative to those above them, creating a spiral current pattern (Figure 1.4). The net motion of the wind-influenced layer, or Ekman layer, is at 90° to the wind direction.

Land masses interfere with the free movement of water, which tends to pile up against downcurrent coasts, creating a slope in the sea surface. The North Atlantic, for example, slopes up from America towards Europe and North Africa, so a true horizontal surface across the ocean will have more water above it, and therefore greater water pressure, on the European side than the American. Water, like wind, attempts to equalize pressure by flowing from high pressure

1

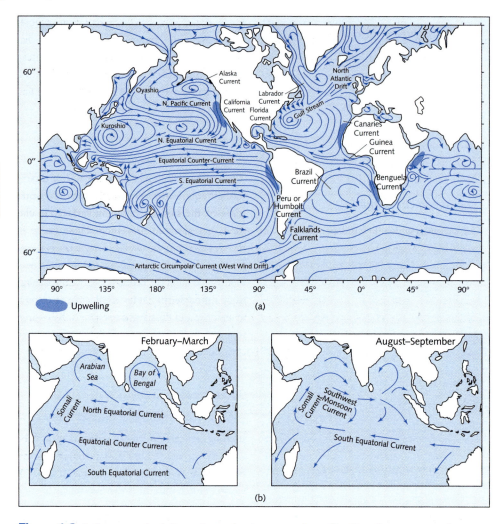

Figure 1.3 Surface ocean circulation patterns, showing zones of upwelling from deep ocean circulation. (a) The global system during February–March. (b) The monsoon (Box 1.1): an example of seasonal changes to oceanic circulation. During February–March, the north-east monsoon drives a small clockwise gyre circulation in the Arabian Sea, a clockwise gyre in the Bay of Bengal, and westerly flows throughout the remainder of the northern Indian Ocean. During August–September, the south-west monsoon creates a clockwise gyre circulation in the northern Indian Ocean and Arabian Sea with an anti-clockwise gyre in the Bay of Bengal. The result is a reversal of the Somali current between the two seasons and the presence of an equatorial counter-current in the north-east monsoon period.

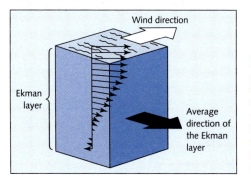

Figure 1.4 The Ekman spiral of water movement generated by wind on the sea surface. Each arrow indicates the direction of movement of that layer, the length of the arrow being proportional to the magnitude of flow. Net movement of the Ekman layer as a whole is 90° to the direction of the wind, the displacement being to the right in the northern hemisphere and to the left in the southern hemisphere.

to low pressure areas, generating a horizontal pressure gradient force, and therefore creating currents which are deflected, by Coriolis force (90° to the right in the northern hemisphere and the left in the southern hemisphere). Currents driven by this pressure gradient are known as geostrophic currents; in the North Atlantic, for example, the east–west pressure gradient force generates a geostrophic flow to the north.

Deep ocean circulation

Deep currents, which flow independently of the wind-driven surface currents, are initiated by surface waters being cooled at high latitudes, which increases their density and causes them to sink to the ocean depths, a process described as downwelling. In most of the world ocean, the water mass in contact with the abyssal sea floor is Antarctic Bottom Water (AABW), which downwells in the Ross and Weddell Seas off Antarctica, although in the northern North Atlantic this is replaced by North Atlantic Deep Water (NADW), composed of Norwegian Sea Deep Water (entering over the Faroes Sill) and Arctic Intermediate Water (AIW: entering from the Denmark Strait). The continual sinking of

these waters drives a slow circulation of water throughout the deep ocean: it is estimated that a water molecule, once moved into the deep ocean, would spend, on average, 510, 250 and 275 years below 1500 m in the Pacific, Indian and Atlantic oceans respectively.

Upwelling

Continual downwelling of water into the deep ocean requires that, eventually, the deep water rises to the surface, rejoining surface circulation in regions of upwelling. These occur in areas where surface water circulation patterns diverge in equatorial regions, leaving a gap which is filled from beneath, and where surface water is forced away from the edge of continents which, in the absence of an extensive continental shelf (as occurs off the western coasts of North and South America and central Africa), forces deep water to rise (Figure 1.3).

Gyres

Localized upwellings or downwellings can be created by gyres, the circular currents produced by cyclonic and anticyclonic winds (Box 1.1).

Box 1.1 The global wind system

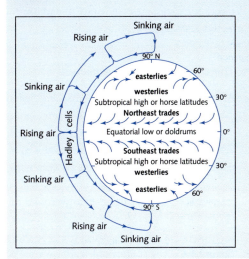

Heat which arrives at the Earth from the Sun is lost through radiation into space. Heat loss is fairly constant with latitude, whereas input of heat is much greater at low latitudes than at high latitudes. Therefore, there is a net gain of energy in the low and mid latitudes and a net loss at the poles.

Assuming that the Earth's temperature is constant, then there must be a mechanism for redistributing this heat, primarily through the circulation of the atmosphere. Hot air rises, creating low pressure at low altitudes, while cool air falls, creating high pressure. Air heated at the equator rises and moves polewards; as it does so, it

1

receives less solar energy, so cools and descends at about 30° latitude, spreading north and south at low altitude. Air at the poles, being cool, descends and moves towards the equator at low altitude; when it reaches temperate latitudes, it meets air moving polewards from the subtropics and this converging air rises.

The movement of air from high to low pressure areas at low altitude and in the opposite direction at high altitude sets up a series of cells of air, the Hadley cells. Air movement is, however, not directly to the north or south, because the rotation of the Earth creates Coriolis force, deflecting air mass fluids to the right in the northern hemisphere and to the left in the southern hemisphere and creating the prevailing wind systems observed on the Earth's surface.

The strength of Coriolis force increases at high latitudes so that, in mid-high latitudes, Hadley cells break down and vortices are produced, creating depressions and anticyclones. Winds circulating around a low pressure system are known as cyclonic (of which true cyclones are extremely low pressure examples), and rotate anticlockwise in the northern hemisphere, clockwise in the southern. High pressure systems are known as anticyclones, their winds rotating clockwise in the northern hemisphere and anticlockwise in the southern.

This theoretical pattern of wind movement is close to that observed over the open oceans, but is broken by the effect of land masses, particularly in the northern hemisphere. The most spectacular example of this is the monsoon winds of the Indian Ocean, created by the presence of the Eurasian land mass. During the summer, air over the land mass heats up and rises, creating a low pressure area. Normally, this would draw in air from all sides, including cooler air from the north which would reduce overall temperatures and raise the pressure. In India, however, the passage of air from the north is blocked by the Himalayan massif, so only warm air from the south is drawn in, which rapidly rises, maintaining low pressure. This sucks in moisture-laden air from the Indian Ocean, the monsoon winds.

An anticyclonic wind forces water into the centre of the gyre, creating an area of downwelling (**Figure 1.5a**), whereas a cyclonic wind pushes surface water out of the gyre, generating upwelling (**Figure 1.5b**). Gyres also occur in large lakes, in which the entire water body slowly rotates around a relatively stagnant centre.

1.2.3 The ocean bed

The ocean bed, or benthic zone, is a relatively small component of the marine environment, but highly diverse. It may be divided into a series of structures (**Table 1.2**).

The coastal zone

The coastal zone is where the sea meets the land, and its area and diversity is, therefore, determined by the topography of the coast.

Despite its small size relative to the rest of the marine environment, the coastal zone supports the highest diversity of marine habitats, including the intertidal (Section 5.2), coral reefs (Section 5.5), estuaries (Chapter 4) and coastal wetlands (Chapter 8).

The continental shelf and slope

At the edge of continents are continental shelves, shallow seas associated with the continental plates (Section 6.8). When the continent and the adjacent ocean are part of the same plate, the continental shelf may be very wide, extending hundreds of kilometres out to sea. In many places, however, particularly around the Pacific Ocean, it is very narrow and deep water is encountered a few kilometres from the shore (**Figure 1.6**). The continental shelf is a zone of deposition for terrigenous sediments, produced

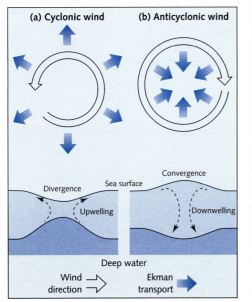

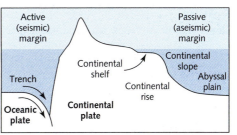

Figure 1.6 Profiles of continental shelf margins. A wide shelf occurs at a seismically passive shelf margin, where deep ocean and continent are part of the same plate. A narrow shelf and trench occur at a seismically active margin, where the oceanic plate is descending beneath the continental margin. Note that the topography of the land mass approximates to that of the adjacent ocean, because the active margin generates uplift of mountain ranges.

Figure 1.5 Ekman currents associated with the circulating patterns of winds around (a) a cyclonic and (b) an anticyclonic atmospheric pressure system, showing how each affects the sea surface shape, the thermocline and localized up- and downwelling. This is the pattern in the northern hemisphere; in the southern hemisphere, the directions of circulation are reversed.

by weathering of continental crust and therefore originating from land via coastal erosion or input from rivers. They consist of boulders, cobbles, pebbles, sands or muds, and dominate coastal regions and the continental shelf, slope

and rise, where they may form deposits 10 km or more thick.

Much of the material that accumulates on the upper continental slope is in an unstable condition and likely to move down slope. These movements are classified according to the degree of deformation of the mass of sediment that has moved. Slides, slumps and debris flows contribute to the formation of the continental rise, the region of increased angle of slope at the base of the continental slope. Slides and slumps result from mechanical failure along inclined planes. Large volumes of sediment move in a way analogous to a landslip, giving rise to a

Table 1.2 Main features of the principal ocean basins				
	Ocean			
	Pacific	**Atlantic**	**Indian**	**World Ocean**
Ocean area (km² × 10⁶)	180	107	74	361
Land area drained (km² × 10⁶)	19	69	13	101
Ocean area/drainage area	9.5	1.6	5.7	3.6
Average depth (m)	3940	3310	3840	3730
Area as % of total:				
shelf and slope	13.1	19.4	9.1	15.3
continental rise	2.7	8.5	5.7	5.3
deep ocean floor	42.9	38.1	49.3	41.9
volcanoes and volcanic ridges*	2.5	2.1	5.4	3.1
ridges†	35.9	31.2	30.2	32.7
trenches	2.9	0.7	0.3	1.7

* Volcanic ridges are those related to volcanic island chains that are not part of constructive plate margins, e.g. the Walvis Ridge in the South Atlantic. They do not include island arcs.
† These are the ridges that correspond to constructive plate margins, e.g. the Mid-Atlantic Ridge.

hummocky topography, although there is relatively little deformation of the original mass and the sedimentary layers can still be traced. They can occur on slopes as little as 2° and may be triggered by earthquakes or simply after the rapid accumulation of sediments. Debris flows are caused in the same way, but are the sluggish movement of a mixture of sediments, down slopes as little as 0.1°, which obliterate the original layering of the deposit.

Turbidity currents are triggered by the same mechanisms as slides and debris flows, but can travel at up to 90 km h^{-1} and transport 300 kg m^{-3} of sediment in suspension. They can move material up to 1000 km and are probably responsible for scouring out submarine canyons and the deposition of sand layers on the abyssal plain (see below). Some canyon systems were probably cut by river flows when sea levels were lower, and the deposition of sediments from major estuaries at their head means that turbidity currents are fairly frequent events.

Ocean ridges

Ocean ridges, which currently occupy about 33% of the ocean floor, are the site of production of new sea floor (Section 6.8). Basaltic magma is extruded from the centre of the ridge as the plates separate, so seafloor spreading tends to be symmetrical about, and perpendicular to, the ridge. The mid-Atlantic ridge is a slow-spreading ridge (each plate is moving away from the ridge at 1–2 cm yr^{-1}, widening the Ocean by 2–4 cm yr^{-1}), producing a well-defined valley 25–30 km wide and 1–2 km deep, with the sea floor sloping away from the ridge at an angle of about 1°. The East Pacific Rise, in contrast, is a rapidly spreading ridge (6–8 cm yr^{-1}), so the median valley is poorly developed, and even absent in places, while the slope is typically 0.2°.

The deep sea floor

Between the continental margins and the mid-ocean ridges lies the abyssal plain, with a slope of less than 0.05°, accounting for 42% of the Earth's surface. This is covered by pelagic sediments, which settle from suspension in the open ocean and are often biological in origin (Section 6.3). These were formerly known as oozes, which gives some idea of their properties. They aggregate very slowly, so dominate only where terrigenous inputs are low. Rising rapidly from the surface of the abyssal plain, suggesting their bases are buried in sediment, are abyssal hills (rising less than 1 km above the plain) and sea mounts (rising more than 1 km). Abyssal hills are most common near ridge systems, where there has been little time for pelagic sediment to build up and mask topographic relief produced by the spreading centre. Most sea mounts, which are most abundant in the Pacific, arise as sub-sea volcanoes, forming islands if they reach the surface. Guyots are sea mounts with a flat top, created by wave erosion of the sea mount when it broke the sea surface prior to sea level rise, subsidence of the mount, or a combination of the two.

Deep trenches are formed where an oceanic plate is being subsumed beneath a continental plate and so occur relatively close to land (Figure 1.6). They are steep-sided and can reach great depths (Figure 1.1).

1.2.4 Inland waters

Inland waters are patchily distributed across the Earth's land surface. Glacial ice is confined to high latitudes and, in relatively small amounts, high altitudes, by its requirement for low temperatures. The very highest latitudes are cold deserts, with very little precipitation, and ice persists not because it is continually replenished, but because what little snow falls is unlikely to melt. Groundwater can be found under most of the Earth's surface, but only where it is close to the surface, usually in association with rivers, is it a significant biological habitat.

Rivers, lakes and wetlands are dynamic systems with a high turnover of water, their distribution and abundance being determined by precipitation and morphology of the landscape. Lakes and wetlands occur most abundantly where precipitation is high; rivers originate in these areas but can flow through much more arid areas.

Rivers

Rivers are drainage channels for the excess of precipitation over evaporation and, as such, are the main conduit for the return of water from land to the sea. Their formation requires enough of a slope to enable water to flow downhill easily, but inputs from steep headwaters can maintain flow in the lower reaches of a river even over low gradient terrain. Furthermore, the constant mobility of water, along with the sediments it carries, will rapidly create a discrete channel within which water is normally confined.

River systems originate most abundantly where rainfall is high, but their nature as channels for the transport of water means that they extend beyond the source region, normally into areas whose rainfall is inadequate to maintain the flow without inputs from upstream. In arid zones, river channels may be common but water flows only rarely, as run-off from occasional storms. Rivers rising in uplands or adjacent arid regions may flow perennially in their upper reaches but evaporate away, or be lost to groundwater, when they reach lower-lying areas. Most of the rain which falls or flows into arid regions never reaches the sea: only the largest rivers, such as the Nile and the Euphrates, have enough volume to withstand the high rates of evaporation.

Lakes

Lakes, like rivers, are unevenly distributed, but their distribution correlates less clearly with precipitation than do the headwaters of river systems, because they require a basin within which to form (Box 7.1) and, in high mountain ranges, such basins may be rare. Most of the world's lakes are in the recently deglaciated

areas of northern North America and northern Europe. Equatorial regions, despite their precipitation, support few large permanent lakes, because they are generally low-lying and flat, although features such as temporary ponds and oxbow lakes (Box 7.1) are very common; an exception is in East Africa, where tectonic activity has created a number of large basins now occupied by lakes.

Arid zones support saline lakes, either temporary or permanent (Section 7.2). The importance of saline lakes is often overlooked, despite the large volume of water they contain (Table 1.1). Admittedly, over 70% of the entire volume of inland saline water is in the Caspian Sea (itself a remnant of a former marine system), but freshwaters, too, are concentrated: about half of all fresh water occurs in just seven lakes: Baikal, Tanganyika and the Great Lakes of North America.

Wetlands

Wetlands occur where inputs of water exceed evaporation, but outflow is impeded by flat topography. Their distribution is discussed in Section 8.7. The vast majority of the world's wetlands occur in areas with the highest precipitation, but some of the largest individual wetlands, including the Okavango and Inner Niger Delta in Africa and the mouths of the Euphrates and Tigris in south-western Asia, occur in arid zones with very little direct rainfall, and are entirely dependent for water upon the large rivers which feed them. Coastal saline wetlands are maintained by the sea so they, too, require little or no direct precipitation.

1.3 Chemistry of water

Naturally occurring water is never pure, but contains a range of components, both suspended and in solution. The composition and concentrations of these substances, particularly those in solution, determine the quality of water.

1.3.1 Dissolved inorganic matter

Salinity

Dissolved ions in water are expressed as units of mass (mg l^{-1}), parts per million (ppm), or sometimes as chemical equivalents (meq l^{-1}), while the total amount of dissolved material in a sample is referred to as its salinity. Salinity may be determined by evaporating off the water and weighing the residual salts. This has problems, however, as the drying process may alter some salts, for example the loss of hydrogen chloride (HCl) from hydrous magnesium chloride (MgCl), or combustion of carbonates. The constancy of seawater composition (see below) allows determination of salinity by measuring the concentration of a single constituent, such as chlorinity (the concentration of chloride [Cl^-]), where salinity = 1.80655 [Cl^-]. Salinity can then be expressed as parts per thousand (ppt) or ‰. Chlorinity can easily be determined by titration, and this was the routine method for estimating seawater salinity until the 1960s, when it was superseded by electrical measurement. Salinity is now defined and measured according to an internationally agreed Practical Salinity Scale, which specifies it as a dimensionless (and therefore unitless) property which is measured electronically, by determining the conductivity of water.

Whereas the ionic composition of sea water is remarkably consistent, that of saline lakes differs considerably. Most have a high concentration of sodium chloride (NaCl), albeit with other ions in very different proportions to sea water or each other, but some, known as soda lakes, are dominated by carbonates (CO_3^{2-}) and hydrocarbonates (HCO_3^-). Therefore, the Practical Salinity Scale, being based on chlorinity, is only applicable to water of marine origin. The salinity of saline lakes is defined as the sum total of all ion concentrations, or total ion concentration, and should always include an indication of units used. These units should ideally be g kg^{-1} or ppt (parts per thousand), although g l^{-1} are widely quoted. The two are widely considered to be equivalent and indeed at low

salinities they are. However, in the range of salinities recorded from salt lakes there is a clear divergence: as the density of the medium increases, the g l^{-1} reading becomes higher than ppt. For sodium chloride solution, 70 ppt is equivalent to 73.4 g l^{-1}, while 140 ppt is 154.1 g l^{-1} (Williams and Sherwood 1994).

Conductivity

Conductivity is a measure of the ability of water to conduct electricity, as it is dissolved ions, rather than water molecules themselves, which are the conductors. Conductivity is expressed in Siemens per centimetre (S cm^{-1}) or, more normally for fresh waters, mS cm^{-1} or μS cm^{-1}, measured at 25°C. An increasing conductivity reading signifies increasing concentration of dissolved ions. Distilled water has a conductivity of 2–4 μS cm^{-1}, clean tap water is normally in the range 60–100 μS cm^{-1}, a clean upland river or lake on hard, erosion-resistant geology will typically give a reading of less than 300 μS cm^{-1}, whereas an organically enriched river's conductivity is usually in excess of 500 μS cm^{-1}. Increasing conductivity in fresh waters is often used as an indicator of pollution, although there are many natural sources of ions, and rivers become naturally more enriched with distance from source (Section 3.2), while ponds may have very high conductivities. As an extreme example of natural enrichment, of course, the conductivity of the sea is 35,000 μS cm^{-1}!

A useful feature of conductivity is that it approximates to ion concentration; a conductivity of 60 μS cm^{-1} will be more or less equivalent to a total dissolved ion concentration of 60 mg l^{-1}.

Dissolved matter in the sea

Sea water contains a high proportion of dissolved components, generally around 35 g kg^{-1}, giving a salinity of 35. This concentration varies little, except in estuaries or semi-enclosed seas with a high volume of fresh water inputs (Section 4.2) or inlets in arid zones where evaporation rates are very high.

Components of sea water can be divided into major constituents, present at concentrations of 1 ppm or more, which account for 99.9% of the salt in the oceans, and minor or trace constituents, the remaining, non-biologically active salts, which are often used as chemical tracers of the source of waters. Most of the 92 naturally occurring elements have been isolated from sea water (Table 1.3), and those which are missing will probably be identified once appropriate analytical techniques have been developed. Chemical composition of the dissolved

1

Table 1.3 The composition of sea water			
Element		**Concentration (mg l^{-1})**	**Some probable dissolved species**	**Total amount in the ocean (tonnes)**
Chlorine	Cl	1.95×10^4	Cl^-	2.57×10^{16}
Sodium	Na	1.077×10^4	Na^+	1.42×10^{16}
Magnesium	Mg	1.290×10^3	Mg^{2+}	1.71×10^{15}
Sulphur	S	9.05×10^2	SO_4^{2-}, $NaSO_4^-$	1.2×10^{15}
Calcium	Ca	4.12×10^2	Ca^{2+}	5.45×10^{14}
Potassium	K	3.80×10^2	K^+	5.02×10^{14}
Bromine	Br	67	Br^-	8.86×10^{13}
Carbon	C	28	HCO_3^-, CO_3^{2-}, CO_2	3.7×10^{13}
Nitrogen	N	11.5	N_2 gas, NO_3^-, NH_4^+	1.5×10^{13}
Strontium	Sr	8	Sr^{2+}	1.06×10^{13}
Oxygen	O	6	O_2 gas	7.93×10^{12}
Boron	B	4.4	$B(OH)_3$, $B(OH)_4^-$, $H_2BO_3^-$	5.82×10^{12}
Silicon	Si	2	$Si(OH)_4$	2.64×10^{12}
Fluorine	F	1.3	F^-, MgF^+	1.72×10^{12}
Argon	Ar	0.43	Ar gas	5.68×10^{11}
Lithium	Li	0.18	Li^+	2.38×10^{11}
Rubidium	Rb	0.12	Rb^+	1.59×10^{11}
Phosphorus	P	6×10^{-2}	HPO_4^{2-}, PO_4^{3-}, $H_2PO_4^-$	7.93×10^{10}
Iodine	I	6×10^{-2}	IO_3^-, I^-	7.93×10^{10}
Barium	Ba	2×10^{-2}	Ba^{2+}	2.64×10^{10}
Molybdenum	Mo	1×10^{-2}	MoO_4^{2-}	1.32×10^{10}
Arsenic	As	3.7×10^{-3}	$HAsO_4^{2-}$ $H_2AsO_4^-$	4.89×10^9
Uranium	U	3.2×10^{-3}	$UO_2(CO_3)_2^{4-}$	4.23×10^9
Vanadium	V	2.5×10^{-3}	$H_2VO_4^-$, HVO_4^{2-}	3.31×10^9
Aluminium	Al	2×10^{-3}	$Al(OH)_4^-$	2.64×10^9
Iron	Fe	2×10^{-3}	$Fe(OH)_2^+$, $Fe(OH)_4^-$	2.64×10^9
Nickel	Ni	1.7×10^{-3}	Ni^{2+}	2.25×10^9
Titanium	Ti	1×10^{-3}	$Ti(OH)^4$	1.32×10^9
Zinc	Zn	5×10^{-4}	$ZnOH^+$, Zn^{2+}, $ZnCO_3$	6.61×10^8
Caesium	Sc	4×10^{-4}	Cs^+	5.29×10^8
Chromium	Cr	3×10^{-4}	$Cr(OH)_3$, CrO_4^{2-}	3.97×10^8
Antimony	Sb	2.4×10^{-4}	$Sb(OH)_6^-$	3.17×10^8
Manganese	Mn	2×10^{-4}	Mn^{2+}, $MnCl^+$	2.64×10^8
Krypton	Kr	2×10^{-4}	Kr gas	2.64×10^8
Selenium	Se	2×10^{-4}	SeO_3^{2-}	2.64×10^8
Neon	Ne	1.2×10^{-4}	Ne gas	1.59×10^8
Cadmium	Cd	1×10^{-4}	$CdCl_2$	1.32×10^8
Copper	Cu	1×10^{-4}	$CuCO_3$, $CuOH^+$	1.32×10^8

Table 1.3 Continued			
Element		**Concentration (mg l^{-1})**	**Some probable dissolved species**	**Total amount in the ocean (tonnes)**
Tungsten	W	1×10^{-4}	WO_4^{2-}	1.32×10^{8}
Germanium	Ge	5×10^{-5}	$Ge(OH)_4$	6.61×10^{7}
Xenon	Xe	5×10^{-5}	Xe gas	6.61×10^{7}
Mercury	Hg	3×10^{-5}	$HgCl_4^{2}$, $HgCl_2$	3.97×10^{7}
Zirconium	Zr	3×10^{-5}	$Zr(OH)_4$	3.97×10^{7}
Bismuth	Bi	2×10^{-5}	BiO^+, $Bi(OH)_2^+$	2.64×10^{7}
Niobium	Nb	1×10^{-5}	Not known	1.32×10^{7}
Tin	Sn	1×10^{-5}	$SnO(OH)_3^-$	1.32×10^{7}
Thallium	Tl	1×10^{-5}	Tl^+	1.32×10^{7}
Thorium	Th	1×10^{-5}	$Th(OH)_4$	1.32×10^{7}
Hafnium	Hf	7×10^{-6}	Not known	9.25×10^{6}
Helium	He	6.8×10^{-6}	He gas	8.99×10^{6}
Beryllium	Be	5.6×10^{-6}	$BeOH^+$	7.40×10^{6}
Gold	Au	4×10^{-6}	$AuCl_2^-$	5.29×10^{6}
Rhenium	Re	4×10^{-6}	ReO_4^-	5.29×10^{6}
Cobalt	Co	3×10^{-6}	Co^{2+}	3.97×10^{6}
Lanthanum	La	3×10^{-6}	$La(OH)_2$	3.97×10^{6}
Neodymium	Nd	3×10^{-6}	$Nd(OH)_3$	3.97×10^{6}
Silver	Ag	2×10^{-6}	$AgCl_2^-$	2.64×10^{6}
Tantalum	Ta	2×10^{-6}	Not known	2.64×10^{6}
Gallium	Ga	2×10^{-6}	$Ga(OH)_4^-$	2.64×10^{6}
Yttrium	Y	1.3×10^{-6}	$Y(OH)_3$	1.73×10^{6}
Cerium	Ce	1×10^{-6}	$Ce(OH)_3$	1.32×10^{6}
Dysprosium	Dy	9×10^{-7}	$Dy(OH)_3$	1.19×10^{6}
Erbium	Er	8×10^{-7}	$Er(OH)_3$	1.06×10^{6}
Ytterbium	Yb	8×10^{-7}	$Yb(OH)_3$	1.06×10^{6}
Gadolinium	Gd	7×10^{-7}	$Gd(OH)_3$	9.25×10^{5}
Praseodymium	Pr	6×10^{-7}	$Pr(OH)_3$	7.93×10^{5}
Scandium	Sc	6×10^{-7}	$Sc(OH)_3$	7.93×10^{5}
Lead	Pb	5×10^{-7}	$PbCO_3$, $Pb(CO_3)_2^{2-}$	6.61×10^{5}
Holmium	Ho	2×10^{-7}	$Ho(OH)_3$	2.64×10^{5}
Lutetium	Lu	2×10^{-7}	$Lu(OH)$	2.64×10^{5}
Thulium	Tm	2×10^{-7}	$Tm(OH)_3$	2.64×10^{5}
Indium	In	1×10^{-7}	$In(OH)_2$	1.32×10^{5}
Terbium	Tb	1×10^{-7}	$Tb(OH)_3$	1.32×10^{5}
Tellurium	Te	1×10^{-7}	$Te(OH)_6$	1.32×10^{5}
Samarium	Sm	5×10^{-8}	$Sm(OH)_3$	6.61×10^{4}
Europium	Eu	1×10^{-8}	$Eu(OH)_3$	1.32×10^{4}
Radium	Ra	7×10^{-11}	Ra^{2+}	92.5
Protactinium	Pa	5×10^{-11}	Not known	66.1
Radon	Rn	6×10^{-16}	Rn gas	7.93×10^{-4}
Polonium	Po	Not known	PoO_3^{2-}, $Po(OH)_2$	Not known

Table 1.3 does not represent the last word on seawater composition. Even for the more abundant constituents, compilations from different sources differ in detail. For the rarer elements, many of the entries in the table will be subject to revision, as analytical methods improve. Moreover, most constituents behave non-conservatively, making averages less meaningful.

components is remarkably constant, so while the total salinity may vary in response to local evaporation, precipitation and mixing, this has no effect on the relative proportions of the main constituents. Exceptions to this occur in enclosed seas and estuaries (Section 4.2), where input processes may affect the composition, and in geologically or biologically active areas, including shallow tropical seas where extensive reefs may be removing calcium carbonate ($CaCO_3$) (Section 5.5), or oceanic hydrothermal vent systems whose outflows are chemically very different from normal sea water (Box 6.3).

Sources and sinks of ocean salt

The most obvious source of salts for the sea is the weathering of rocks by rainwater and hence additions from river water to the sea (Table 1.4). However, some salts abundant in river water are relatively uncommon in crustal rocks. An important example is chloride (Cl^-), the dominant ion in oceanic waters, which is common in rock only where evaporites have been deposited by the sea. The Cl^- in rivers, therefore, originates mainly from the sea. It leaves the ocean as an aerosol and is carried in the atmosphere to fall as rain and return to the sea via rivers. Volcanoes produce large quantities of hydrogen chloride (HCl) gas, and the high level of volcanic activity early in the Earth's his-

tory is probably the source of the majority of the chloride present in today's oceans.

The time that the average molecule spends in the ocean, its residence time (RT), may be estimated by dividing the total mass of a given substance dissolved in the oceans by its rate of supply. As weathering of rocks is continuing but composition of sea water remains relatively constant, there must be sinks for the material brought into the ocean by rivers. Much of it is precipitated out as sediment, particularly bicarbonate, calcium (Ca^{2+}) and silicate (SiO_2), which are proportionally less common in sea water than river water. Removal mechanisms for these salts include inorganic precipitation, reaction of dissolved substances with particles and biological processes such as reef building and construction of silicate and carbonate skeletons (Section 6.2).

Calcium carbonate in oceans

Calcium carbonate is used by many marine organisms as a skeletal material. Upon death, these organisms sink and release the calcium and carbonate ions back into solution.

$$Ca^{2+} (aq) + CO_3^{2-} (aq) \leftrightarrow CaCO_3 (s)$$

Surface ocean waters are supersaturated with $CaCO_3$, but spontaneous precipitation is rare, as most of the carbonate ions are weakly bound

Table 1.4 Chemical constituents of rainwater and rivers			
	Rainfall (continental)	Rainfall (marine and coastal)	Rivers*
Ca^{2+}	0.2–4.0	0.2–1.4	5.3–24.2 (13.4)
Mg^{2+}	0.05–0.5	0.4–1.5	1.4–5.2 (3.4)
Na^+	0.2–1.0	1.0–5.0	3.2–7.0 (5.2)
K^+	0.1–0.5	0.2–0.6	1.0–1.6 (1.3)
Cl^-	0.2–2.0	1.0–10.0	3.4–7.0 (5.8)
SO_4^{2-}	1.0–3.0	1.0–3.0	3.5–15.1 (8.3)
HCO_3^-	0	0	26.7–80.1 (52.0)
SiO_2	0	0	6.8–16.3 (10.4)

* Range of mean values for different continents, with the world average in parentheses. All figures are estimates of natural values, from which anthropogenic pollutants have been excluded.
Adapted from Allan (1995) after Berner and Berner (1987). All concentrations are in mg l^{-1}.

1

to magnesium ions, forming magnesium carbonate ($MgCO_3$).

$CaCO_3$ solubility increases as depth (and therefore pressure) increases and as temperature drops. As skeletal material sinks, therefore, it enters the region in which $CaCO_3$ will begin to dissolve, the lysocline region. The depth below which all $CaCO_3$ has entered solution is the carbonate compensation depth (CCD), which is typically 4–5 km deep. Variations in the depth of the lysocline are controlled by the chemistry of the sea water, while variations in the CCD are the result of both water chemistry and the rate of supply of carbonate debris; it is, for example, normally shallower near the continental margins as the higher biological activity in the sediment aids its dissolution.

Dissolved matter in inland waters

Inland waters, in contrast to the sea, show a wide variation in chemical constituents. While most are classed as fresh water, which means salinity lower than 3 mg l^{-1}, there is a large number of lakes, classed as saline, in which salinity may be many times higher than that in the sea and very different in composition. Furthermore, chemical composition can vary greatly even within a single water body. The concentration of solutes generally increases in rivers with distance from the source and can vary in response to discharge (Section 3.2), while freshwater lakes become more solute-enriched with time (Section 7.6) and saline lakes may experience rapid fluctuations in salinity, determined by rainfall and volume of water (Section 7.2).

The chemical composition of inland waters is determined particularly by the catchment. Rain brings some inputs but, except in areas with very hard rocks and thin soils, or adjacent to the coast, is relatively unimportant. On average, about 40% of rain becomes run-off and enters water courses, the rest evaporating, so chemical constituents of rainwater, which will not evaporate, will be concentrated 2.5 times in run-off water. Many constituents of fresh water are, however, at much higher concentrations

than these (Table 1.4), demonstrating the importance of catchment weathering.

1.3.2 Dissolved gases

The most important gases in water are oxygen, for respiration, and carbon dioxide, for photosynthesis. Both originate mainly from the atmosphere, but diffusion mixes gases very slowly and physical mixing is required to ensure that they penetrate beyond the surface layers. Physical mixing in wave-swept water bodies and turbulent rivers will maintain high concentrations, but in less turbulent water bodies, concentrations will be determined to a large extent by biological activity. A large biomass of photosynthesizers can raise oxygen levels and lower carbon dioxide levels by day and, through respiration, produce the opposite effect at night.

Oxygen

Table 1.5 shows oxygen (O_2) concentrations at saturation, in which the amount in solution in pure water, at the given temperature and pressure, is in equilibrium with that in the air. High rates of photosynthesis create conditions of supersaturation (saturation in excess of 100%), in which oxygen concentrations are so high that bubbles of undissolved gas may be produced on plant surfaces. In contrast, high rates of respiration reduce concentrations well below saturation.

The concentration of oxygen in solution, even in pure water, is determined by temperature (Table 1.5) and pressure, a 100 m increase in altitude reducing solubility by about 1.4%. Solubility is also reduced by salinity because the more solutes it contains, the less oxygen water can hold. Sea water therefore contains approximately 20% less oxygen than fresh water at an equivalent temperature and pressure.

Carbon dioxide

Carbon dioxide (CO_2) is subject to the same effects of temperature, pressure and salinity as

Table 1.5 Oxygen concentrations in water at saturation		
Temperature (°C)	Concentration (mg l^{-1})	Concentration (ml l^{-1})
0	14.63	10.45
5	12.77	9.12
10	11.28	8.06
15	10.07	7.20
20	9.08	6.49
25	8.26	5.90
30	7.57	5.41
35	6.98	4.99
40	6.47	4.62

Values given are for pure water at sea level; contaminants and changes in pressure will reduce these (see text). Concentrations are normally expressed as mg l^{-1}; volumetric concentrations (ml l^{-1}) are also given here, to allow easy comparison with the atmospheric concentration: 210 ml l^{-1}.

oxygen, but is considerably more soluble in water because it combines with alkali metals to form bicarbonates and carbonates according to the reaction:

$$CO_2 + H_2O \leftrightarrow H_2CO_3 \leftrightarrow H^+ + HCO_3^- \leftrightarrow 2H^+ + CO_3^{2-}$$

Therefore, very little CO_2 gas is present in water. Total dissolved carbon in the oceans decreases with depth to about 1000 m and then is fairly constant to the bottom. This is the result of carbon uptake in the surface zone by biological activity (photosynthesis and skeletal production) while, in deep waters, lower rates of carbon uptake are balanced by regeneration from decomposition of organic particles.

1.3.3 Acidity and alkalinity

Acidity

Acidity, the ability of water to neutralize alkalinity, is normally determined by the pH (which has no units), a measure of the concentration of hydrogen ions in solution ([H^+]):

$$pH = -\log_{10}[H^+]$$

All water contains some hydrogen ions, because a small number of water molecules will dissociate:

$$H_2O \leftrightarrow H^+ + OH^-$$

Pure water at 24°C has a [H^+] of 10^{-7} mol l^{-1}, giving a pH of 7. This is, therefore, classed as neutral, a greater pH signifying alkaline water while a lower pH is acid. Note that, as it is a logarithmic scale, a change of one pH unit is equivalent to a tenfold change in [H^+], and that increasing [H^+] is signified by lower pH values.

Alkalinity and buffering capacity

Rainwater is slightly acid (pH around 5.6), because it dissolves CO_2 from the atmosphere to produce carbonic acid:

$$H_2O + CO_2 \rightarrow H_2CO_3$$

Carbonic acid, in turn, dissolves carbonates, particularly $CaCO_3$ and $MgCO_3$, which act as a buffer, neutralizing excess acidity. The alkalinity of a solution is a measure of this buffering capacity, and is effectively the measure of the concentration of carbonate and bicarbonate in solution, although, in some lakes, borate (BO_4^-) may be important. Note that the term alkalinity can be misleading, because it does not refer to how alkaline a solution is; even water with a low pH (i.e. acid) can have a high alkalinity.

Acid waters

For freshwater organisms, the pH range 5.5–8.5, within which most natural water bodies fall, is effectively neutral (and often referred to as circumneutral). Acidification as a pollution problem occurs where two criteria are met. First, rain is acidified by anthropogenic release of sulphates and nitrogen oxides, which form sulphuric and nitric acid in the atmosphere, reducing rainfall pH below 5.0 or even 4.0 (10–100 times more acid than natural rainfall). Secondly, hard geology, with little weathering of buffering agents, limits the buffering capacity, so the excess acidity is not neutralized. It is important to remember, however, that many water bodies are naturally acidic. Volcanic activity contributes to acidification of rainfall by emitting sulphates. Lakes in active volcanic craters can be very acid, with a pH of 2.0 or even less. Very slow decomposition of plant material in peat-producing wetlands (Section 8.1) produces humic acids, acidifying pools and water bodies in such environments. *Sphagnum* moss (Section 8.6, Box 8.3), which is well adapted to such environments, contributes to acidification of water by stripping Ca^{2+} ions from water and replacing them with H^+.

Alkaline waters

Highly alkaline water bodies are relatively uncommon, but include soda lakes, whose pH can be in excess of 11.0, which contain high concentrations of sodium carbonate (Na_2CO_3) and sodium bicarbonate ($NaHCO_3$) derived from lava.

Sea water

Sea water, in contrast to inland waters, has a pH typically 8.0 ± 0.2, with very little variation, because it is extremely well buffered with a high alkalinity generated by the high concentration of carbonates. It is difficult to measure the pH of sea water accurately, so in practice it is usual to measure the alkalinity and total dissolved carbon. It is then possible to use empirical chemical relationships to determine concentrations of CO_3^{2-} and HCO_3^- and hence calculate the pH.

With increasing atmospheric concentrations of CO_2, resulting from the burning of fossil fuels, more CO_2 has been dissolving into the seas and oceans. This drives the carbonate equilibrium system to produce more carbonate, resulting in the release of more H^+ ions and therefore increasing acidity. There is now a firm basis for believing that human activities will not just warm the planet but also turn the oceans more acid. The oceans absorb about one-third of the carbon dioxide produced by

the burning of fossil fuels and as a result are currently around 0.1 pH units more acidic than in pre-industrial times (Orr *et al.* 2005). While a slight change in pH may not be a problem for many taxa it will significantly impact the ability of marine organisms to lay down shells and skeletons derived from calcium carbonate. Thus it might lead, in addition to any direct effects of global warming, to impacts on coral reefs, shell-fish (clams, scallops, winkles) and plankton (particularly coccolithophorids) (Doney 2006).

Global models have tended to suggest that significant effects may only arise over periods measured in centuries, but more detailed modelling suggest that effects in high latitude systems, such as the Southern Ocean, may occur within as little as 50 years (Orr *et al.* 2005).

1.4 Energy inputs in aquatic systems

1.4.1 Photosynthesis

Primary production is the creation of organic molecules from inorganic components, using an external energy source. By far the most common mechanism is photosynthesis, using sunlight to synthesize sugars from CO_2:

$$6H_2O + 6CO_2 \rightarrow C_6H_{12}O_6 + 6O_2$$

This mechanism is used by plants, algae and Cyanobacteria ('blue-green algae', formerly known as Cyanophyta). Other requirements include nitrogen, normally as nitrate (NO_3^-), and phosphate (PO_4^-) for protein synthesis. Of these, CO_2 is rarely limiting but the others, along with biotic processes, can have a major effect upon rates of production.

Light penetration

Light intensity in water is always very much lower than that in the atmosphere immediately overhead. Much light never manages to penetrate water, being reflected from the surface,

Box 1.2 Light attenuation

Light attenuation can be estimated according to the following formula:

$$I_d = I_0\, e^{-kd}$$

where I_d is the light at depth d, I_0 is the light at the surface (depth 0) and k is the extinction coefficient. We can measure k directly with a light meter or it can be estimated as:

$$k' = 1.7 D_s$$

where D_s is the Secchi depth (**Box 7.3**).

Attenuation varies according to the wavelength of light. In clear waters, blue-green light (wavelength around 480 nm) penetrates to the greatest depth, while, in turbid waters, blue light is selectively scattered and the wavelength of maximum penetration moves towards the red, ending up in the green part of the spectrum at about 550 nm.

while light which does enter the water column is absorbed or scattered (reflected) by water itself, as well as by dissolved organic molecules, phytoplankton and suspended particulate matter. Obviously the more turbid the water, the more light is scattered and absorbed.

The decrease in light intensity with depth is called attenuation (**Box 1.2**). Many aquatic macrophytes grow on the water surface or have emergent components extending into the air, and intertidal algae can photosynthesize during low tide, so light attenuation is not a problem, but submerged photosynthesizers will be light limited.

Compensation point and critical depth

Phytoplankton (Section 2.2) are floating photosynthesizers which are mixed vertically by water movements. **Figure 1.7** illustrates the relationship between photosynthesis, respiration and depth in a water column. Rate of photosynthesis is dependent upon the intensity of light and so, apart from a small amount of photoinhibition near the surface, it declines with increasing depth, whereas respiration rate remains relatively constant, so that eventually a

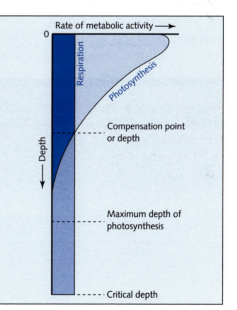

Figure 1.7 Change in photosynthetic rate by phytoplankton with depth. Photosynthesis declines with increasing light attenuation (Box 1.2) whereas respiration remains relatively constant. The depth at which production exactly matches consumption is the compensation point. The critical depth marks the point above which total respiration and total production are equal; note that this point is below the maximum depth at which photosynthesis occurs. The actual depths will vary according to species, water clarity, temperature, season and, of course, time of day.

1

depth is reached where the energy assimilated by photosynthesis is exactly matched by the energy requirements for photosynthesis, a point known as the compensation point or depth. Above this point, net production can occur, while below it, there is an excess of respiration over production. The point at which total photosynthesis matches total respiration (i.e. the two shaded components in **Figure 1.7** cover the same area) is the critical depth.

Sverdrup's model

Obviously, a phytoplankter which remains beneath the compensation point will be unable to sustain itself but, in practice, vertical mixing moves phytoplankton through different depths in the water column. The mixing depth is the depth to which phytoplankton are mixed, and is generally the depth of the thermocline (**Box 1.3**, but see Section 7.3 for exceptions). If the mixing depth is greater than the critical depth,

then there can be no net production, but if it is less than the critical depth, net production occurs. This is known as Sverdrup's model and links primary production to the critical depth versus mixing depth relationship.

The model makes some important assumptions:

1 There is a uniform depth distribution of phytoplankton in the mixed layer.

2 The extinction coefficient of light (**Box 1.2**) is a constant (this is rarely true as a result of self-shading by the phytoplankton).

3 Nutrients are not limiting.

4 Phytoplankton production is proportional to radiation.

5 Respiration is constant with depth.

The layer of water from the surface to the compensation depth is often referred to as the photic (or euphotic) zone; in practice, this is very difficult to measure accurately and is

---

Box 1.3   Stratification and the thermocline

Increasing the temperature of water reduces its density, so warm water floats on top of cool water. Solar energy heats up the surface of a water body, and this warm surface water is mixed, by wave action, with deeper water. If, however, there is little wave action, the depth of mixing will be shallow and heated water will remain close to the surface, as it is of a lower density than deeper water. Therefore, it receives yet more solar energy and becomes even warmer.

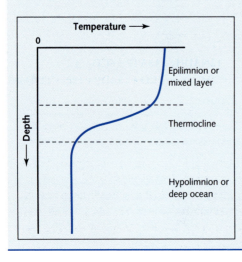

This leads to stratification of the water column into distinct horizontal layers: warm, low-density water on top and cooler, higher density water beneath. Within each layer there is free mixing and little temperature variation, but between layers there is little mixing and therefore a clear temperature difference.

Three vertical layers are created in this way: the warm upper layer (epilimnion to limnologists; mixed layer to oceanographers), the cool lower layer (hypolimnion or deep ocean) and, between the two, the zone of rapid temperature change, called the thermocline (occasionally referred to as the metalimnion by limnologists).

generally considered to be down to a depth where the light value is about 1% of surface light.

Measuring primary production

Primary production by phytoplankton is generally measured by following either CO_2 uptake or O_2 production. The former uses radio-labelled $^{14}CO_2$ (the most widely used technique) while the latter measures change in O_2 either by electrode, titration or using 'heavy water': $H_2^{18}O$.

As respiration occurs simultaneously, an adjustment for the respiration rate is applied by making measurements in the dark. In order to gain realistic estimates, samples are incubated under conditions which are as natural as possible in both the light and the dark for 4–12 h. This is either done in the controlled environment of an incubator or *in situ*, by suspending sample bottles from a buoy. This technique assumes that, excepting photosynthesis, the processes occurring in the light and the dark are the same, which may not always be the case. These techniques also tend to underestimate photosynthesis, as the soluble products exuded by phytoplankton (Section 2.2) are not readily measured. Therefore, some care must be taken in interpreting the results.

Nutrient limitation

Although light is clearly limiting at great depths, there are many environments where light intensities are high and yet primary production is low, even close to the surface. This is normally because one or more nutrients are limiting. As in terrestrial systems, limiting nutrients in aquatic systems are normally nitrates and phosphates, although nitrate is the most common limiting nutrient in marine systems, whereas phosphate is most commonly limiting in freshwaters. Silica commonly limits production by diatoms (Sections 6.4 and 7.4), while iron has recently been identified as a limiting nutrient in some parts of the ocean (Section 6.4).

Nutrient limitation in open waters is often a result of water stratification (Box 1.3); nutrients are depleted in the upper, illuminated layers while concentrations remain high in the depths, beneath the photic zone (Sections 5.4, 6.3, 7.3).

Biotic limitation

Primary production may be limited by the activities of grazers consuming the photosynthetic organisms. This is considered in detail in Sections 5.4, 6.5 and 7.4.

1.4.2 Chemosynthesis

In the absence of light, certain bacteria can generate organic matter using the chemical energy in reduced compounds to drive their metabolism according to the following general equation, where X signifies one of a number of reduced substances.

$$XH_2 + H_2O \rightarrow XO + 4[H^+ + e^-]$$

reduced substances   oxidised form   reducing power

The reducing power is used to generate energy via the cytochrome system, a process which is most efficient if free oxygen is available.

$$4[H^+ + e^-] + mADP + mPi + O_2 \rightarrow 2H_2O + mATP$$

$$2[H^+ + e^-] + NAD \rightarrow NADH_2$$

These can then be used to assimilate sugars using carbon dioxide:

$$12NADH_2 + 18ATP + 6CO_2 \rightarrow C_6H_{12}O_6 + 6H_2O + 18ADP + 18Pi + 12NAD$$

Chemoautotrophic bacteria include nitrogen-fixing species (Box 8.1), which derive their energy from ammonia (NH_3) or nitrite (NO_2^-), while others use sulphur compounds, such as sulphides (S^{2-}) or methane (CH_4).

The most celebrated sites for the occurrence of chemoautotrophy are deep ocean vents (Section 6.7, Box 6.3). Reduced compounds are normally oxygenated in the presence of oxygen, and the interface between vent waters and oxygenated deep ocean waters is one of the few situations where reduced compounds and oxygen occur together.

1.4.3 Detritus

Detritus is of fundamental importance to aquatic systems and the dominant source of energy for many (Sections 3.4, 4.4, 5.4). Even in environments where primary production occurs, certain primary producers are rarely grazed as living organisms, but are consumed after death, as detritus. The terms autochthonous and allochthonous detritus are often applied, the former referring to detritus derived from primary production *in situ*, the latter to detritus imported from elsewhere, including the terrestrial environment. A range of different types can be identified, the main distinction being between particulate and dissolved components.

Particulate organic matter

The terms particulate organic matter (POM) and particulate organic carbon (POC) refer to non-living particulate matter in water bodies, the detritus. These terms are sometimes used interchangeably, although the distinction should be what was measured, the entire mass of organic material or just the estimated carbon content. Freshwater biologists generally distinguish between coarse particulate organic matter (CPOM) and fine particulate organic matter (FPOM). CPOM has a diameter of a least 1 mm, and will include recognizable plant fragments, including coarse woody debris (CWD), up to the size of entire trees, along with carcasses of macrofauna; it is often relatively heavy and will sink to the bed unless resuspended by turbulence or a strong current. FPOM, with a diameter less than 1 mm, will settle out only in sheltered waters and much may remain in suspension. While plankton (Section 2.2) are living particles of organic matter, they are not normally classed as POM, although the distinction is artificial for some filter-feeding organisms which consume both living and non-living particles. The non-living POM suspended in association with plankton is occasionally referred to as tripton; tripton and plankton together constitute the seston.

Dissolved organic carbon

Dissolved organic matter (DOM) or dissolved organic carbon (DOC) is defined as that which passes through a 0.45 μm mesh, although it is now known that some colloidal organic matter and even very small bacteria and viruses will pass through this mesh size.

Bulleted summary

- The Earth's aquatic habitats are part of a single hydrological cycle. The vast majority of water is contained in the open ocean, but coasts and inland waters, despite their very small relative volume, are structurally diverse.

- Within the open oceans there are two separate water circulation systems: surface water circulation, driven by the wind, and deep water circulation, driven by cool water sinking at high latitudes. The two are connected only in regions of downwelling and upwelling.

- Salinity and proportions of dissolved constituents are very constant in the sea, in contrast to the wide variation encountered in inland waters, and even within a single water body at different times or places.

- Photosynthesis is limited to shallow waters and, additionally, is often nutrient-limited. Chemosynthesis can be locally important but the dominant energy source for many aquatic systems is detritus, mainly from external sources.

1

Further reading

Open University Course Team (1989). *The ocean basins: Their structure and evolution*. Oxford, Pergamon Press.

Open University Course Team (1989). *Seawater: Its composition, properties and behaviour*. Oxford, Pergamon Press.

Open University Course Team (1989). *Ocean circulation*. Oxford, Pergamon Press.

Open University Course Team (1989). *Waves, tides and shallow-water processes*. Oxford, Pergamon Press.

Open University Course Team (1989). *Ocean chemistry and deep-sea sediments*. Oxford, Pergamon Press.

Assessment questions

1. What processes may cause natural variation in freshwater conductivity?

2. What are the principal methods for measuring primary production in aquatic systems?

3. What chemical equilibrium provides sea waters with their buffering characteristics?

4. Briefly describe the combination of physical, chemical and biological conditions that allows a phytoplankton bloom to be initiated in a deep lake or the open ocean.

2

Living in Aquatic Systems

2.1 Water as a medium for life

2.1.1 Sea water versus fresh water

The conditions under which life originally evolved are the subject of much debate, but what is beyond reasonable doubt is that the sea was the cradle for multicellular development. The development and radiation of marine forms was well advanced before the land, or even fresh waters, were colonized at all. All water provides buoyancy and protection from desiccation and extremes of temperature. The fundamental difference between marine and freshwater systems is salinity: sea water, unlike fresh water, provides a ready supply of dissolved salts, essential for aquatic organisms, which need to maintain high ionic concentrations in their tissues and internal fluids. The

high concentration of salts in the sea and, crucially, its constancy, allow marine organisms to have an internal salt concentration equal to external concentrations. Most marine species therefore have internal sodium and chloride concentrations very similar to those of the external medium, although concentrations of other ions such as calcium and magnesium may be lower, requiring active ionic regulation. If they are transferred to fresh water and they lack effective osmoregulatory abilities, salts will diffuse from their bodies and be lost through excretion without being replaced, while excess water may be absorbed by osmosis, leading to cell damage. Therefore, to colonize fresh water, these problems must be overcome by developing an efficient excretory mechanism to remove excess water or by sealing tissues and internal fluids from the water outside, and by active maintenance of internal salt concentrations. A third option—reducing internal salt concentrations—has been adopted by most freshwater organisms. However, although their internal ion

2

concentrations are normally lower than those of marine species, they are always considerably higher than the fresh water in which they live, because these salts are essential in metabolism.

Osmoregulators and osmoconformers

Organisms whose body fluids follow changes in external osmotic conditions are known as osmoconformers, whereas those which maintain constant internal concentrations are called osmoregulators. This process is considered further in Section 4.3.

Osmoregulation requires one or both of two adaptations: an impermeable external membrane and ability actively to remove excess water or excess salt. Freshwater animals need to be efficient osmoregulators, and one consequence is that they produce considerably more urine than their marine relatives. Urine production by marine animals is typically less than 10% of their body weight per day, whereas freshwater shrimps of the genus *Gammarus* produce around 40% per day, while water fleas (*Daphnia*) can produce more than 200% and the freshwater bivalve *Anodonta* more than 400%. Simple excretion of excess water would, however, result in loss of essential ions, so these are selectively removed from the urine before excretion.

A consequence of active osmoregulation is that freshwater species have significantly higher metabolic rates than closely related marine forms. Freshwater shrimps of the genus *Gammarus*, for example, have a rate of oxygen uptake up to 65% higher than marine species, and at least 11% of the total metabolism of the freshwater species *G. pulex* is devoted to active ion transport to maintain internal concentrations. Acidification exacerbates the problem of ion exchange: not only does the low buffering capacity of acid water reduce concentrations of ions such as sodium (Na^+), but high concentrations of H^+ ions compete with Na^+ for active transport sites across gill membranes. Therefore, rather than absorbing Na^+, animals that are not adapted to acid waters absorb H^+ and become ion stressed.

2.1.2 Diversity of aquatic life

Diversity of marine life

Of approximately 40 phyla in the animal kingdom, 38 occur in the sea, and 20 of these are confined to marine environments. Amongst other kingdoms, marine diversity is less extensive, so that, for example, only 14 of the 27 phyla in the kingdom Protoctista (algae and single-celled eukaryotes, commonly known as protists) occur in the sea, although three of these—Phaeophyta (brown seaweeds), Rhodophyta (red seaweeds) and Foraminifera—are amongst the most familiar of non-animal marine organisms.

Major groups of marine animals include Echinodermata (starfish, sea urchins etc.), which are confined to sea water, and Porifera (sponges) and Cnidaria (jellyfish, corals, sea anemones etc.), which are almost exclusively marine, while Mollusca (bivalves, snails, octopus etc.) and Nematoda are at their most diverse in the sea. Arthropoda are represented almost exclusively by crustaceans, including crabs, shrimps and barnacles, and arachnids (sea spiders). Where light penetrates to the sea bed, or in the intertidal zone (Section 5.2), macrophytes (multicellular photosynthesizers) are dominated by algae, generally described as seaweeds: Rhodophyta and Phaeophyta, which are both almost exclusively marine, along with some multicellular Chlorophyta (green algae). Single-celled algae are dominated by the mainly marine Dinoflagellata, Haptophyta (coccolithophores) and Actinopoda, along with the exclusively marine Foraminifera. The only abundant algae which are common in both marine and fresh waters are diatoms (Bacillariophyta).

Diversity of freshwater life

Maintaining the correct ionic concentration is the fundamental stumbling block that restricts many groups of organisms to the sea. Therefore fresh waters support only 16 animal phyla, several of which are mainly marine groups for which the freshwater environment is peripheral. There is however one phylum, the Micrognathozoa, confined to fresh water, although it is currently represented by a single species collected in 1994 from a cold spring in Greenland (Kristensen 2002). Protists are relatively diverse in fresh waters, with 20 phyla recorded. The other two multicellular kingdoms, the fungi and the true plants, are largely terrestrial, but the latter clearly originated in fresh waters and includes the Bryophyta (mosses and liverworts) and Spherophytes (horsetails) which, although not truly aquatic, live in very damp environments. One group of true plants, the Angiosperms (flowering plants), has successfully re-invaded fresh waters to the extent that it dominates aquatic marginal habitats, and even includes several marine representatives in coastal areas.

The freshwater macrofauna is dominated by insects, particularly orders such as Ephemeroptera (mayflies), Plecoptera (stoneflies), Odonata (dragonflies), Trichoptera (caddis flies) and Diptera (true flies), whose adult phase is a terrestrial winged form; excepting Diptera, these orders are almost exclusively freshwater as larvae. Insects with aquatic adult stages include Hemiptera (waterboatmen, pond skaters etc.) and Coleoptera (beetles). Annelids (true worms and leeches), bivalve molluscs and crustaceans may be locally abundant, although their diversity rarely matches that in the sea; among the smaller fauna, however, the crustacean class Branchiopoda, including copepods and cladocerans (water fleas), achieves similar diversity in both the sea and fresh waters. Around 40% of fish species are freshwater, with centres of high regional diversity (see Section 7.5). Macrophytes are dominated by vascular plants, although filamentous Chlorophyta (e.g. *Cladophora*) can be locally abundant, while single-celled algae, apart from diatoms, are dominated by unicellular Chlorophyta and a series of phyla which are mainly or exclusively freshwater, including Chrysophyta, Euglenophyta (e.g. *Euglena*) and Gamophyta (e.g. *Spirogyra*).

2.1.3 Colonization of fresh waters

Direct colonization

Representatives of predominantly marine groups entered fresh waters directly from the marine environment, without an intermediate terrestrial stage. Freshwater bivalve molluscs clearly originated in this way, as their filter-feeding habits require immersion in water. Similarly, hydroids, the only freshwater cnidarians, and the relatively few freshwater sponges, are obligate aquatic species. Of the main groups of Annelida, Polychaeta have a poor system of osmoregulation and are mainly confined to marine environments, whereas Oligochaeta (true worms) and Hirudinea (leeches) possess an effective osmoregulatory system and have successfully colonized fresh waters and, in a few cases, have secondarily colonized marine or brackish environments.

Secondary colonization

Many organisms colonized fresh waters secondarily, via terrestrial ancestors. Insects exploded across the land after development of their waterproof cuticle; the same development allowed colonization of fresh waters as it effectively sealed body fluids from the outside world, so ion exchange ceased to be a major problem. Fungi are almost exclusively terrestrial, with few species in aquatic environments, but they include specialist freshwater forms, the aquatic hyphomycetes (phylum Deuteromycota), which play a crucial role in aquatic decomposition processes (Section 3.4).

Among freshwater snails, prosobranchs (e.g. limpets) have gills and colonized directly from the sea, while pulmonates (e.g. pond snails) have lungs, and include terrestrial land snails, suggesting recolonization from the land. Crustacea probably colonized from the sea; there are shrimps today in estuaries which have a wide salinity tolerance (**Figure 4.14**), but equally some crustaceans, such as woodlice, have colonized damp terrestrial habitats.

2.1.4 Recolonizing the sea

Although there are many examples of organisms colonizing fresh waters from the sea, there are relatively few cases where freshwater organisms have successfully invaded the sea.

Insects

Insects are easily the most diverse group of animals, accounting for the vast majority of all known species, and dominate terrestrial and freshwater environments alike, yet only a few hundred species are known from marine systems, almost all of them being confined to the intertidal or to saline wetlands. The only truly oceanic species are hemipterans of the genus *Halobates*, wingless species which live on the surface of the sea and lay their eggs on floating debris and which can, therefore, persist many hundreds of kilometres from land.

Many theories have been proposed for the general absence of insects from marine environments, including an inability to overcome the physiological constraints of respiration and osmoregulation or the problems associated with physical barriers, such as wave action and surface tension. Clearly, these are inadequate explanations because most of them apply also to fresh waters, while the problem of osmoregulation has been overcome by saltmarsh and rockpool species, as well as by those species occurring in saline lakes. More plausible is the competition theory of Usinger (1957): insects attempting to colonize the sea would have faced competition from established groups, particularly crustaceans. It may also be a consequence of their ability to fly; insects' wings do not work in water and aquatic species require a terrestrial environment in which to emerge. A possible reason, therefore, for the absence of insects beyond the sea coast is the lack of land or exposed surface on which to metamorphose into the adult phase. Van der Hage (1996) has proposed an alternative explanation: insects coevolved with angiosperm plants, themselves almost absent from the sea because their pollination mechanisms, whether passive or through the activity of pollinating animals, do not work under water. Therefore, angiosperms dominated the land and, secondarily, freshwaters, providing resources for insects. Had angiosperms invaded the sea, then insects would have followed. This theory is compelling but, as Van der Hage concedes, does not explain the absence of predatory and detritivorous insects in the marine environment.

Teleost fish

The major exception to the general absence of organisms colonizing the sea from freshwaters is the teleost, or ray finned, fish (Actinopterygii). These originated in late Silurian, around 410 million years ago (Ma), apparently in fresh water. While some early teleosts did enter sea water, none of these lineages presented. The first teleosts with modern descendants colonized the sea during the Jurassic period (208–144 Ma), but only after the appearance of the Acanthomorpha (Acanthopterygii), which comprise the majority of modern marine fish did the modern diversity of marine fish develop. Acanthomorphs first evolved in the Cretaceous period (144–65 Ma), but diversified in the sea from around 55 Ma.

Teleost fish are hypo-osmotic to sea water, meaning that their internal water fluids contain a lower concentration of dissolved components than the surrounding water, so that without an impermeable barrier they would lose water by osmosis. The fish themselves have osmoregulatory organs to ensure that the correct osmotic pressure is maintained, but their eggs, whose osmolarity is similar to that of the parents, do not, and therefore need a large reservoir of water. Without such a reservoir, eggs will not survive, and more primitive teleost fishes, including salmonids, sturgeon and lampreys, are unable to spawn in the sea and must return to fresh waters to do so (see Section 4.3).

Teleosts were able to colonize the sea because they developed a gene that gives instructions for breakdown of a yolk protein, producing a large pool of amino acids which serve to draw water into the egg while still within the ovary, thereby ensuring a large reservoir of water which keeps the embryo alive until it develops its own osmoregulatory mechanisms. Once released, the water is sealed within the egg by chemical reactions in its outer membranes. The egg is therefore able to develop with an adequate reservoir of water but, equally, being less dense than sea water it floats. This secondary consequence of a large reservoir of water ensures that eggs are widely dispersed by surface currents and increases the possibility of creating isolated populations. Therefore, it may be the reason why teleost fish speciated so rapidly within the marine environment (Finn and Kristoffersen 2007).

2.2 Ecological groupings of aquatic organisms

Taxonomically, organisms inhabiting marine and freshwater environments are very different. Ecologically, however, there are clear similarities. Ecological classifications group organisms according to common functions and, within the aquatic environment, there are two basic ways in which this is normally done: classification according to habitat or according to functional feeding mechanism.

2.2.1 Classification by habitat

Aquatic organisms can be categorized into four major groups—pelagic, benthic, neuston, and fringing—according to the part of the water body which they inhabit.

Pelagic organisms

Pelagic organisms are those which live within the water column, and can be further subdivided into plankton and nekton. Plankton, which include algae, bacteria and a variety of animals, are either completely passive, or with powers of locomotion too weak to swim against horizontal currents, and therefore drift in the water column. Traditionally, plankton have been divided into phytoplankton (algae) and zooplankton (animals), terms which reflect their perceived ecological function, rather than any rigid taxonomic divisions. The nekton, in contrast, are active swimmers capable of independent motion; all are animals, particularly fish, but including also whales, turtles, squid, and, as temporary members, aquatic birds such as penguins, gannets and grebes, and mammals such as otters.

Plankton may be further subdivided. Holoplankton are permanent members, represented by many taxa in the sea but dominated in lakes by branchiopod crustaceans, including cladocerans and copepods. Meroplankton are temporary members, spending only part of their life cycle in the plankton. They include larvae of anemones, barnacles, crabs and even fish, which, later in life, will join the nekton. Meroplankton are very much a feature of the sea, particularly coastal waters, as the often sedentary adult forms of coastal species use their planktonic stage for dispersal. They are less diverse in fresh waters, but can be individually very abundant, including certain dipteran (midge) larvae.

Benthic organisms

The benthic community, or benthos, comprise organisms on the bed of the water body. Animals attached to or living on the bottom are referred to as epifauna, while those which burrow into soft sediments or live in spaces between sediment particles are described as infauna. Hyporheos are those species which

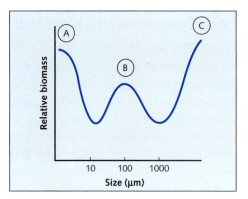

Figure 2.1 Biomass–size spectra for benthic organisms occurring in soft coastal and estuarine sediments, showing the peaks in frequency associated with (A) the microbial component; (B) the meiofauna; and (C) the macrofauna.

2

inhabit sediments where surface waters and groundwaters interact (the hyporheic zone). Attached multicellular plants and algae are referred to as macrophytes, while single-celled or filamentous algae are called periphyton, benthic microphytes, or microphytobenthos. Epiphytic algae are those which grow on macrophytes.

Benthic consumers can be divided by size into macrofauna (>500 μm [0.5 mm] in length), meiofauna (10–500 μm) and microorganisms (<10 μm). This is not simply a classification of convenience, because most benthic species do fall clearly into one of these size categories (**Figure 2.1**).

Neuston

Neuston are those organisms associated with the water surface, where they are supported by surface tension. They include small Acari (mites) and Collembola (springtails), along with some larger insects. Most neuston require a very still water surface and this component of the fauna is, therefore, taxonomically very restricted in rivers and the sea, although by no means absent; it includes, in marine environments, the insect *Halobates* (Section 2.1) and floating cnidarians such as the Portuguese man-of-war (*Physalia physalis*).

Fringing communities

Fringing communities are floral communities that occur where the water is shallow enough for plentiful light to reach the bottom, allowing the growth of attached photosynthesizers, which may be entirely submerged or emergent into the air. Freshwater fringing communities are dominated by higher plants such as water lilies and reeds, though usually with a high biomass of small attached algae, while marine communities are composed almost entirely of algal seaweeds. These communities will merge in space with local plankton and also are doubtfully distinct from benthic communities, as the majority of macrophytes are attached to the substrate. Wetlands (Chapter 8) are composed of this type of vegetation, often supporting significant terrestrial as well as aquatic components.

2.2.2 Classification by functional group—benthos

The differentiation of organisms into guilds or functional groups, based upon what they eat and how they eat it, has been pursued particularly with respect to benthic assemblages. Cummins (1974) proposed a classification which differentiated five groups of benthic invertebrates, based upon their trophic role in freshwaters. Grazer-scrapers, in this classification, feed upon attached algae; shredders eat CPOM, whereas collectors feed upon FPOM, collector-filterers removing them from the water column while collector-gatherers pick them off the river bed. Finally, there are predators which, like predators in every system, consume other living animals. Such a system is also commonly used in marine environments, in which similar categories are recognized, albeit with slightly different names. The guilds originally proposed by Fauchald and Jumars (1979) are widely used: carnivores, filter feeders, herbivores, surface deposit feeders and subsurface deposit feeders.

In practice, this distinction, often considered to be into clear-cut groups, is imprecise.

Periphyton, for example, normally occur as part of a biofilm (the *aufwuchs* of river ecologists), also containing fine particulate detritus and bacteria, held together by bacterial secretions, and most grazer-scrapers will scrape and consume the entire film, whether or not it contains algal cells. Shredders and collector-gatherers are often the same species at different stages in their life cycles and some species, normally considered to be detritivores, will become predators under certain circumstances. These include freshwater shrimps and caddis larvae which, if the opportunity arises, will become predatory and even cannibalistic.

Despite these problems, however, the classification of macroinvertebrates into functional feeding groups is useful, so long as it is understood that they should not be too rigidly applied. It is useful when the taxonomy of the organisms sampled is poorly resolved and is valuable in attempts to use aspects of an organism's biology, its biological traits, as a means of assessing ecological functioning (Section 2.5).

2.2.3 Classification by functional group—pelagic communities

In pelagic communities, attempts to classify on the basis of type of food eaten are less successful because, typically, consumers are opportunistic and will eat anything that is the correct size for their mouthparts to deal with.

The pelagic food web

Traditionally, pelagic food webs were considered to be straightforward: small phytoplankton carried out primary production, and this energy was consumed by progressively larger classes of organisms (Figure 2.2a). Phytoplankton were considered to be dominated by algae, mainly because sampling was done by nets of a mesh size which only caught these relatively large cells (Figure 2.3). It is now recognized that, in addition to this 'net-phytoplankton', there is also a range of very

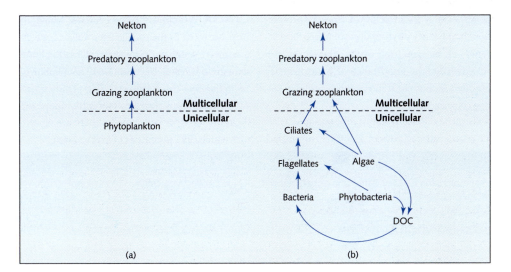

Figure 2.2 (a) The traditional food web in pelagic systems. (b) Incorporation of microbial components makes this food web more complex. All unicellular organisms would formerly have been grouped together as 'phytoplankton'. Curved arrows show the direction of the microbial loop.

2

Large pelagic species, particularly the nekton, are almost exclusively predatory but, among smaller species, there is much overlap in size between the various photosynthetic groups and zooplankters, such that consumers at these levels are both predators and grazers. The traditional distinction between phytoplankton (algae) and zooplankton (animals) is upset by the role of phagocytic (literally: cell-eating) protists, which act as grazers and predators at very small scales, and the distinction between grazers and predators is also untenable. Plankton are therefore classified by size (Table 2.1), although again there is overlap as many species will change to a larger size class as they grow older.

The microbial loop

Many of the bacteria in pelagic systems are photosynthetic, but others specialize in feeding upon DOC, derived from much larger algal cells. Up to 15% of sugars synthesized by algae diffuse into the surrounding water, while much more is added by the messy feeding of zooplankton, which rip algal cells apart, shedding yet more DOC into the water column. Estimates of the loss of DOC vary from 20 to 60% of total primary production.

This DOC is consumed by bacteria, which, in turn, are eaten by microflagellates. Ciliates, which are small enough to eat microflagellates but large enough to be eaten by 'traditional' zooplankton, complete the link (Figure 2.2b). Azam *et al.* (1983) proposed the term 'microbial loop' for

Figure 2.3 Retrieval of a plankton net from a research vessel. The size of the net and the way it is deployed influence the species captured and whenever such samples are analysed these constraints must be borne in mind. Photo by C. Frid.

small phytoplankton, consisting of photosynthetic bacteria and microflagellates, which may be responsible for more of the primary production than the net-phytoplankton. More modern views of pelagic food webs, therefore, still acknowledge the importance of basal primary production, but put it into a more complex setting (Figure 2.2b).

Table 2.1   Types of plankton as classified by size
Femtoplankton (0.02–0.2 μm) and picoplankton (0.2–2 μm), collectively referred to as ultraplankton. Mainly bacteria.
Nanoplankton (2–20 μm). Microflagellates.
Microplankton (20–200 μm). The 'classical' phytoplankton (algae) and grazing zooplankton.
Macroplankton (200–2000 μm [0.2–2 mm]). Predatory zooplankton and some large grazers.
Megaplankton (>2 mm). Includes cnidarians—jellyfish—which are amongst the largest planktonic organisms. Jellyfish were virtually ignored in the 'traditional' food web, because sampling techniques missed them: their gelatinous nature ensured that nets would shred them rather than sample them. The importance of these organisms is still unclear.

this bacterial portion of the food web; it is by this microbial activity that DOC, which would otherwise be lost to the pelagic system, re-enters the food web (Sherr and Sherr 1991).

2.3 Aquatic life cycles

2.3.1 General features of life cycles

Simple and complex life cycles

Reproduction and subsequent development follows either a simple or a complex life cycle. Simple life cycles are shown by organisms such as mammals and birds, in which the newly born offspring is morphologically similar to the adult and develops by gradual modification over time. Complex life cycles involve two or more morphologically very dissimilar stages with an abrupt change—a metamorphosis—between them, and are typical among most invertebrate groups.

Asexual and sexual reproduction

Almost all aquatic organisms have the ability to reproduce sexually, and for most this is the only option available. Amongst protists, the commonest method is conjugation, in which two cells merge and exchange genetic material, before separating once more. Multicellular animals and some algae reproduce by producing haploid cells known as gametes, two of which merge to create a diploid cell which then develops into a new individual. Typically the organisms are divided into two sexes, the male producing small mobile gametes called sperm while the female produces larger, immobile gametes called eggs. Higher plants and most algae alternate between a haploid and a diploid multicellular phase.

Asexual reproduction occurs by two mechanisms. Fission involves the physical separation of part of an individual to create a new individual. Simple cell division is typical of unicel-lular protists, while budding is common in many invertebrate groups, particularly colonial organisms such as corals, tunicates and hydroids. Parthenogenesis is the development from unfertilized eggs; it occurs among several disparate phyla of animals.

Sexual and asexual reproduction each have advantages and disadvantages. Asexual reproduction has the advantage that rates of population growth can be very high, but the disadvantage that development of genetic adaptations to changing conditions is not possible, so probability of extinction is high. A trade-off adopted by some animals is to reproduce sexually when conditions deteriorate, producing a resistant form which remains dormant until conditions improve, following which reproduction is asexual until conditions deteriorate once more. This strategy is common among freshwater taxa associated with temporary water bodies, such as rotifers and cladoceran crustaceans, both of which practise cyclical parthenogenesis, but it is relatively rare in the marine environment, presumably because conditions remain relatively stable.

Among emergent macrophytes, growth from rhizomes is the main form of reproduction in established beds. However, sexually produced seeds can survive, often for many years, as a seed bank in the sediment, ready to germinate under conditions indicative of creation of new habitat, such as rewetting after an extended dry period.

2.3.2 Ensuring fertilization

A major disadvantage of sexual reproduction is that gametes from the different sexes need to come into contact with each other. Some species are hermaphrodite, both male and female gametes being produced within the same organism, and many of these individuals are able to self-fertilize, but as the primary advantage of sexual reproduction is genetic exchange, mixing of gametes between individuals is required occasionally. Most animals are in any case either one sex or the other, so strategies to ensure gametes come into contact with each other are essential.

External fertilization

Typically, aquatic species take advantage of the buoyancy and mobility of the fluid medium to disperse gametes, so direct contact between male and female is not necessary. Gametes are released by both sexes into the water, in which fertilization takes place. The chance of a single pair of gametes coming into contact is, however, remote, so large numbers need to be produced. Furthermore, it is essential that sperm and eggs are released at the same time, and therefore specific environmental cues are required. In coral reefs, the corals themselves and various other sedentary species undergo mass spawning. The main cue for these events appears to be subtle changes in day length and the lunar cycle.

An interesting question is what cues are used by deep sea organisms, beyond the range of light penetration or seasonality of water temperatures. It is probable that they respond to very subtle changes in temperature, or that release of gametes by one individual disperses pheromones which induce other individuals to do the same. It is also possible that benthic detritivores respond to seasonal changes in the 'marine snow', the flux of detritus from the surface waters. On the abyssal plain both the quantity and the quality of the inputs of organic matter settling from above can influence reproductive and other behaviour (Hudson *et al.* 2003).

Internal fertilization

Synchronous release of gametes can be avoided by internal fertilization, for which the male and female must physically come together. This requires either that the two sexes join together before adulthood or that adults are mobile. It is a relatively rare strategy among aquatic organisms.

Internal fertilization requires production of fewer gametes than external fertilization, but still requires both sexes to be present at the same time. Semelparous species only reproduce once and therefore normally have a relatively short adult stage to the life cycle, in contrast to iteroparous species which reproduce more than once during their adult life. Most aquatic insects are semelparous and it is important therefore that not only are gametes produced simultaneously but that the adult stage is synchronized among all individuals within a species. Aquatic insects in temperate zones are typically univoltine, producing one generation per year; emergence of adults is therefore synchronized to a specific time of year to ensure that reproduction occurs, determined by photoperiod. Semivoltine species, in which the life cycle extends over two years, also emerge as adults at a specific, predictable time of year. Multivoltine species, which have several life cycles per year, are more likely to respond to specific water temperature levels as cues to emerge. In the tropics, high temperatures and absence of clear seasonality allow rapid development and continuous reproduction, so cues for synchronous adult emergence other than temperature or photoperiod, which vary little, need to be found. As in tropical seas, therefore, many emergence patterns are based on lunar cycles (Figure 2.4).

2.3.3 Dispersal

Most benthic organisms have a dispersal phase and a relatively sedentary phase, and the different phases are reflected in different stages within a complex life cycle.

Sedentary benthic species generally produce a planktonic larval form which acts as the species' dispersal phase. This stage is planktotrophic, deriving its nourishment by eating other planktonic organisms. However, here there is a trade-off between distance carried and predation risk. The longer a planktonic stage lasts, the more likely it is to be eaten, but the further it can be carried; distant dispersal, in

2

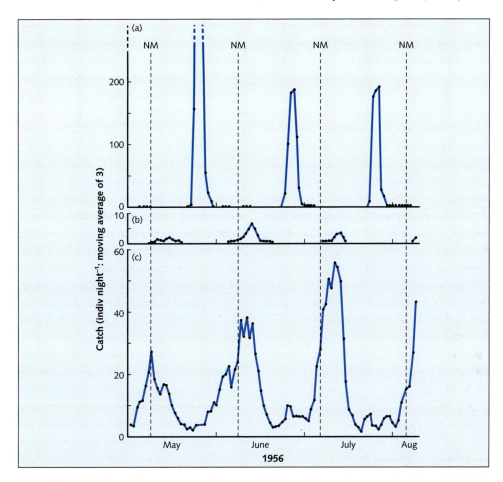

Figure 2.4 Examples of lunar phase-related emergence of tropical aquatic insects, based on night catches in light traps on Lake Victoria, Uganda. (a) *Povilla adusta* (Ephemeroptera); (b) *Clinotanypus claripennis* (Chironomidae); (c) *Tanytarsus balteatus* (Chironomidae). NM, new moon. From Corbet (1964).

turn, is risky because if planktonic larvae are dispersed too far they may no longer be in the vicinity of suitable habitat for the adult. Along marine coasts, tidal currents are normally stronger than residual currents, so a larval stage will be transported a relatively long distance during the first tide, but after this the rate of dispersal will decline, so further advantages to dispersal will only be achieved by having an extended larval phase. Furthermore, reproduction involves a trade-off between number of offspring and the energy invested in each one. Even among animals which do not practice parental care, either many large or few small eggs can be produced. Larger eggs contain more food for the developing embryo, but require a longer period of time before hatching. The optimum length of time for the larval phase is therefore determined by the frequency and predictability of suitable colonization sites for adults. If these are predictably located close to the point of release, then relatively few, large offspring with a short larval phase is advantageous, but if they are unpredictably located, then a large number of small offspring with a long larval phase is more beneficial.

An alternative dispersal strategy adopted by many freshwater taxa is to produce an aquatic larval form and a terrestrial adult form, the adult stage being the time when dispersal occurs and is associated with active oviposition site choice. Most aquatic insects have retained their flying adult stage, and all but Hemiptera (water boatmen and their relatives) and some Coleoptera (beetles) are fully terrestrial as adults, returning to the water only to lay eggs. Amphibians, too, have a fully aquatic larval phase (tadpoles) and then an adult phase that may remain closely associated with the water (some newts, for example) but equally may become fully terrestrial (such as many frogs and toads).

Movement between freshwater and marine habitats is rare, but some fish species spend their main period of feeding in the sea and then migrate to fresh waters to spawn. Conversely, while many crustacean species are permanently freshwater, others spend most of their lives in fresh water but need to return to the sea periodically to reproduce; these include grapsoid crabs such as the Chinese mitten crab (*Eriocheir sinensis*), and atyid shrimps in the Caribbean. One group of fish, the eels, also follows this pattern (see Section 4.3).

2.3.4 Marine versus freshwater strategies

Contrasting conditions between marine and freshwater environments mean that related taxa may adopt very different reproductive strategies in each environment. Whereas most

marine benthos produce widely dispersing planktonic larvae, which are released as eggs or even as gametes into the surrounding water, this is an inappropriate strategy for river benthos, as their offspring would be carried downstream and lost.

Crabs

Crabs practice internal fertilization, the male transferring sperm which the female will often retain for an extended period until she is ready to fertilize her eggs. However, once eggs have been fertilized, the strategies adopted by marine and freshwater species diverge.

Most marine crabs produce many thousands of eggs (**Figure 2.5a**). Following fertilization, these are released into the water, where they hatch to release a larval stage called a zoea. This moults several times, increasing in size and then metamorphoses into a megalops, which is relatively short lived, soon metamorphosing into a juvenile crab (**Figure 2.6**). In some cases the timing of reproduction is important to ensure that larvae are in the appropriate area when they are ready to settle (see Section 4.3), but the basic mechanism is the same.

Freshwater crabs provide an interesting contrast with marine crabs. They produce relatively

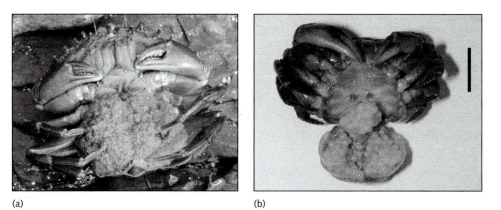

(a)   (b)

Figure 2.5 (a) A marine crab (*Carcinus maenas*, the shore crab), showing the large number of eggs carried; these will be released into the plankton, and their subsequent life cycle is shown in **Figure 2.6**. Reproduced with the kind permission of Steve Trewhella. (b) East African freshwater crab (*Potamonautes* sp.) showing the small number of large eggs carried; these are cared for by their mother throughout their development and for several weeks after the fully formed crabs hatch. Photo M. Dobson.

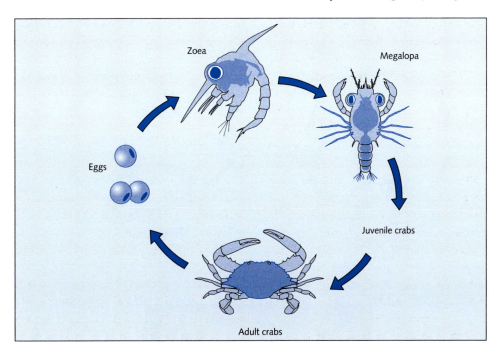

Figure 2.6 Marine crab life cycle. Marine crabs undergo a series of life stages during a planktonic phase, before settling on the bed as juvenile crabs.

few eggs, but each is considerably larger than those of marine species (**Figure 2.5b**). The female continues to carry these eggs in the brood pouch under her abdomen even after fertilization. The larval stage is passed entirely within the egg, which hatches to release a fully formed, albeit small, crab. This too stays in the care of its mother, often for several weeks, thereby avoiding a passive dispersal stage. A typical brood of a crab in the African genus *Potamonautes* will contain 60–120 eggs; information on survival rates is not available, but maternal care will certainly enhance this.

The reason for the difference in life cycles is that the passive planktonic stage would be disastrous for a freshwater species living in a river, as zoeae would be washed downstream and lost from the population. Therefore, maternal care is adopted, the much smaller number of offspring being compensated for by the improved survival rate to adulthood. Other freshwater crustaceans, including crayfish and gammarid shrimps, also brood their young in this way.

Bivalve molluscs

The blue mussel (*Mytilus edulis*), a common species in the North Atlantic, illustrates a typical marine bivalve life cycle (**Figure 2.7a**). This species practices external fertilization, in which eggs and sperm are released simultaneously into the water. Each female will release several million eggs. The free-living larva enters the plankton as a trochophore, before metamorphosing into a veliger, characterized by a ciliated feeding and swimming organelle called a velum. It remains in the plankton for 3 weeks to 6 months, before settling onto a solid substrate and metamorphosing into a juvenile mussel.

Freshwater mussels, in contrast, show three different types of life cycle. The first, typified by the zebra mussel (*Dreissena polymorpha*), is similar to that of marine species. Annual production per adult can be more than 1 million eggs or 10 billion sperm. Following external fertilization the larva passes through the trochophore and veliger stages. Between 18 and

2

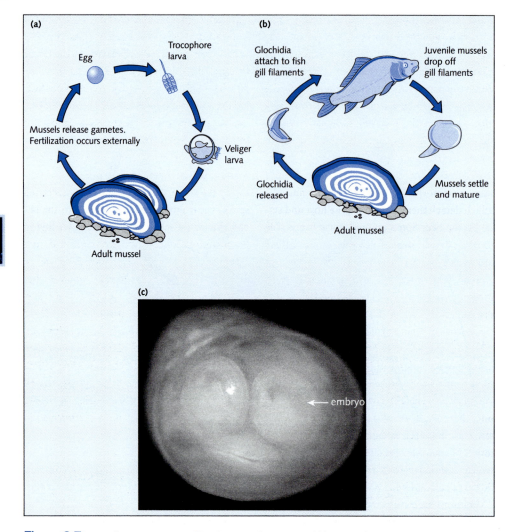

Figure 2.7 Reproductive strategies of bivalve mussels. (a) Typical life cycle of a marine mussel, in which a series of free-living planktonic stages precedes the final static adult state; freshwater zebra mussels also follow this pattern. (b) Life cycle of a freshwater unionid mussel, showing the larval stage during which a vertebrate host is parasitized. (c) Juvenile sphaeriid mussels developing within the gills of the adult (from Guralnick 2004).

90 days after fertilization, the veliger settles on a solid surface and crawls on it until appropriate cues induce it to attach itself. From this it develops into the plantigrade stage, feeding using gills, and finally metamorphoses into a juvenile mussel.

The second strategy, adopted by unionid mussels, involves larval parasitism of a vertebrate host (Figure 2.7b), usually a fish, although at least one species, the eastern North American salamander mussel (*Simpsonaias*

ambigua), parasitizes an amphibian, the mudpuppy (*Necturus maculosus*). Males release sperm into the water but, in contrast to marine bivalves, the female retains her eggs in pouches within her gills. When the female draws sperm through her gills during respiration, the eggs are fertilized and retained until they develop into larvae called glochidia. These are then transferred to the host by one of two mechanisms. Some species have fleshy lures, resembling prey, to attract fish; when a fish attacks the lure, thousands of glochidia are released, some of which will be drawn into the gills of the fish. Other species release large numbers of glochidia together in packets called conglutinates; when a fish attempts to eat the conglutinate, its membrane ruptures, releasing glochidia into the host's mouth, from where they find their way into the gill. Glochidia attach themselves to gill filaments and remain there, feeding on the host's gill tissues, until they metamorphose into free-living mussels, which detach themselves from the host and settle on the bed. Some unionid mussels are able to parasitize a range of fish species, but others are specific to a single species or a group of closely related species. The European pearl mussel (*Margaritifera margaritifera*), for example, requires salmonids such as brown trout (*Salmo trutta*) to complete its life cycle.

The third strategy, employed by pea mussels (Sphaeriidae), is for fertilization to occur in the same way as unionids, but then to brood the young within the gill cavity until they develop into the adult form **(Figure 2.7c)**. The larval stage is therefore completely suppressed. A female can carry a clutch of up to 20 individuals, which collectively can comprise as much as 90% of the adult body weight.

The reproductive strategies adopted by unionids and sphaeriids have both dispensed with a free-living larval stage, and involve some degree of maternal brooding. As with freshwater crabs, these are strategies that reduce the probability of loss of larvae in running water environments.

Intertidal gastropods

Given the similarity in diet and the harsh environmental conditions encountered on rocky shores one might expect the various species to show convergence in their life history regimes, but there is a great diversity of reproductive and life history strategies. Among coexisting species of periwinkles on northern European shores, the following life histories are found. The edible periwinkle *Littorina littorea* releases her

eggs after internal fertilization as a number of gelatinous egg masses (each containing 2–3 eggs). The egg masses are pelagic and the larvae hatch after a few days as planktonic swimmers, which remain in the plankton feeding and drifting for up to six weeks before metamorphosing and settling on a shore. Two other species also retain a planktonic larva—the upper shore *L. arcana* and the small periwinkle *Melarhaphe neritoides*. The latter species lives at the very top of the shore, often in a zone only receiving spray, yet it must release its eggs into the sea and the larvae spend several weeks feeding in the plankton. In contrast, two upper and mid-shore species, *L. saxatilis*, and *L. neglecta*, are ovoviviparous, the female retaining her eggs in an internal brood chamber until they hatch and emerge as miniature adults. Three other species—*L. obtusata*, *L. mariae* and *L. nigrolineata*—have an intermediate strategy of producing a benthic egg mass from which miniature adults emerge. Why closely related species inhabiting similar environments should adopt such different life history strategies in this way is unclear.

2.3.5 Life cycles of aquatic parasites

The mussels of the family Unionidae illustrate an example of a group that adopts a parasitic stage for part of its life cycle. Several other aquatic groups adopt the same strategy, in which the parasitic larval stage is protected and nourished by its host, which also facilitates wide dispersal, of which horsehair worms and digeneans are interesting examples.

Horsehair worms

Horsehair worms or Gordian worms (phylum Nematomorpha) are a mainly freshwater group as adults, but the larval stage parasitizes terrestrial arthropods such as grasshoppers. Therefore, it requires two stages to reach its host. Eggs are laid in still water, from which larvae hatch and sink to the bottom and seek a

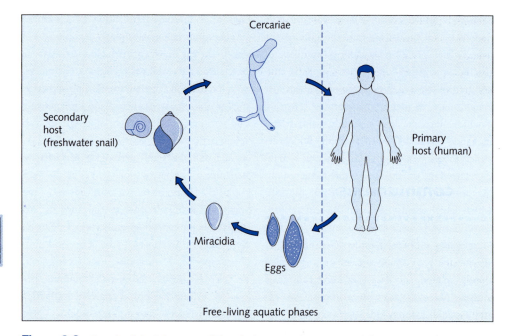

Figure 2.8 Life cycle of the *Schistosoma* fluke which causes schistosomiasis (bilharzia) in humans. This organism occupies hosts at different stages of its life cycle, a freshwater snail as its secondary host and humans as its primary (reproductive phase) host, with short but crucial free-living aquatic stages to allow transfer between hosts.

host. Having entered the body of an aquatic host they remain dormant; if the host is a larval insect, which emerges as a flying adult but then is eaten by a terrestrial invertebrate predator, the horsehair worm larva penetrates the gut of its new host and continues its development. It consumes its host's body and by the time it reaches the adult stage it occupies most of the host's abdomen. It now faces a second problem, that of how to return from a terrestrial host to the water, and Thomas *et al.* (2002) have demonstrated that it can manipulate its host's behaviour to seek out and enter water. The adult phase is short-lived and does not feed. Recently, evidence has come to light that not only can the adult horsehair worm *Paragordius tricuspidatus* leave its host's body, but it can also escape if the host is eaten by a vertebrate predator (Ponton *et al.* 2006).

Digeneans

Digeneans are members of the phylum Platyhelminthes, a group which includes free-living flatworms as well as many parasitic forms, including flukes and tapeworms. Each stage of the digenean life cycle requires a host, and usually the complete life cycle passes through two or three hosts. However, they are classed as aquatic because they include active free-living larval stages.

Schistosomes (*Schistosoma* spp.) have typical life cycle for digeneans, alternating between humans as the primary host and freshwater planorbiid snails as the secondary hosts (**Figure 2.8**); the importance of this group is that they cause the debilitating human disease schistosomiasis (also known as bilharzia). An egg released from an adult within a human hatches in fresh water to release the free-swimming miracidium. This seeks out a snail and enters its body, where it undergoes metamorphosis into the second swimming stage, the cercaria. This stage emerges from the snail and actively seeks a human host, which it enters by penetrating the skin.

Digeneans with three hosts have two intermediate hosts. The cercaria enters the second intermediate host and develops into a resting stage, or metacercaria, enclosed within a cyst. It remains this way until the second intermediate host is eaten by the definitive host following which it reactivates and develops to adulthood.

2.4 The organization of aquatic communities

The fundamental determinant of an organism's distribution is its physiological tolerance of the abiotic environment. Every species has a range of physicochemical conditions under which it can live, and these tolerances ultimately control where a species can be found, although it requires conditions close to its optimum before it can breed successfully (**Figure 2.9a**). The most obvious facet of the aquatic environment is salinity (Section 1.3), to which every species will have its own tolerance limitations (**Figure 2.9b**), although many other factors, such as temperature, pH and light intensity, will also have effects and all will, of course, interact with each other. The actual distribution of a species rarely, however, matches that which would be predicted from laboratory measurements of its physiological requirements, due to the modifying effect of biotic interactions with co-occurring species and the consequences of the spatial and temporal variability of the physical environment.

2.4.1 The role of physical disturbance

Water in almost all aquatic systems is moving. Movement is most obvious when it is relative to a solid substratum, as in rivers and coastal seas, and we can identify current and wave action, respectively, as the main physical processes. These are apparently harsh environments, structured entirely by the extreme physical forces, but this is not necessarily the case. Aquatic organisms in mobile environments are

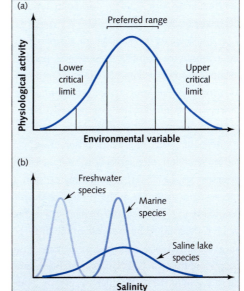

Figure 2.9 (a) A hypothetical tolerance curve of an organism, showing the range within which life is possible, the critical limits within which it can function properly and the preferred range. The centre of the preferred range marks the physiological optimum. (b) Salinity tolerance curves for a freshwater species and a marine species, whose ranges are very narrow (i.e. they are stenohaline) and a species adapted to a fluctuating saline lake (i.e. euryhaline), which has to be tolerant both of very high salinity and a wide variation.

adapted to, or even dependent upon, regular or continuous movement of this sort, because levels of variation in the physical environment which occur within the average life span of an organism will provide an evolutionary drive to which the species will respond. When the environment fluctuates outside this range—examples being strong currents associated with a 50-year flood event, waves associated with a severe storm or unusually high summer temperatures—this will result in removal or mortality of individuals and can be classed as disturbance. It is important to recognize that an event which is a disturbance to one community will not be for another; for example, ice scour on the shore of a warm temperate lake is a major disturbance event, but in polar regions it is a common occurrence to which the local biota are well adapted. The crucial parameter is the frequency with which a particular level of the factor occurs.

At intermediate frequencies of disturbance, the community becomes a series of patches forming a mosaic. Some patches are recently disturbed and support a pioneer community, others are occupied by the competitive dominant, and others are at some point in between. So, depending on our scale of observation, we see in the community as a whole an increase in diversity, as the disturbance frequency allows species from all stages in the successional process to coexist (Connell 1978). The size of the patch is determined by the scale of the disturbance, and patch size will tend to increase with decreasing frequency of disturbance. For example, a storm of the scale which occurs, on average, every year may overturn a relatively small proportion of the stones on the bed of a river or coastal sea, creating patchiness on stone surfaces the size of individual stones, while a storm of the scale encountered every 20 years overturns all the stones, creating disturbance patches as large as the entire stretch of river or shoreline affected.

The most widely studied disturbances are generated by the vagaries of the physical environment (Sousa 1984), but it is also possible to consider various biological factors as disturbance events. Predation not only removes individuals, depressing the population below the carrying capacity of the environment, but effectively acts as a small-scale disturbance, in that resources are freed in a patch where the predator was foraging. Similarly outbreaks of parasitism, disease etc., can be viewed in this patch dynamics way.

2.4.2 The role of biotic interactions

Hairston *et al.* (1960) identified biotic interactions as the primary determinants of gross community structure in ecological systems. Observing that the Earth's terrestrial surface is dominated by plants, they reasoned that these organisms, the primary producers, must

be limited by competition for space, water or light. As herbivores were unable to limit the growth of green plants to any great extent, this suggested that they, in turn, were kept below their theoretical carrying capacity by predation. Therefore, if herbivores are limited by predation, it follows that predators are limited by competition for a limited food resource (Figure 2.10a). This model has been criticized, most notably on the grounds that plants persist not through predator control of grazers but because they possess physical and chemical defences, but the basic premise, that communities are biotically structured, holds.

The model of Hairston *et al.* (1960) is, by their own admission, applicable only to terrestrial systems, where physical disturbance is of only local importance. Aquatic systems are, however, affected by disturbance and physical heterogeneity, both spatial and temporal, to the extent that it cannot be overlooked in describing their community structure. Menge and Sutherland (1987), therefore, developed a model of community organization which takes into account physical as well as biological factors (Figure 2.10b). This acknowledges that, although predation and competition are important components of most aquatic systems, their effects are tempered by the underlying abiotic (and particularly physical) environment and that furthermore the physical environment in aquatic systems continually creates heterogeneity and disturbance, across the entire range of spatial and temporal scales. Biotic interactions are therefore less likely to be the primary determinants of community structure than they are in terrestrial systems.

2.5 Ecological functioning

Ecological functioning (or function) is a concept that refers to keeping systems working through the activities of living organisms

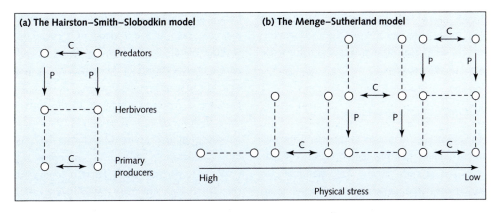

Figure 2.10 Alternative models of community structure. (a) The Hairston–Smith–Slobodkin model, showing competition operating at top and bottom levels, while predation structures the intermediate level. (b) The Menge–Sutherland model, in which the relative importance of competition and predation are modified by physical stress. At the highest stress levels, it acts as a continual disturbance, allowing no species the opportunity to establish but, as physical stress is reduced, the community becomes more complex until, at low stress levels, it is equivalent to that envisaged by Hairston *et al.* (1960). C, competition; P, predation; solid arrow, strong effect; dashed line, weak or no effect.

2

(Bolger 2001). It has been defined variously as referring to ecosystem processes or biogeochemical activities, the sum total of these processes, the flow of energy and materials through the biotic/abiotic components of ecosystems and as a combination of ecosystem processes and ecosystem stability. A useful practical approach comes from Naeem *et al.* (2004) who define ecosystem functioning as referring to 'the activities, processes or properties of ecosystems that are influenced by its biota'.

The main ecological functions delivered by aquatic ecosystems are as follows.

Energy and elemental cycling

Organisms excrete a range of inorganic and organic substances. These vary from the liberation of organic molecules by microbes and phytoplankton/benthos through to the release of undigested food remains by larger organisms. Ultimately however, all organic material is broken down to inorganic forms. A variety of organisms can take these inorganic forms and, using energy, synthesize them into organic matter. In many cases in both the water column and the sediments this process is short-circuited

with simple organic material being used in the synthesis of complex molecules rather than inorganic forms. While microbes are the main chemical engineers in the decomposition process and a range of autotrophs (including microbial photo- and chemoautotrophs) are the main players in the synthesis pathways, macroorganisms play an important role. Many decomposers are only able to work on molecules in solution or on very small particles. The physical ingestion, digestion and egestion by macrofauna is therefore important in liberating dissolved and small particulate organic matter that the microbes can actually utilize.

Modification of physical processes

Biological structures often interact with physical aspects of the environment and so alter the dynamics of physical processes. Amongst the most dramatic examples of this are the offshore coral reefs that provide protection of the inshore lagoons and coasts from the action of oceanic waves. Less dramatic are the effects of dense stands of aquatic macrophytes on waves and currents, producing refugia from high flow and areas of sediment deposition. On smaller scales, beds of tube-building polychaete worms can modify the local flow and sedimentary environment, making it more suitable for settlement of some species.

Habitat/refugia provision

The physical structure of living organisms such as reed beds, floating macrophytes, biogenic reefs and, in open oceans large mammals and sargassum weed, provide a number of ecological roles, including increasing the surface area available for colonization, providing cryptic habitat for juveniles, and reducing the efficiency of predators through provision of refugia.

Productivity

The rate of production of new organic matter varies between environments, and at a range of spatial and temporal scales. In a number of aquatic environments the system is based on allochthonous material and chemoautotrophy. This ecosystem functioning in aquatic systems is markedly different to that in terrestrial systems.

Energy redistribution

Movement of organisms over large spatial scales will redistribute energy from its point of origin in the aquatic system. In most environments, at least some predators range over large distances and between habitat units. In large water bodies there may also be extensive horizontal or vertical movement by primary producers and consumers. Thus the 'food' generated, either from primary production or by exploitation of detritus, is made available for use within the habitat and is also exported.

Community resistance and resilience

A key aspect of biological systems is their resilience, defined as their tendency to return to their 'normal' state after a perturbation. In contrast, resistance is the degree to which a system does not alter its state in the face of a perturba-

tion. Thus biological systems that 'absorb' a pressure, such as the impacts of a pollutant, without changing are showing resistance. If, once a critical threshold is reached, the system switches to a new configuration, but once the pressure is removed it reverts to the former state, then it can be said to be resilient. This is a rather simplistic account but forms the basis for consideration of system variability and response to perturbations. There is some evidence that high diversity systems are more resistant and more resilient than low diversity ones. The ability of systems to absorb human insults and recover is a key concept for environmental management and the ability to deliver sustainability.

Propagule export

In many aquatic systems propagules arrive from some considerable distance away, for example marine fish larvae may drift hundreds of kilometres in the plankton, and stream insect larvae are replenished by adults dispersing via the aerial phase. Thus a key function of any aquatic habitat is to supply larvae for export to other areas.

Adult immigration and emigration

In the same way that many areas are exchanging larvae, so adults also move between areas. These movements can include foraging excursions by predators that range over large areas, but also include regular migratory movements, as seen in many fish and aquatic bird species.

All habitats and the communities in them probably deliver all these functions to some extent. However, the amount of each function delivered per unit area/volume is likely to vary between systems. For example, a structurally complex reef will provide more habitats and refugia than a mobile sand bank, but the latter, depending on location, may be more important for modifying physical processes, such as absorbing wave energy and so reducing the physical stress on the coast.

Bulleted summary

- Sea water is the optimum medium for life, which achieves a high taxonomic diversity in marine environments. The low salinity of fresh waters is the main reason for their restricted diversity, relative to the sea. Most higher taxa now found in fresh waters originally colonized directly from the sea, although some, most notably insects and higher plants, entered via the land. Few taxa have colonized the sea from fresh waters, a notable absence from marine environments being insects, but teleost fishes are a major exception, having originated in fresh waters but now dominating marine habitats.

- Aquatic organisms can be classified into ecological groups, either on the basis of habitat or by functional group. Habitat-based classifications differentiate between pelagic organisms, floating or swimming in the water column, and benthic species, living on or in the bed of the water body. Neuston, occupying the surface film, and fringing communities can also be distinguished. Functional classifications differentiate organisms on the basis of what they eat and how they eat it. Most consumers are, however, opportunistic and consume particles according to size, without distinguishing between living and non-living components. These classifications work equally well in marine and freshwater systems.

- Most aquatic organisms have complex life cycles, and have to overcome problems associated with synchronous release of gametes or synchronous emergence of adults. Dispersal mechanisms are determined by the habitat occupied. Parasitism of other animals by one or more life cycle stages has developed in several groups of aquatic animals.

- Aquatic food webs can be complex at the microbial scale. The microbial loop is important in transferring dissolved organic carbon (DOC) into a particulate form that is then available for larger consumers.

- The distribution and relative abundance of aquatic organisms is determined, to a large extent, by disturbance which, at a low frequency, can enhance species diversity by inducing patchiness. Whereas terrestrial communities are structured primarily by biotic processes, the fundamental determinant of aquatic community structure is frequency and intensity of physical disturbance.

- Aquatic organisms contribute to ecological functioning, the mechanisms that keep ecological systems working. They are responsible for food supply, nutrient recycling, structural habitat and modification of physical processes, among others.

Further reading

Barnes R.S.K. and Hughes R.N. (1990) *An introduction to marine ecology*, 2nd Edn. Oxford, Blackwell Scientific Publications.

Barnes R.S.K., Calow P., Olive P.J.W., Golding D.W. and Spicer J. (2000) *The invertebrates—a synthesis*, 3rd Edn. Oxford, Blackwell Publishing.

Cheng L. (ed.) (1976) *Marine insects*. Amsterdam, North-Holland Publishing Company.

Frid C. and Dobson M. (2002) *Ecology of aquatic management*. Harlow, Pearson Education.

Margulis L. and Schwarz K.V. (1988) *Five kingdoms. An illustrated guide to the phyla of life on Earth*, 2nd Edn. New York, W.H. Freeman and Co.

Williams D.D. and Feltmate B.W. (1992) *Aquatic insects*. Wallingford, CAB International.

Assessment questions

1   What difficulties must marine organisms overcome in order to colonize fresh water?

2   How does a complex life cycle facilitate dispersal?

3   Why do the life cycles of freshwater crabs differ from those of their marine relatives?

4   How does disturbance influence predatory and competitive interactions?

3

Rivers

■■■■■■■■■■■■■■■■■■■■■■■■■■■■■■■■■

3.1 Introduction

The tiny fraction of the world's water that is contained in rivers (Table 1.1) belies their enormous significance as geomorphological processes on the Earth's terrestrial surface. At any one time, there are approximately 2000 km^3 of water contained in river channels worldwide, but the volume of water discharged annually into the sea by rivers is around 41,000 km^3, demonstrating that the average river, even discounting evaporation, replaces its water about 20 times per year. A river is a channel of flowing water, whose movement is determined by gravity and is therefore downhill. Some rivers may cease to flow, and may even dry up completely, if inputs of water are low; indeed, rivers in regions with clearly defined wet and dry seasons may be seasonal (intermittent) and in arid zones they can be unpredictably ephemeral,

only flowing during or immediately after rain. The majority of rivers are, however, continuously flowing (perennial), though discharge and rate of flow will be variable. Their linearity and unidirectional flow are of fundamental importance in determining their structural and biological features, as are the volume of water present and its quality.

The nature of a river is determined by its catchment. Geology and topography, in association with the hydrological activity of the river itself, determine the physical form of the channel and the nature of the substratum, while water chemistry is determined principally by the terrestrial environment from which its water originates, although the quality of water added through precipitation can be locally important.

3.2 The abiotic environment

3.2.1 Flow

The defining feature of a river is the continual downstream movement of its water. However, this movement is by no means uniform, and the complexities and fluctuations in flow define the physical environment of a river. Flow patterns are determined by the volume of water and its velocity (which collectively define its discharge) and also by channel morphology. The velocity of water movement, or current, varies across the river channel. At the interface of the solid substratum and the moving water, friction creates a viscous sublayer, the boundary layer. A vertical profile shows velocity to be lowest adjacent to the bed, in the boundary layer, and to increase with increasing depth. A horizontal profile, too, shows that velocity varies across the channel, in response to depth, channel topography and the presence of obstacles. Some areas have rapid flow, including the thalweg, a narrow channel within the main body of water in which flow is greatest. To the side of the thalweg are areas of lower flow, including those where the water may be stationary, or even flowing upstream.

A common feature of most rivers is that flow varies in response to short- and long-term weather and climatic patterns. This is best illustrated by small headwater streams, in which recent rainfall events can be picked up as peaks in flow, with a rapid rise in discharge followed by a slow decline (Figure 3.1a). Between these events, the river will return to its baseflow, the normal low flow encountered when there are no direct inputs from the catchment. Baseflow is maintained by constant export from water storage areas, such as upland wetlands or, more normally, groundwater. Where a river runs above the water table, it will lose water to groundwater, and so will dry up at these times. Intermittent rivers flow during seasons when

their channels are below the water table; ephemeral rivers are always above the water table and fed entirely by surface inputs, hence their short periods of flow.

Whereas peaks in discharge are normally frequent and short-lived in headwater streams, a different pattern is shown by large rivers with many tributaries. These tend to show seasonal changes in discharge, as the peaks from individual tributaries merge (Figure 3.1b), and a predictable flood cycle caused by such a discharge pattern will support associated wetlands (Section 3.6 and Chapter 8).

3.2.2 River channel form

Rivers normally run in clearly defined channels. The size of the channel is determined by discharge, but as this can fluctuate widely in response to precipitation or inputs from upstream, rivers generally flow at somewhat less than full channel capacity (Figure 3.2a). Channel size and structure is determined by occasional high-flow events rather than baseflow, so natural river channels have a capacity many times that normally encountered. Occasional high discharge events will however cause the river to overflow, flooding the adjacent land (Figure 3.2b). In its lower reaches, a river running through a floodplain (see below) may have a poorly defined channel, constantly changing its form, and may become 'braided': split into a series of interconnected channels.

Natural river channels are never homogeneous in form, but contain a variety of structural features, producing a range of current velocities even within a short stretch. These may be conveniently divided into whole-channel features and partial-channel features.

Whole channel features

A low order stream (Box 3.1) displays an approximately regular alternation between riffles and pools (Figure 3.3). Riffles are relatively shallow, high gradient stretches, where high-velocity water passes over a substratum of

3

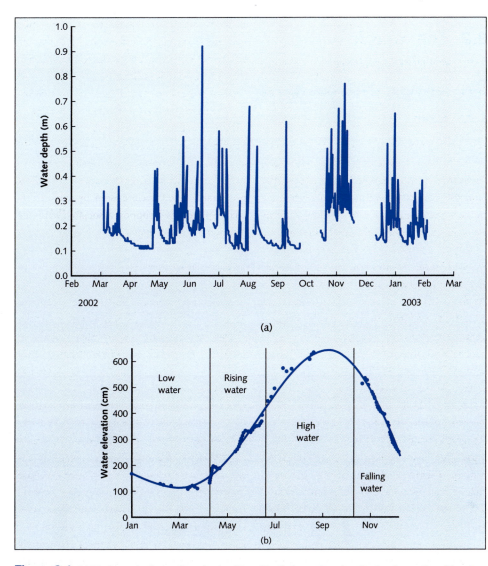

Figure 3.1 (a) Discharge hydrograph and annual trend in discharge for a headwater stream (Arnold and Dobson, unpublished data). (b) Annual pattern of discharge for a large river, the Cinaruco River, Venezuela. A clear cycle of low, rising, high and falling water can be identified, determined by seasonal fluctuations in rainfall in its headwaters (from Arrington and Winemuller 2006).

large cobbles or boulders, many of which break the surface and cause turbulence. Pools are relatively deep, low gradient stretches, with slow-flowing water and relatively fine substratum. Between the two may occur glides (or runs), where water flows rapidly but smoothly, as the substratum does not break the surface.

The pool-riffle sequence occurs where bed substratum is coarse and gradient is shallow. Where the channel is steep the stream will tend to form a pool-step sequence, where water falls over accumulation of large boulders or runs over bedrock into deeply scoured pools. In a small stream, a relatively large piece of organic matter, such as a fallen tree, can also create a step (Figure 3.8).

At low flow, riffles and steps are areas of high water velocity, whereas pools are calm and slow-flowing. As discharge increases, however, flow increases in pools more rapidly than in

(a)

(b)

Figure 3.2 (a) During periods of low discharge, a large river will flow considerably below its channel capacity, and in extreme cases, bed sediments may be exposed in certain parts of the channel. Note that the banks of this river have been artificially increased in height to control flooding (River Loire, northern France). (b) Very high discharge may exceed the capacity of the river channel, causing flooding in the adjacent area. The river here is now constrained by a dam upstream, but heavy rain required release of more water than normal, inundating the formerly regularly flooded floodplain (Guadeloupe River, Texas). Photos M. Dobson.

Box 3.1 Stream order

Stream ordering is a method of assigning a number to each river stretch, this number being an indication of its relative size within a drainage basin. Two of the commonest methods for applying ordering are considered here.

The system most widely used by river ecologists is the Strahler Method. All headwaters without tributaries are assigned to 1st order. When two 1st order streams converge, they form a 2nd order stream, two 2nd order streams form a 3rd order stream, and so on.

A lower order tributary does not change the order of the main stream so that, for example, a 4th order stream can be joined by 1st, 2nd and 3rd order tributaries without changing its order; only merging with a 4th order stream will create a 5th order channel.

In this method, the smaller, headwater streams rapidly change their order as they merge with each other, but the rate of change slows down as channels become larger. Large rivers such as the Garonne or the Rhône in France are only 9th order at their mouths, and even the Amazon barely gets

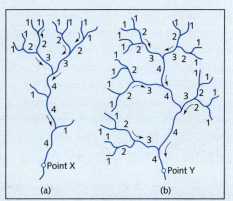

(a)   (b)

into double figures. The disadvantage of this system, therefore, is that stream order can bear no relation to discharge as smaller tributaries, however many of them, will not affect stream ordering. Note that the two outlet streams in the illustration above are both 4th order but, assuming the same initial discharge for all the 1st order streams, point Y in system (b) will have twice the discharge of point X in system (a).

An alternative, the Shreve Method, overcomes this problem by being additive: the order numbers are summed whenever two branches meet.

The problem with this system is that the numbers can quickly become very large.

In principle, whenever stream order is quoted, the system by which it has been calculated should be given. In practice, this is not normally a problem, because the vast majority of research into river ecology is on smaller streams, whose order

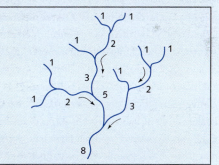

rarely exceeds three, and is often one or two, in both of these classifications. Indeed, many publications refer simply to 'low-order' streams, which can be taken to mean three or less.

Stream order is quoted to give the reader an opportunity to visualize the size of the river being studied. This works because there is consistency within a geographical area: a reference to a 2nd order stream in eastern North America will be meaningful to anybody who has experience of streams in that part of the world. Unfortunately, it might not be meaningful to those unfamiliar with eastern North America and, worse, those with experience of rivers in very different places, such as the humid tropics or the Australian outback, may get the wrong impression of the type of stream being discussed. Stream ordering, despite its value, suffers the same limitations as ecological classifications, and should be interpreted accordingly.

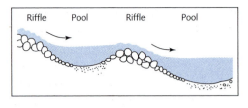

Figure 3.3 The riffle–pool sequence of a river channel, showing variations in substratum composition and surface turbulence that it generates.

riffles until during flood conditions it can be similar in both features. These high flow events maintain pools by scouring them during these periods, whereas the coarse sediment in riffles is much less susceptible to movement. Even at low flow, areas of scour occur on the outside of meander bends and immediately downstream of waterfalls; in both cases a pool is eroded and maintained by continual water movement.

Partial channel features

Within the main riffle pool sequence, a range of other features may be present. Some, such as gravel bars, undercut banks and slow-flowing sections on the inside of a meander, are associated with the riffle pool sequence or with the sinuosity of the river channel. Others are more unpredictable in their occurrence, including features such as clumps of aquatic vegetation, trailing bankside vegetation, fallen trees, roots and large boulders. Gravel bars on the edge of a relatively straight section of channel (side bars), on the inside of a bend (point bars) or in the middle of the channel (mid-channel bars) are all indicators of a dynamic river channel that is continually eroding and re-depositing sediment. At a smaller scale, minor variations in conditions can be found in different areas of a single riffle, upstream or downstream of channel features such as clumps of vegetation, or even on different sides of stones.

3.2.3 River zonation

In addition to the localized variations in channel form, longitudinal trends can be identified along a river system, such that it can typically be divided into three stylized zones: the headwater, or 'erosion' zone, the middle-order or 'sediment transfer' zone and the lowland or 'sediment deposition' zone.

These three zones can be defined in terms of supply of water and of sediments. The erosion zone is the major source of water and of sediment. As channel slope is relatively steep, deposition of sediment, if it occurs, is localized or ephemeral, the channel being essentially eroding in nature. Therefore substratum particle size is large (cobbles and boulders) and, occasionally, the river may have eroded to the bedrock. Discharge is highly variable and responsive to local precipitation. The steep channel slope and coarse substratum may produce turbulent flow, in which riffles, rapids and even waterfalls will be present. In the sediment transfer zone, river gradient is reduced so that water and sediment are transported with little net loss or gain. Any deposition of sediment is balanced by erosion elsewhere as the river cuts a new channel. The sediment deposition zone is low gradient, with stable and often predictable discharge. Here the river deposits its sediment load, typically as it approaches the sea and develops a delta or an estuary. Even in areas of high rainfall, the immediate catchments in the middle-order and lowland zones will not receive enough precipitation to maintain the river, which therefore derives most of its water via discharge from upstream.

This idealized river structure can be modified by local geology. For example, the river may be divided into distinct floodplain reaches by bands of hard or less eroded bedrock. Even large rivers can pass through stages where they are constricted by topography. The River Danube has a zone of deposition downstream of Vienna (Austria), beyond which it passes through a relatively constricted gorge in what is effectively a sediment-carrying zone once more; the process is repeated further downstream in Romania, as it passes through the Carpathian Mountains. The River Nile passes over cataracts in southern Egypt, well below the stretch in which such rapids would occur in an idealized river, while the River Niger in Mali has a very large delta (Section 8.5), below which it becomes a relatively constrained river channel once more.

River classification

The way that a river changes along its length has inspired many attempts to classify the various stretches in a way that allows direct comparison between river systems. River classifications based upon purely biological parameters, such as the presence of particular species, are of only limited value, mainly because they are restricted to geographical regions in which those species occur. European rivers, for example, have been divided, moving in a downstream direction, into trout zone, grayling zone, barbel zone and bream zone, determined in each case by the presence of typical fish species; this system, devised for central Europe, fails even in Britain, where the grayling is a rare species. Nowadays, therefore, river ecologists, particularly those working in the upper reaches of river systems, use the stream ordering system, borrowed from hydrologists (Box 3.1).

3.2.4 Groundwater

River water percolates into the sediments, creating a store of fluvial groundwater which flows, albeit much more slowly than the river itself, in a downstream direction. This groundwater can be important in regulating the flow of a river. Groundwater inputs are low in upstream sections, where there is little alluvium, although groundwater may be the source if the river is spring fed. Further downstream, groundwater will enter the river if the water table is high, and will be responsible for ensuring continual flow during periods of low precipitation.

3.2.5 Floodplains

A river in a sediment transfer or deposition zone generally has a floodplain, a flat valley bottom constructed of loose or unconsolidated alluvial sediment deposited when the river

floods or shifts its channel. Geomorphologically, a floodplain is an area flooded by the river with a predictable—though by no means regular—frequency; a five-year floodplain, for example, is that part of the valley which will be inundated by a flood of a magnitude occurring, on average, once every five years. Ecologically, however, the floodplain is better restricted to the zone close to the river which receives regular, seasonal flooding. The difference in flooding between low-order streams and larger rivers (Figure 3.1) is of fundamental ecological importance, as organisms in low-order streams are adapted to surviving the rigours of flood events, whereas those in higher-order rivers may be adapted to benefit from the predictable inundation of adjacent floodplains.

3.2.6 Changes in water quality

Rainwater contains relatively few impurities (Table 1.4) but, as it flows through or over the ground, it picks up solutes and suspended matter. Once these enter the river channel, they normally remain there, becoming concentrated by further inputs and by evaporation. Therefore, proceeding from its source towards its mouth, the concentration of solutes and suspended particles in a river gradually increases, although rapid changes over a short distance can occur when a river channel passes through areas of contrasting catchment geology, vegetation or land use.

Major fluctuations in the concentration of impurities result from high discharge, which resuspends sediment from the bed and if it causes flooding will input matter from the floodplain. During heavy rain, excess water from the catchment passes rapidly into the river channel, bringing with it a pulse of sediment, detritus and solutes. As an example, much of a heavy fall of acid rain may pass straight into water courses via overland flow and therefore not interact with the soil, so that even if the buffering capacity of the catchment soil is high, a pulse of acid water reaches the river without being neutralized, and can lower its pH considerably for a short period of time.

3.3 Ecological effects of the abiotic environment

3.3.1 Pelagic and benthic assemblages

The movement of water in a river is the primary physical process acting upon living organisms, but its effects will be determined by the size and structure of the organism, its activity and its position within the water column (Box 3.2). The current in particular has the effect of carrying downstream any organisms living within the water column. For this reason, pelagic communities in running waters are normally poorly developed, or even absent, and fish are amongst the few permanent residents of the water column in rapidly flowing waters. The benthos is, therefore, the dominant ecological group in most rivers.

Adaptation to microhabitats

Most benthic species are able to swim very weakly, if at all, and endeavour to remain on or in the substratum. Epibenthos have adopted a range of morphological strategies to withstand dislodgement by the current, including hooks, claws and suckers, while behavioural strategies include orientation of the body parallel to the direction of flow to reduce the surface area impacted by the current. The boundary layer is normally so thin that only the very smallest organisms (microorganisms and the smaller meiofauna) are unaffected by current.

Flow is not simply a stressor, with a negative impact on aquatic organisms. Constant movement of water replenishes food resources and oxygen, and carries away waste products. Organisms that attach themselves or cling to the river bed in rapidly flowing areas are safe from many predators, whose mobility makes them susceptible to dislodgement. Organisms or their propagules can use flow as a mechanism for dispersal, although inevitably this is mainly one-way.

Each species has specific habitat requirements and will inhabit only those patches of river bed to which it is best suited. For example, species which need high concentrations of oxygen, or feed directly from the water column, live in exposed, fast-flowing riffles. Other groups are more abundant in pools, backwaters and other patches of reduced flow, either because they require the relatively fine sediments or organic matter which aggregate in such environments or because they are intolerant of flow, including, for example, active predators requiring pond-like habitats in which to swim.

Patchiness and species diversity

The diversity of the biota of rivers is enhanced by the structural diversity of the channel, which leads, in turn, to a diversity of flow regimes even within a short stretch of river. Obviously, the greater the number of microhabitats represented, the greater the overall diversity of the river. The most diverse of all are those beds containing a mosaic of habitats, comprising a series of discrete entities, or 'patches'. These include the major channel types and whole channel features, but also partial channel features, which act as specialist microhabitats (Section 3.2). Any single patch may support very few species, but a variety of patches within a single stretch of river will ensure that overall species diversity is high.

As a river becomes larger, the bed becomes more homogeneous and much of the structural diversity is lost, but pelagic species assemblages may become more varied and marginal wetland habitats add diversity. Artificial channelization, in which a river is straightened and its channel made uniformly wide and deep, drastically reduces the range of microhabitats present, both by removing channel features and by cutting the river off from its marginal habitats, with a consequent reduction in species diversity.

Box 3.2   The physics of flow

The flow of water within a channel may be smooth (laminar) or turbulent. In laminar flow, water moves as a series of parallel layers, travelling at different speeds but with no interference between layers. However, uneven substratum or high velocity can upset this smooth movement, causing interference among the different layers and creating turbulent flow. Laminar flow is in theory ideal for benthic organisms, because it is predictable and unlikely to cause structural damage, but turbulent flow is important in maintaining passive floating particles, such as diatoms and FPOM, in suspension, and also in mixing nutrients, heat and oxygen throughout the water column.

The interaction between viscosity and inertia is an important determinant of habitat for river organisms. Viscosity is the resistance to deformation, while inertia is the resistance to changes in velocity. If viscous forces are high, then laminar flow is promoted, whereas high inertial forces promote turbulence. The ratio of inertial to viscous forces produces the Reynolds number (Re).

$$\mathrm{Re} = \frac{VL}{v}$$

where: V = velocity (m/s)
L = a characteristic length, such as that of an organism under study (m)
v = kinematic viscosity (m^2/s), determined by dividing dynamic viscosity by fluid density. This is a constant for a given water temperature: at 10°C $v = 1.308 \times 10^{-6}$ and at 20°C $v = 1.007 \times 10^{-6}$.

Re is a dimensionless measurement. Values <500 signify laminar flow, whereas those >3000 signify turbulent conditions, with the intermediate values signifying transitional conditions. For very small organisms, such as bacteria, whose L is very low, Re is tiny (typically 10^{-4}–10^{-5}). This protects the organism from turbulence, as it is coated in a highly viscous fluid. However, this slows down movement of resources to the organism and movement of waste products from it. For invertebrates, Re ranges from around 1 for small individuals to 1000 or more for larger individuals. For large fish, Re can be in the millions, its value increasing further among actively swimming animals.

Re can also be measured for the river channel itself, by using channel depth as L, and the mean flow from a vertical profile as V. Laminar flow only occurs if flow is well below 0.1 m/s, and therefore most river flow is turbulent.

Another commonly used indicator of flow conditions is the Froude number (Fr).

$$\mathrm{Fr} = \frac{V}{\sqrt{gd}}$$

where: V = velocity (m/s)
d = water depth (m)
g = acceleration due to gravity (m/s^2).

Like Re, Fr is a dimensionless parameter. Fr < 1 indicates slow or smooth flow, while Fr > 1 shows that flow is fast and turbulent.

Groundwater connections

Lateral and vertical connections exist between the river and its associated groundwater in allu-vial sediments. Many benthic invertebrates will enter the sediment on the river bed, even if for brief periods, but some species, known as amphibites, have a life cycle which involves

3

both surface water and groundwater systems. The most celebrated amphibites are various species of stoneflies associated with the Flathead River, Montana: the nymphs spend one or two years in the sediments before migrating to the main river to emerge as flying adults; these then lay their eggs in the river and the newly hatched nymphs migrate back into the sediments. Movements in this system are generally lateral, and stonefly nymphs have been extracted from the broad alluvial terrace of the Flathead River up to 3 km from the river channel itself. The Flathead River is perhaps exceptional in the extent of these movements, but such lateral connections probably occur wherever conditions are suitable, the degree to which organisms penetrate from the river being related to the strength of hydrological connection between the river and its alluvial sediments. A high turnover of water in the interstitial spaces enhances availability of food and oxygen and encourages lateral movements of river species, but will also lead to passive dispersal of organisms within the sediments, such as diatoms in the Flathead River floodplain which have been recorded in complete darkness up to 350 m from the river. A lower flushing rate by

river water will reduce numbers of amphibites and of species which become part of the hyporheos only occasionally or accidentally. Groundwaters thus hydrologically separated from surface waters will, however, offer stable environmental conditions and will be typified by an increased importance of the permanent inhabitants of groundwater—the stygobites.

3.3.2 Longitudinal distribution patterns

Zonation

River biota respond to changes in the physico-chemical environment that occurs along a river's course, the precise physicochemical requirements of the majority of benthic species resulting in a downstream zonation. This is well illustrated by net-spinning caddis flies of the family Hydropsychidae (**Figure 3.4**), which are amongst the dominant benthic invertebrates of many European rivers. Hydropsychids in the Welsh River Usk show a clear zonation of species from headwaters to lower reaches (**Figure 3.4a**): *Hydropsyche siltalai* frequents rapidly flowing, turbulent conditions, whereas

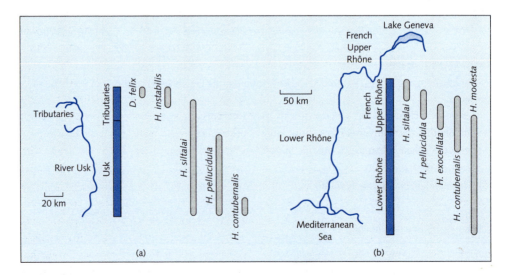

(a)   (b)

Figure 3.4 Longitudinal distribution patterns of hydropsychid caddis larvae in (a) the Welsh River Usk (after Hildrew and Edington 1979), and (b) the French River Rhône (from Bravard *et al.* 1992). The species present change in response to different physicochemical conditions along the rivers' lengths. D, *Diplectrona*; H, *Hydropsyche*.

H. pellucidula becomes dominant as flow is reduced, so these two species can coexist along much of the river by virtue of the complex nature of its physical environment, each showing a microdistribution related to the patchiness of its preferred current regime. The other species are not, however, responding simply to current: *H. contubernalis*, for example, is tolerant of low oxygen saturation and moderate organic enrichment (which rarely occur separately in rivers); *Diplectrona felix* and *H. instabilis*, on the other hand, each require relatively cool water and have precise, but slightly different, temperature tolerances. On a larger scale, a zonation of five species of *Hydropsyche* occurs in the French River Rhône, downstream of Lake Geneva. Here, *H. siltalai* and *H. pellucidula* are confined to the upper reaches, several species overlap in the middle reaches of the river, but only one, *H. modesta*, occurs in the Lower Rhône **(Figure 3.4b)**.

3.3.3 Effect of water chemistry

Changes in water chemistry can have a profound effect on river species assemblages. Most freshwater organisms require at least some dissolved impurities, without which they are unable to maintain the correct internal ionic balance for survival (Section 2.1). Generally, however, as the concentration of solutes in fresh water increases and in human terms the water becomes dirtier, the diversity of species present will decline.

This is well illustrated by organic inputs to water. Benthic macroinvertebrates show large differences in their tolerance to organic matter loading, such that the species assemblage in a given river can be used to predict the water quality, and vice versa, with reasonable accuracy. Indeed, there are many taxa which are known to be good biological indicators of the quality of water and whose absence may be used as evidence for organic pollution. Where organic loading is very high, macroinvertebrates are limited to the few most tolerant species, particularly oligochaete worms and

certain chironomid larvae, although these may be present in very large numbers. A slightly lower organic loading allows pollution-tolerant molluscs and leeches to persist, along with baetid mayflies and crustaceans such as hoglouse (*Asellus* sp.). As water quality improves, so species diversity increases, but only the cleanest water will support stoneflies and the majority of mayfly and caddisfly taxa.

Abrupt changes in water quality can cause marked discontinuities in species distributions, for which the River Goyt, a major tributary of the River Mersey in northern England, provides a well documented example. A single point source of organically rich effluent changed the water quality of the river substantially, with an immediate loss of stoneflies, mayflies and other pollution-sensitive species. The more hardy of these groups returned to the river in smaller numbers downstream of a tributary, the River Sett, which diluted the effluent in the River Goyt.

Biological patterns such as these are widely exploited in the development of biotic indices of water quality, in which the differential sensitivity of organisms is taken into account. Each species present is given a score, reflecting its sensitivity to organic pollution, so that the more sensitive taxa such as stoneflies are allocated high scores, whilst very tolerant species have low scores. Summing the scores from a site produces an index number: the higher this number, the cleaner the water. Most biotic index systems are based upon macroinvertebrate species, for reasons of practicality, but other taxa, including bacteria, can be used.

Similar trends have been observed with increasing acidification, as sensitivity to low pH is a common feature among freshwater organisms. Thus, spatial studies of a range of sites will generally show species richness declining as pH falls and, within a single site, acidification of water can drastically reduce the number of species present. Ironically, stoneflies, which are among the species most sensitive to organic loading, are in many cases tolerant of very acid waters and may be amongst the few

3

macroinvertebrates present in highly acidified streams. River pH can change rapidly, particularly if the catchment is susceptible to acid pulses (see Section 3.2), and low rainfall during summer may result in a seasonal reduction in acidity, so a direct reading of water pH may not pick up short-term detrimental falls in pH. For this reason, differential sensitivity of invertebrate taxa to low pH has been used to develop an acidification index, whereby the presence or absence of indicator taxa gives an indication of susceptibility to acidification (Davy-Bowker *et al.* 2005).

3.3.4 Drift

The water column of any river will contain large numbers of 'drift': individuals of normally benthic species suspended in the water column and being carried downstream by the current. Drifting invertebrates may enter the water column after accidental dislodgement, but also as a deliberate behavioural tactic, to find new food resources or to avoid inclement conditions, including predation. Actively entering the water column in this way can be a dangerous strategy, because once part of the drift, most benthic invertebrates have little or no control over where the current will carry and deposit them, and they are at the mercy of predators such as salmonid fish and net-spinning caddis flies. It is possibly for this reason that drift rates are generally higher at night than during daylight, as darkness reduces the chance of being caught by a visual predator. The nocturnal increase in drift, which is triggered by light intensity, is probably a mixture of active dispersal and increased levels of accidental dislodgement, as individuals that spend the day under stones emerge at night to forage on upper surfaces and thus become susceptible to dislodgement.

Catastrophic drift

The number of individuals drifting at any one time, even during the night, is normally only a small proportion of the total population of benthic invertebrates. Accidental drift on a large scale may, however, be triggered by floods or other adverse events, including inputs of a pollutant. This 'catastrophic drift' can occur at such high densities that it significantly lowers benthic densities; if in response to pollution, it may be either an active attempt to avoid the undesirable conditions or a passive response, drifting organisms having been disabled by the pollutant and therefore no longer able to resist the current.

Hall *et al.* (1980) provide a very clear example of the effect of a pollution episode upon both drift rates and benthic densities. Following artificial acidification of a small stream in the Hubbard Brook Experimental Forest, New Hampshire, there was a major increase in drift densities over a four-day period (**Figure 3.5**). The organisms drifting could be divided into three groups: (a) acid-tolerant species, whose drift rates remained constant despite acidification; (b) species which, prior to acidification, had not been recorded in the drift, and which, therefore, drift only in extreme circumstances; (c) species whose drift rates increased enormously in response to acidification. After the first pulse, drift rates settled down (**Figure 3.5**), despite continual addition of acid for a further five months, but remained high relative to the control stretch because benthic densities in the acidified zone had been reduced to around one quarter of those in the control.

3.3.5 Flooding and flood refugia

Organisms are adapted to the typical flow regime of their environment, but may have difficulty if the flow changes rapidly. Periods of relatively high or low flow may be predictable and regular if, for example, they relate to snow melt, but most small rivers suffer irregular spates, which can have major effects upon organisms living on the river bed, or even upon the river channel structure itself. Giller *et al.* (1991) provide an example of a flood at a scale expected once every hundred years, appreciably

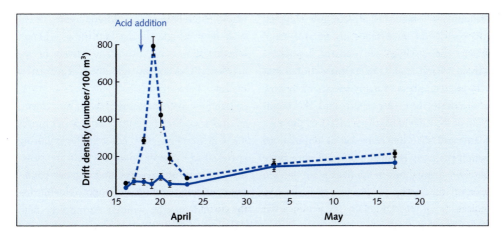

Figure 3.5 Catastrophic drift in response to acidification of a stream. The numbers drifting per day in the acidified stretch (dashed line) were, for several days, many times those drifting in the reference stretch (solid line). From Hall *et al.* (1980).

3

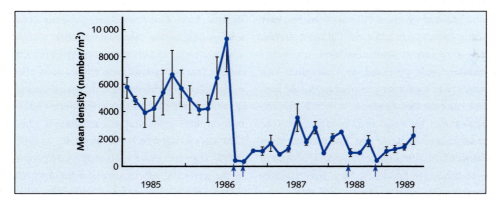

Figure 3.6 Effect of a major flood on benthic macroinvertebrate density in the Glenfinnish River, Ireland. The mean number of individuals recorded per m² is plotted. Arrows along the horizontal axis indicate major (1986) and minor (1988) floods. From Giller *et al.* (1991).

longer than the generation time of even the longest-lived inhabitants of the system. This flood, in the Glenfinnish River in southern Ireland, reduced populations of benthic invertebrates so severely that even several years later there was little sign of recovery (**Figure 3.6**). Most spates are rather less disruptive than this example, however, and what is remarkable is the speed at which most communities will recover, frequently more rapidly than an enhanced level of reproduction would allow. Clearly, therefore, these organisms are able to persist through all but the most severe floods in appreciable numbers, by exploiting places within the river system where disturbance is reduced relative to that in surrounding areas. There are three types of these flood refugia available during high discharge events: the floodplain, the hyporheic zone and the channel itself.

The floodplain

If a river overtops its banks, water spills out over the floodplain, carrying large numbers of dislodged organisms. As the water enters the flood plain, however, its velocity is reduced and the organisms will be deposited. That this process occurs is evidenced by pools left on floodplains after floodwater has receded: these contain river organisms, which, now trapped, will perish. Its importance as an area from which recolonization can occur is less clear, although Badri *et al.* (1987) demonstrated that very large numbers of many species of benthic invertebrates were deposited onto the flood-plain of the Rdat in northern Morocco during a spate, and that the recovery of populations within the river channel coincided with large numbers drifting out of the floodplain while it was still inundated.

The hyporheic zone

The hyporheic zone is a layer of saturated sediments and subsurface water beneath the river channel, within which invertebrates can occur in substantial numbers. Spates, other than very severe ones which scour the sediment deeply, have little impact upon this zone, which has therefore been proposed as a refugium into which river organisms can migrate during high flow. Against this, Palmer *et al.* (1992) demonstrated that the hyporheic zone may not be enough; they showed that the meiofauna in Goose Creek, Virginia, suffered severe losses in even moderate floods, as they were concentrated in the upper layers of sediment. Furthermore, vertical migration in response to raised flow was not rapid enough to protect them from being washed away with the sediment.

The river channel

The river channel itself may seem to be an odd place to look for flow refugia during a flood. The very heterogeneity of the channel, however, ensures that there will be 'dead zones' where flow remains low, even during spates, allowing otherwise vulnerable benthic or swimming species to persist. Dead zones occur wherever physical features, such as backwaters or boulders, provide patches where the flow of the river is interrupted. They may experience some flow, but this will fluctuate little relative to the rest of the channel. Evidence from Broadstone Stream in southern England suggests that some

macroinvertebrates do, indeed, aggregate in such flow refugia during high flows, though whether they move into them actively or are accidentally dislodged and then deposited is unclear.

High flow alters patterns of water movement, so that areas that experience little water movement at baseflow may not be refugia during spates. For example, pools, which are slow-flowing deposition areas under most conditions, undergo an increase in flow rate during floods and may be scoured during very high flow events (Section 3.2).

3.3.6 Refugia through complex life cycles

Refugia from disturbance are generally considered only at the microhabitat scale within a river stretch, but Lancaster and Belyea (1997) suggest that between-habitat refugia must also be considered. Species with complex life cycles, in which different stages occupy different habitats, can use one habitat as a refugium if other habitats are disturbed. For example, aquatic insects with terrestrial adult stages may persist in the terrestrial habitat during a flood which destroys the larval population within the river, the adults then being able to facilitate recolonization of the larval habitat. In order to function, however, the two life stages must be present at the same time so that there are adults which persist after the loss of larvae, and each life stage must be relatively long-lasting, whereas most aquatic insects have short-lived adult stages with little temporal overlap with larval stages. A more likely use of different habitats as refugia is shown by cased caddis of the genus *Sericostoma* in Oberer Seebach, Austria, which, in addition to an adult stage, has a spatial separation of larvae: young larvae live in the hyporheic zone, up to 1 m deep, and are more likely to persist during spates than older larvae, which live close to the sediment surface.

A second large-scale refugium described by Lancaster and Belyea (1997) is the same habitat at a different site, so that if a population is devastated by disturbance, recolonization can occur from a different river. This is related to the complex life cycle mechanism in that organisms require a means to move between rivers. Again, aquatic insects, with their flying adult stage, are good examples, although their dispersal abilities as adults are limited, hence the slow recovery of streams such as that illustrated in Figure 3.6.

3.3.7 Drought

Persistent periods of low flow or drying can be equally disruptive to stream organisms as floods. Unlike floods, which are rapidly occurring disturbance events, drought as a stress occurs with increasing intensity over a long period of time. All rivers and streams undergo occasional events in which flow is severely reduced, but with increased water abstraction and diversion for human use, the prevalence of such events is increasing in many parts of the world.

Drought has a number of effects on the stream environment (Lake 2003). Most obvious is loss of water. As water levels drop, surface flow will eventually cease and, with further water loss, shallow environments such as pools may dry up, reducing the river to a series of disconnected pools although possibly with subsurface water movement. Reduced water volume results in biota becoming stranded in drying remnants, and mortality can be high. Biota will also become concentrated into remaining pools, increasing population densities but also attracting predators, including aerially mobile aquatic taxa such as water beetles but also terrestrial predators, exploiting the aquatic organisms' inability to escape. In addition to this direct impact, however, water temperatures may rise, increasing the risk of deoxygenation. Cessation of flow causes detritus, DOM and nutrients to increase in concentration; leachates from detritus may reach toxic levels, while high nutrient levels in unshaded pools can lead to algal blooms. Heavy siltation may also occur.

Whether taxa are lost from a river reach is determined by the intensity of the drought. Although the abiotic effects may increase gradually, the falling water level will cause biota to pass through a series of thresholds, with sudden reductions in species richness (Boulton 2003). For example, as a river becomes disconnected from littoral vegetation, those taxa that need this habitat will be lost; similarly, when riffles dry out, species confined to this microhabitat may disappear rapidly. In environments dominated by intermittent streams, periodic drying is of course a normal feature to which stream organisms are adapted.

Once water returns, recolonization will occur. Taxa that have been able to persist in refugia such as deep pools or subsurface water will return rapidly, whereas those that have been eliminated from the reach may take several generations to return to their original population levels. If a long stretch of a river has dried up completely, then initial colonization may be by fugitive species or those that can recolonize quickly through rapid reproduction. In a coastal stream in South Carolina, Smock *et al.* (1994) recorded that early colonizers following restoration of flow included marine crustaceans that had previously been absent from the stream. Complete recovery can occur within a few weeks and even following severe droughts often takes less than a year.

3.4 Energy inputs

3.4.1 Primary production

Phytoplankton

Primary production in rivers can be very low. Phytoplankton, in particular, face the problems associated with continually being carried downstream and, in larger rivers, insufficient light as a result of being mixed through water of increasing turbidity. There are, however, phytoplankton species which are true river specialists —the 'potamoplankton', which occur most

3

commonly in large rivers. Phytoplankton can become abundant in small or turbulent rivers if there are dead zones, such as backwaters, where the turnover time of water is less than the life cycle of algal species, allowing a continual replenishment into the water column. The Orinoco River and its tributaries in Venezuela show a clear seasonal pattern of planktonic primary production, its maximum coinciding with low discharge during the dry season. The dead zones in this case are secondary channels which, as they stagnate, quickly develop large populations of phytoplankton; minor increases in river discharge will then flush these out into the main channel. There is, therefore, little true potamoplankton, despite the size of the river. The River Meuse in Belgium, in contrast, has few natural dead zones as a consequence of heavy channelization, but supports a high biomass of diatoms with a clear longitudinal zonation. Descy *et al.* (1994) suggest that their source is the middle reaches of the river itself, where high nutrient levels, originating from partially treated sewage inputs, and relatively shallow water contribute to high productivity of a genuine potamoplankton. This, in turn, is exported downstream, where reduced growth and dilution contribute to a gradual reduction in biomass.

Attached photosynthesizers

Attached photosynthesizers, including macrophytes and microphytes, are limited as abundant components of the biota to small streams and the shallower edges or slower flowing reaches of larger rivers; light is typically the limiting factor in deep stretches because of the turbidity caused by transport of suspended sediments and solutes. Most vascular plants are susceptible to damage by flowing water and occur only where current is reduced, but where conditions are good they can become very abundant. In England, some of the most productive rivers for macrophyte growth are chalk streams, where a combination of clear water, high nutrient levels and stable flow allow very high densities to develop. The dominant species is usually stream water crowfoot (*Ranunculus*

penicillatus var. *calcareus*), an annual which will typically achieve a biomass of around $200 \, \text{g m}^{-2}$ within one growing season. Attached algae, the periphyton, are important primary producers, particularly in smaller streams (**Box 3.3**). Single-celled algae growing on stone surfaces are unaffected by current and can become very abundant in clear streams, but their biomass will necessarily remain low. They can, however, be a major energy input to shallow, stony-bedded streams, where they support a specialist guild of grazer-scraper invertebrates. Macrophytic algae can become seasonally abundant in such streams, as they can withstand buffeting by moving water without suffering structural damage.

3.4.2 Detritus

Where primary producers are rare, in river stretches with low light intensity due to turbidity or shading, the primary energy inputs will be in the form of detritus, such as dead leaves from the surrounding catchment, which either fall directly into the water from riparian trees, are blown in laterally off the ground, or are transported from upstream. A river system benefits from excess primary production within its terrestrial catchment or in the riparian zone to support its detritus-based communities. The productivity of a shaded stream will therefore be determined by the productivity of its catchment and the dominant primary consumers will be detritivores. Even in highly productive streams dominated by macrophytes, energy inputs from external sources can be very important. Species which live on macrophytes and apparently feed on them may simply be consuming the epiphytic algae and small particulate detritus which aggregate on their leaves.

Underground rivers, which occur in karstic limestone areas, are completely dark, and their food webs are entirely heterotrophic. Their energy derives from transport of organic matter from the surface and they are dependent, therefore, upon the productivity of a system from which they may be spatially very distant. Most

Box 3.3   The biofilm

Stable solid surfaces on the bed of a river or the littoral zone of a lake will develop a biofilm, a discrete microbial community of bacteria, typically with some algae, protozoa and fungi, within a gelatinous matrix. Biofilms, which may be up to several millimetres thick, are created and maintained by bacteria. A freshly exposed surface will attract bacteria to settle once it has developed a thin layer of organic material, the 'conditioning layer'. These bacteria secrete a signal molecule called acyl homoserine lactone (AHL) into the water medium. In the planktonic phase, the bacteria are at low density and therefore the concentration of AHL is equally low; when they settle on solid surfaces, however, an increase in population density raises the concentration of AHL. Once AHL reaches a critical concentration, in a process known as 'quorum sensing' it triggers AHL-responsive genes, which induce physiological changes. Among the changes stimulated are those associated with the stationary, biofilm-developing phase, including reduced motility and greater secretion of extracellular polysaccharides (EPS), which are the basis of the biofilm matrix.

A mature biofilm is a complex structure, in which the EPS matrix is columnar, with channels between the columns through which water percolates. The bacteria continually secrete EPS, to balance losses through erosion. Quorum sensing helps to maintain stable populations, as high concentrations of AHL induce a stationary phase with reduced rates of cell division. If the population declines, then AHL concentration drops and rates of cell division increase.

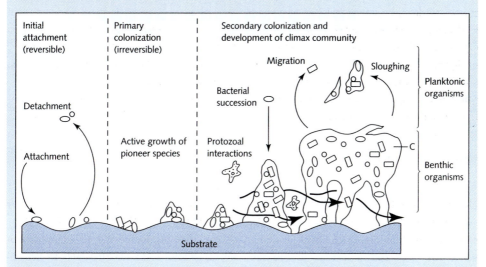

From Sigee (2005).

Within the EPS matrix, algal cells will establish if light levels are adequate, and the intercolumnar spaces will be occupied by grazing ciliates and flagellates. If light levels are high, then diatoms can dominate, leading to development of an algal biofilm.

unpolluted groundwaters, including karstic systems, are characterized by scarce trophic resources, to which subterranean faunas respond by adopting low metabolic rates.

Composition and flux of detritus

Organic matter comes in a variety of forms, from large leaves or pieces of wood to dissolved

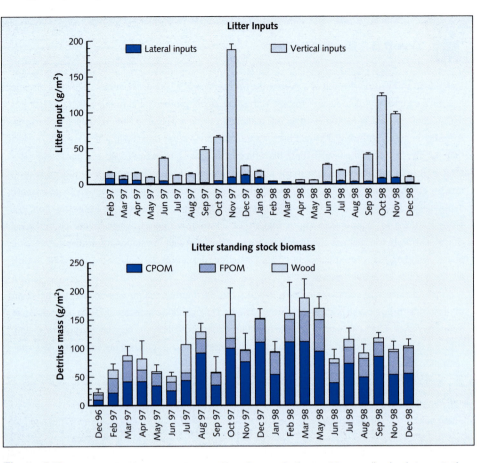

Figure 3.7 Changes in leaf litter inputs (top) and standing stocks (bottom) in a small upland stream in the Peak District, England. There are obvious peaks in inputs during autumn leaf fall, but the high retention ability of the stream channel ensures that standing stocks are consistent throughout the year. From Pretty *et al.* (2005).

components. The main inputs to rivers are in the form of coarse fragments of vegetation from the surrounding catchment, including leaves and twigs and, if leaf-fall is seasonal, litter inputs will show a similar seasonality (Figure 3.7). Other inputs may occur at different times; whereas inputs of leaf litter occur mainly during autumn leaf fall, inputs of flowers and fruits are concentrated in spring and summer and inputs of twigs and bark may be distributed throughout the year. The actual mass of inputs is determined by the tree species; in northern Spain, replacement of native trees by plantations of *Eucalyptus globulus* has not only altered the quality of leaf litter inputs

but also significantly reduced its quantity (Table 3.1).

Leaf litter cannot be consumed by detritivores, or even colonized effectively by decomposers, while moving in the water column; it needs to be retained on the river bed. A river's detritus standing stock is, therefore, determined by how much falls in, but equally by the presence of barriers that hold it. In this respect, woody debris can be of great importance as a retention agent—it decomposes and is consumed very slowly, but acts as a trap, holding large quantities of detritus (Figure 3.8). Removal of such debris will result in a major increase in export of particulate organic matter.

(a)

(b)

Figure 3.8 Two examples of wood in streams acting as a dam: (a) a single large branch has dammed the stream creating a pool (Oaken Clough, northern England); (b) an aggregation of small branches, twigs and leaves (Broadstone Stream, southern England). Photos M. Dobson.

Table 3.1 Leaf litter inputs to two streams in northern Spain		
Litter type	Input to stream running through native vegetation (lateral inputs as percentage of total)	Input to stream running through Eucalyptus plantation (lateral inputs as percentage of total)
Total (g AFDM m^{-2} yr^{-1})	759.3 (19.5)	517.4 (7.7)
Leaf litter (%)	65.7 (23.8)	60.4 (8.8)
Twigs and bark (%)	11.2 (18.0)	20.9 (4.4)
Fruits and flowers (%)	14.3 (6.0)	13.3 (7.8)
Debris (%)	8.8 (18.2)	6.4 (10.1)

From Pozo *et al.* (1997).

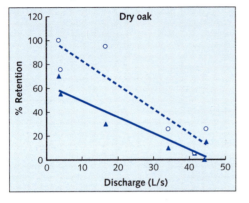

Figure 3.9 Retention of dry oak leaves with (dashed line) and without (solid line) artificial debris dams (logs) in a stream channel, showing the reduction in retention with increasing discharge. From Pretty and Dobson (2004).

Retention is not simply determined by the presence of appropriate physical structures, but also influenced by discharge. During autumn leaf fall, large leaf packs may form in a stream channel during baseflow, but a rise in discharge will entrain and remove most of these, only those leaf packs retained by efficient barriers remaining (**Figure 3.9**).

Conditioning and detritus transformation

Leaf litter breakdown is determined by a combination of biotic and abiotic activities. Invertebrate consumption and microbial decomposition, along with physical attrition and some chemical decomposition, reduce detritus to a variety of breakdown products, including decomposer and detritivore biomass, FPOM, inorganic nutrients and respiration products. Decomposition processes act synergistically. For example, the quality of most plant detritus as a food source for invertebrates is relatively poor; it is low in nitrogenous compounds and much of the carbon is in forms such as cellulose, which macroorganisms are unable to digest. It is, however, readily colonized by microorganisms, particularly hyphomycete fungi, whose filamentous mycelia can attain a biomass in excess of 10% of the mass of detritus itself. The breakdown rate of leaf litter is strongly correlated with fungal activity, which is, in turn, determined by initial lignin content of leaves: species with a low lignin content, such as alder (*Alnus glutinosa*) and ash (*Fraxinus excelsior*), decompose rapidly, whereas evergreen oak (*Quercus ilex*) and beech (*Fagus sylvatica*), whose initial lignin content is high, degrade slowly (**Box 3.4**). Fungal degradation facilitates leaf litter breakdown through assimilation into fungal biomass, but also, in weakening its structure, makes detritus susceptible to fragmentation and leaching. In addition, microbial activity increases the nutritional quality and palatability of detritus for detritivores, a process referred to as conditioning. Invertebrates will selectively feed upon detritus that has been conditioned, gaining nutrition either from the partially decomposed detrital material or from the fungal biomass itself. Invertebrates, in turn, shred leaf litter into smaller particles, thereby increasing its surface area and facilitating bacterial colonization.

Box 3.4   Comparing breakdown rates of leaves—the k value

One way of comparing breakdown rates of different leaves is the single negative exponential model. The figure to the right shows breakdown of leaves of the common African riparian tree *Syzygium cordatum* in Kenya, to which a negative exponential curve has been fitted. This curve is described by the equation:

$$Y = 92.252e^{-0.0223x}$$

In the equation, −0.0223 is the kinetic coefficient of decay, or k value, equivalent to the proportion of mass lost per day. Here, 2.23% of the remaining mass is lost daily.

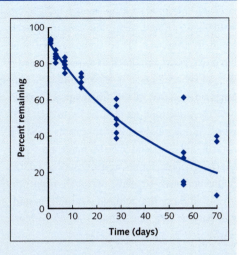

From Dobson *et al.* (2004).

The classic idea, from early work in North America (Petersen and Cummins 1974), is that there are three processing ranges (see the table below), into which different species can be placed. The k value is directly related to palatability to shredders, a higher value indicating a more palatable food resource.

k value	Processing range	Examples
<0.005	slow	beech, pine
0.005–0.010	medium	oak, plane
>0.010	fast	alder, cherry, maple

These values, derived from temperate North American species, are valid for most values recorded in temperate zones, but not in the tropics, where higher temperatures result in faster breakdown rates. Indeed, the example illustrated in the figure above is very rapid, but typical for tropical species, where breakdown rates up to 0.065 have been recorded. Therefore, if using k values to compare breakdown rates in different sites or on different dates, they need to be corrected for temperature. This can be done by quoting k values per degree day rather than absolute day. Degree day is calculated by multiplying the time in days that a breakdown experiment has been running by the mean temperature over the experimental period.

Alone, the k value is a useful first comparison but often unrealistic, as the figure to the right demonstrates. In this example, showing breakdown of litter from alder (*Alnus glutinosa*), a very common European riparian tree species, in a stream in northern England, exponential decay curve fits the actual data very poorly.

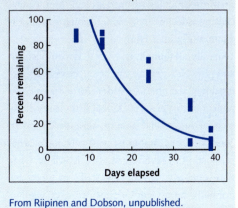

From Riipinen and Dobson, unpublished.

In a highly retentive river, CPOM is processed *in situ* and a higher proportion of FPOM will therefore be exported than is imported. Iversen *et al.* (1982) produced an energy budget for Rold Kilde, a small spring-fed river in a beech wood in Denmark. They demonstrated that input of CPOM amounted to around 716 g m^{-2} yr^{-1}, of which 71% was leaves and 11% was other identifiable plant fragments. Outputs over the same time totalled 535 g m^{-2} yr^{-1}, of which all identifiable plant fragments together contributed 7%, and FPOM made up 92%. Moreover, both inputs and outputs of coarse fragments were strongly seasonal, occurring mainly during the autumn leaf-fall period, whereas fine fragment export was more evenly distributed throughout the year.

Role of invertebrates in leaf litter processing

Terrestrially derived detritus is important to a large proportion of invertebrates in low order streams. Most detritivores in temperate areas respond to the strongly seasonal autumn pulse of leaf litter inputs by synchronizing their life cycles to take advantage of the resource when it is abundant, their larvae hatching in early autumn and then achieving rapid growth rates when the detritus arrives. Wallace *et al.* (1999) demonstrated the importance of detritus inputs to stream invertebrates in a long term exclusion experiment. By constructing a cage around a small stream, they stopped all leaf litter inputs, causing a decline in shredder biomass and production (**Figure 3.10**).

Equally, although microbial decomposition and physical attrition can be effective at breaking down leaf litter, shredders are a major component of this process. As leaf litter is normally of poor nutritional quality, shredders need to consume large amounts, and can eat a large proportion of standing stocks. Much of what they consume is not ingested, as they are normally 'messy eaters', and most of what is ingested is not assimilated but is deposited as faeces, physically fragmented but otherwise little

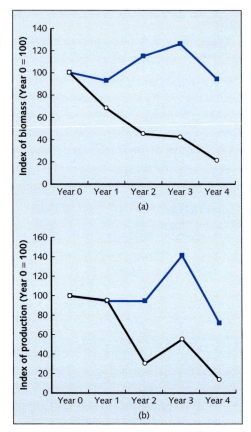

Figure 3.10 Impact of litter exclusion on (a) invertebrate biomass, and (b) shredder production in a small stream at Coweeta, North Carolina. Filled squares, reference stream; open circles, litter exclusion; year 0, pre-treatment; year 4, litter exclusion and wood removal. Adapted from data in Wallace *et al.* (1999).

changed. Therefore, shredding invertebrates have an important role in the transformation process and in providing food for other feeding groups, as FPOM is important as food for collectors, who are unable to process the coarser fractions and are thus dependent upon these transformations.

Feeding and impacts of collectors is considered further in Section 4.4; although this is within an estuarine context, the processes are similar in fresh waters.

The role of species richness

In most streams studied to date, several species of shredders coexist and shredding activity is

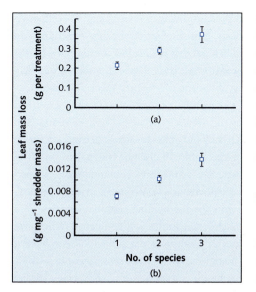

Figure 3.11 Increased shredder richness increases breakdown rate. Mean leaf mass loss (g) at different species numbers, corrected for species-specific differences. (a) Average loss per treatment; (b) average loss per mg shredder biomass; +1 s.e. in each case. From Jonsson and Malmqvist (2000).

sensitive to changes in species richness. Jonsson and Malmqvist (2000) carried out laboratory microcosm studies, placing different numbers of species together and monitoring their progress. When kept alone, different species showed rates of leaf litter consumption that were determined by body mass, but when placed in combination, even after compensating for differences in biomass, more species resulted in higher rates of leaf litter processing (Figure 3.11). Therefore, increased species richness increases the rate of leaf litter processing. This pattern has since been demonstrated in the field, streams with higher diversity of shredders having more rapid breakdown of leaf litter (Huryn *et al.* 2002).

The diversity of leaf litter can also have an impact on breakdown rates. Lecerf *et al.* (2005) demonstrated that shredder activity was enhanced by a high diversity of hyphomycete fungi and that this, in turn, was higher in streams running through mixed forest than through monoculture. Therefore, a greater diversity of trees increases rates of leaf litter breakdown. High tree diversity also increases the probability that inputs to a stream will contain leaf litter with different decomposition rates; in a retentive stream this can benefit shredders by ensuring that food is available for long periods, the invertebrates eating different species in sequence as they become conditioned by microbial activity.

3.4.3 Allochthonous inputs and water quality

Rivers have a biologically mediated ability to absorb and remove organic matter and nutrients, and this has been exploited by mankind for centuries as a mechanism for disposing of sewage and other domestic waste. Untreated sewage contains a large proportion of suspended solids, which cloud the water and clog external organs, such as gills, but its main impact is to cause depletion of dissolved oxygen concentration. Sewage is a very rich source of organic carbon and, by its very nature, breaks into tiny particles with a vast surface area. Onto these particles settle bacteria, which decompose the organic components and in doing so respire, using up the available oxygen in the water. Although this bacterially mediated deoxygenation is detrimental to most river organisms, there are two points to bear in mind. First, in respiring oxygen, bacteria are contributing to decomposition and therefore removal of organic pollutants. Second, all stages of deoxygenation and contamination with organic matter have their equivalents in natural situations. Therefore, for any state of organic pollution, there are aquatic species adapted to the conditions, each of which, through consumption, will contribute to the cleansing process. The biofilm (Box 3.3) is an important element in this cleansing activity. Particles that settle onto the bed are consumed by bacteria which, in turn, provide food for predatory protists living within the biofilm. Excess nutrients are absorbed by algae; if nutrient and light levels are high, the biofilm can change to one dominated by filamentous or multicellular algae such as *Cladophora*, which grow into dense mats but are effective at removing excess nutrients from the water.

Some sewage treatment processes include the use of filter beds, in which the contaminated water percolates through gravel upon which grow bacteria, whose effect is to mineralize and remove organic contaminants. Vervier *et al.* (1993) demonstrated an equivalent process in a gravel bar in the Garonne River, and suggested that these structures are important cleansers of river water, because, as it percolates through the gravel, bacteria extract and remove DOC.

3

3.5 Community structure in rivers

■■■■■■■■■■■■■■■■■■■■■■■■■■■■

3.5.1 Disturbance, competition and predation

Traditionally, the physical environment of rivers has been viewed as the overwhelming determinant of community structure, such that biotic interactions were thought to be of little importance. In recent years, however, many specific examples have been identified which demonstrate the role of biotic interactions, at least at a local scale. Examples of biotic interactions exerting a strong influence on community structure led Hildrew and Giller (1994) to propose three types of river community: disturbance-dominated, competitive and predation-structured. These are closely analogous to the different community types proposed by Menge and Sutherland (1987) from rocky shores, and described in **Figure 2.10**.

Disturbance-dominated communities are composed of species among which there are no influential interactions. Physical disturbance in such systems is so intense and frequent that communities never become stable enough for biotic interactions to become important, and species dynamics are essentially independent.

A predator living in such a situation will, obviously, have a major detrimental effect upon an individual that it eats, but consumption by the predator species will have an insignificant effect on abundance of the prey species as a whole.

Competitive communities are those governed by a resource which is both limiting and patchily distributed, such as detritus occurring in the form of discrete leaf packs. In some ('shuffled competitive communities'), competing species are very similar in their requirements for the limiting resource, and within a patch, competition will be intense. Whichever species colonizes a patch first will dominate it ('founder control'), but coexistence is possible because it is a matter of chance which species will arrive first. In other communities ('partitioning competitive communities'), competitors have slightly different requirements and partition the resource accordingly. In both cases, a high degree of mobility is required to facilitate colonization or to enable an individual to find a patch that provides its specific requirements. Interspecific competition has been identified in a number of situations, particularly among sedentary species which require an area of stream bed in which to forage. The North American caddis larva *Leucotrichia*, for example, is an algal grazer which attaches its retreat to stone surfaces and uses it as a base from which it aggressively defends its foraging area from all intruders, thereby affecting distribution and numbers of other grazers, both sedentary and mobile. The European net-spinning caddis *Plectrocnemia conspersa* will compete intraspecifically for a similar resource: a space in which to construct its net. If two individuals occupy a vacant site at the same time, an aggressive encounter will ensue and will continue until the loser concedes defeat by launching itself into the water column and drifting away.

In predation-structured communities, the activity of predators has a significant effect on numbers of their prey. In some, predators may aggregate on certain patches of prey—those where prey density is highest or the habitat is preferable, for example—leaving other areas as refugia from predation. In others, spates may increase predator pressure by concentrating both predators and their prey into restricted flow refugia. In New Zealand, the crayfish *Paranephrops zealandicus* is an important omnivore whose feeding activities can significantly reduce biomass of leaf litter, benthic algae and most invertebrate taxa. However, chironomid midge larvae in the subfamily Chironominae increase in the presence of the crayfish (Usio and Townsend 2002). Chironominae are relatively sedentary, living in tubes, so will have reduced susceptibility to direct bioturbation from crayfish. Furthermore, the presence of crayfish may benefit them indirectly by removal of insect predators, relaxation of competition caused by removal of other primary consumers, and by improved food availability caused by crayfish feeding and defecation.

In Puerto Rico, March *et al.* (2002) demonstrated the effect of shrimps on algae and on benthic invertebrates. Diatoms were the dominant type of benthic algae in the presence of shrimps, whereas when they were excluded, filamentous algae became dominant (Figure 3.12). Shrimp grazing kept filamentous algae in check, and their predatory activities also kept biomass of chironomid midge larvae low. In contrast, they appeared to have a positive effect on mayflies; shrimp foraging activities reduced sediment deposition, which probably facilitated mayfly grazing. The impact of shrimp exclusion was clear at high and mid altitude sites, but not at low altitude sites, where filamentous algae were absent; this was because snails, which were abundant at low altitudes, had more of an effect on algal biomass than shrimps.

The examples of the crayfish in New Zealand and the shrimp in Puerto Rico demonstrate an impact of a large crustacean on the stream community. Interestingly, both illustrate an important feature of many stream systems: the omnivorous nature of dominant species.

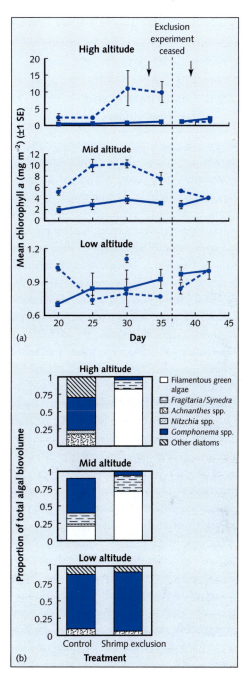

Figure 3.12 Impact of shrimps on algal growth on tiles in different reaches of a stream in Puerto Rico. (a) Chlorophyll *a* biomass in control quadrats (solid lines) and those from which grazing shrimps have been excluded (dashed line). (b) Algal community composition; all except filamentous green algae are single-celled diatoms. From March *et al.* (2002).

Trophic levels appear to be relatively poorly defined in streams, and omnivory is commonly observed.

3.5.2 Disturbance and species diversity

Which of the three community types—disturbance-controlled, competition-controlled or predation-controlled—will prevail in a given area is determined by a complex mix of inter-acting forces, but the intensity of disturbance will be the overriding feature (see Section 2.4). If environmental stress, such as discharge or frequency of floods, is high, then this disturbance will structure the community. As environmental stress is reduced, however, competition will start to occur, but even relatively low levels of disturbance will maintain the system in a non-equilibrium state, by removing patches of competitively superior species, thus ensuring that they do not dominate the entire system. Predation will only be a major determinant at low levels of environmental stress, as most predators need to be mobile to seek their prey and are therefore more susceptible to dislodge-ment. It must be emphasized, however, that the communities are not mutually exclusive: extreme disturbance to epilithic algae may be tolerable perturbation to an otter.

Neither are they fixed. McAuliffe (1984) demonstrated that the competitive dominance of *Leucotrichia* over larvae of the moth *Parargyractis* was tempered by physical disturb-ance. The moth constructs a net which, upon emergence of the adult, is destroyed, creating space for a new colonizer of either species, whereas the *Leucotrichia* larvae construct stone cases which persist after emergence of the adults and, in the majority of cases, are reoccupied by the subsequent generation. The expectation, therefore, is that, over time, *Leucotrichia* would take over completely. This does not occur because occasional spates over-turn stones, destroying the caddis retreats and, temporarily, giving equal advantage to both

species on the newly exposed stone surfaces. A second type of physical disturbance, reduced summer discharge, allows a third species, the chironomid *Eukiefferiella*, to coexist. During low summer flow the upper surface of many stones will be exposed, causing *Parargyractis* to abandon its net but killing *Leucotrichia*. These species, with a single generation per year, must wait for the following summer and a new cohort before recolonization can take place. *Eukiefferiella*, however, has several generations per year, so can take advantage of the tem-porary absence of the competitively superior species.

Continual disturbance at a very small scale contributes to the patchiness of the river bed environment and therefore to overall species diversity of the benthos. Without physical dis-turbance, *Leucotrichia* would be the only graz-ing species present, but localized events allow two other species to coexist, each in different patches. Thus, for benthic communities, mod-erate disturbance results in the highest species diversity (Section 2.4).

At a geographical scale, however, even sea-sonally predictable disturbance can serve to reduce diversity. A fluctuating physical envir-onment requires long-lived species such as fish to be adapted to a range of conditions. Swim-ming species, as opposed to those able to live on the river bed or in the sediment, show little habitat specialization because no habitat is pre-dictably long-lived. The effect of this on species diversity is well illustrated by the larger rivers of tropical Africa. The River Congo catchment is a low gradient basin with rich habitat diver-sity but, more significantly, is unique in Africa in that it straddles the equator, with tributaries in both the northern and southern hemispheres. The equatorial basin experiences little seasonal variation in rainfall; seasonal rainfall patterns occur in its tributaries, but peaks on one side of the equator are countered by dry conditions on the other. Therefore, fluctuations in water level and discharge are small in the main basin. In contrast, the other main rivers in Africa are each confined to a single hemisphere, so the entire catchment experiences the same seasonal variation in rainfall, resulting in discharge and water levels being high during the wet season and low during the dry season. A consequence of this is that, whereas the Rivers Niger, Nile and Zambezi support, respectively, 134, 115 and 110 species of fish, the Congo has nearly 700.

3.5.3 Community structure and traits

Comparing ecological communities among different rivers is complicated by the differing species assemblages often encountered over small geographical scales. One way of overcoming this is to classify organisms in terms of traits: characteristics, whether morphological, behavioural or relating to life history, that can be quantified and therefore compared among species. Within streams, the first application of traits was the classification of benthic invertebrates by functional feeding group (Section 2.2). More recently, the trait concept has been expanded to cover the full range of attributes of organisms, and attempts have been made to summarize all the available biological information about taxa in a tabular form, thereby allowing easy comparison among species.

Use of traits allows clear predictions to be made about community responses to a given environment. For example, Townsend and Hildrew (1994) proposed that flow-related disturbance would have an impact on traits, species inhabiting disturbed systems having traits relating to resistance to floods—such as clinging behaviour, streamlined morphology, terrestrial life stages—or to resilience—such as adult mobility, habitat generalism, rapid population growth. Probably the most straightforward trait to measure is body size. This is also the trait that can tell us most about an organism's response to its environment, as body size relates to so many features of its life. For example, size directly determines predation risk,

risk of dislodgement from the stream bed, and availability of refugia such as interstitial spaces. It is also indirectly related to other processes, large size in invertebrates being typical, for example, of shredders and predators, whose food resources are generally large particles, while species that have rapid growth and short generation times will inevitably be small-bodied.

3.5.4 Are tropical streams and temperate streams different?

The tropics cover the largest single area of the world's land mass, but their rivers have been relatively poorly studied, and most of our understanding of rivers, and particularly of small streams, derives from research carried out in the north temperate zone. In recent years, however, evidence has come to light suggesting that there may be important differences in community structure, and therefore in function, between temperate and tropical streams.

Many of the perceived differences relate to detritus processing. The shredding invertebrate guild, so important in leaf litter consumption in temperate streams (Section 3.4), appears to be uncommon in many widely dispersed tropical systems. It is unlikely that leaves from tropical tree species are less nutritious than their temperate counterparts (Box 3.4), and a possible explanation for this difference is that of competitive exclusion, the continuous high temperatures giving bacteria a competitive advantage over invertebrates. Alternatively, shredders may indeed be common in tropical streams, but under-recorded. There is evidence that large crustaceans and even fish occupy the shredding niche in many tropical areas, in contrast to temperate regions in which the role is carried out primarily by insect larvae and smaller crustaceans. The fundamental difference between tropical and temperate streams may not, therefore, be in the abundance and functional importance of shredders, but simply in their taxonomic composition.

3.6 Ecological functioning

3.6.1 Longitudinal and lateral linkages

The flow regime in rivers introduces a downstream movement to processes such as energy flow and nutrient cycling. The changing conditions that can be expected as one moves downstream along a river channel are integrated, in that upstream processes, by virtue of the unidirectional nature of the current, have a knock-on effect downstream. Thus, the eroding nature of high gradient streams provides the sediment which is then deposited in the low gradient river further downstream. Equally, build up of sediment and other particulate or dissolved matter modifies light penetration and oxygen concentration. Just as there are physicochemical links, so are biological processes connected in this way. Production of plankton in the middle reaches of the River Meuse (Section 3.4), for example, supports its high biomass further downstream and, in turn, the production of consumers in the lower reaches. Similarly, the export of fine particles of detritus from low order streams suggests that breakdown processes in the headwaters are not only supplying food to collectors *in situ*, but also to those that live further downstream, so upstream production and downstream consumption of FPOM are coupled. Several models have been proposed in recent decades in an attempt to provide a simple mechanism for explaining biological connections between the various stages of a river system; three of the most influential are described in **Box 3.5**.

A consequence of this continual movement is that nutrient cycling processes in rivers differ from those in other systems (**Figure 3.13**). Nutrient cycling is the transfer of an element from a dissolved inorganic form—during which it is available as a nutrient for an autotroph—through incorporation into living tissue and its

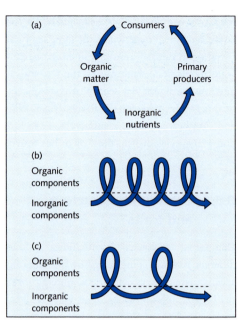

Figure 3.13 Diagrammatic representations of (a) a nutrient cycle in a terrestrial ecosystem; (b) a nutrient spiral in a retentive river, in which downstream movement of materials is slow; and (c) a nutrient spiral in a non-retentive river, in which downstream movement of materials is rapid.

subsequent transformation back into the inorganic form by decomposition processes. In most terrestrial systems, nutrient cycling takes place within the same geographical area, with relatively little input or output of elements (**Figure 3.13a**), but in rivers, a given element will be continually transported downstream, either in solution or as part of a living organism, and the cycle is broken, becoming more of a spiral (**Figure 3.13b, c**). Losses of nutrients downstream are, therefore, high and must be matched by inputs from upstream.

3.6.2 Other ecological functions

In low-order streams, frequent fluctuations in discharge ensure that habitat modification by organisms is small scale and short lived. Patches of vegetation may develop in sheltered backwaters or during season when flow is low, but these are rarely permanent. The main exception

Box 3.5 Models of longitudinal and lateral linkages within stream systems

Among the many models of linkage within river systems, three of the most influential are presented here.

The River Continuum Concept or RCC (Vannote et al. 1980)

This proposes that energy inputs will change in a longitudinal direction, in relation to channel size, degree of shading, light penetration etc., and relative importance of functional feeding groups will change in tandem. One major prediction of the RCC is that downstream stretches, with little primary production, support detritus-based communities which are heavily dependent upon breakdown of plant litter in upstream reaches, and therefore exploit the excess production of such upstream sites in the same way that the river as a whole exploits excess terrestrial production in its catchment. Different streams enter the continuum at different points so that, in the example illustrated, the forest stream is shaded by trees and has a large initial input of terrestrially derived coarse particulate detritus, primarily leaf litter, whereas the desert stream, with fewer trees and less shading, is dominated initially by photosynthetic algae.

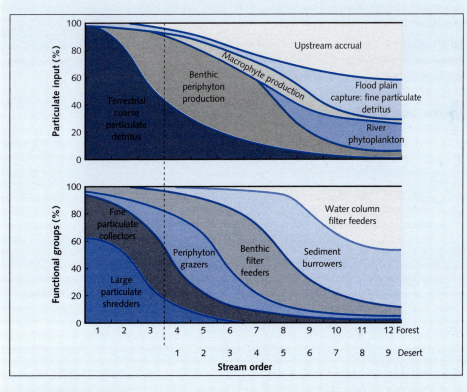

From Minshall *et al.* (1985).

Box 3.5   Continued

Despite its value in conceptualizing longitudinal linkages along a river system, the RCC fails to explain adequately processes in very large rivers, whose requirements for organic carbon are apparently not met by upstream production (Sedell *et al.* 1989). It also treats floods as unpredictable or catastrophic, failing to recognize their predictable seasonality in larger rivers.

The Flood Pulse Concept or FPC (Junk et al. 1989)

This model overcomes the difficulties faced by the RCC. Its basis is that the pulsing of river discharge—the flood pulse—is the major force controlling biota in river floodplains, and that lateral exchange between the river channel and its floodplains is more important in determining nutrient and carbon supply in lower reaches than longitudinal connections.

The key component of the FPC is the 'aquatic/terrestrial transition zone' (ATTZ), which is the ecological floodplain as defined in Section 3.2. Organisms use the floodplain for feeding and reproduction, and the main channel to move to other parts of the floodplain or as a refugium during low flow. Among fish, the river is often the main spawning site, but eggs or fry are transported into the floodplain, which acts as a nursery. Such movements have been demonstrated from several floodplain systems in the tropics, including the Inner Delta of the Niger River in Mali (Section 8.5).

The main limitation of the FPC is that active movement between the channel and the floodplain has so far been identified only among fish and a few other vertebrates. It may, therefore, be limited in value to consideration of taxa which can actively move. Extensive floodplains contain permanently flooded lakes and channels, which act as a source for less mobile organisms to populate the floodplain during inundation without requirement for immigration from the river channel. Certainly the inundated floodplain of the Orinoco River in Venezuela (Section 3.4) supports a planktonic system independent of that in the river channel itself and exports relatively little phytoplankton biomass to the river, despite its high productivity; zooplankton and benthos may show the same pattern, with the only inputs to the river being excess individuals.

The Riverine Productivity Model or RPM (Thorp and Delong 2002)

This model was developed to explain processes in large rivers with relatively constricted channels, in which neither allochthonous inputs nor floodplain export can produce enough carbon to support the observed productivity. In its original form it proposed that the major source of organic matter assimilated by animals in large rivers is derived from a combination of allochthonous inputs from the riparian zone (as predicted by the RCC), but also autochthonous autotrophic production in the river channel. It therefore downplayed the importance of the floodplain as a source of energy. It has since been modified to place more emphasis on primary production and less on constricted channels, stating that the primary energy source supporting overall metazoan production and species diversity in mid- to higher-trophic levels of most rivers (at least 4th order) is autochthonous primary production entering food webs either directly via algal grazers or indirectly via decomposers. The RPM acknowledges that most carbon in rivers is allochthonous in origin, but contends that the majority of these inputs are poor quality and therefore difficult to assimilate. Algal primary production, in contrast, is more labile and therefore preferred by animal consumers, whether grazers or detritivores.

Figure 3.14 A beaver dam in Colorado, showing the pool that develops behind it. Photo M. Dobson.

3

to this is the influence of large vertebrates. Pre-eminent amongst these is beaver (*Castor canadensis*), whose dam-building activities can fundamentally alter river channels **(Figure 3.14)**.

The constant flow in rivers ensures that migration or dispersal of propagules is limited. Indeed organisms such as freshwater crabs and crayfish, whose main marine relatives exploit water movement to disperse planktonic larvae, have dispensed with the planktonic life stage in rivers and care for their offspring until they are large enough to function independently as benthos (Section 2.3). However, some migration does occur. Most aquatic insects have a terrestrial adult phase, during which flight allows them to move between systems; this probably occurs very rarely, but is frequent enough to ensure genetic mixing (Wilcock *et al.* 2003). River catchments usually contain a variety of other freshwater habitats such as lakes and wetlands, between which accidental movement often occurs. For example, large numbers of plankton washed out from lakes or backwaters may support a potamoplankton that is unable to sustain itself (Section 3.4), or that is the main food resource for filter feeders at the lake outlet. In other cases, such as the lateral movement of fish into floodplains to reproduce, it is deliberate. Despite most rivers being connected to the sea, however, there is little movement between these environments. In the absence of a suitable freshwater competitor, estuarine species may invade fresh waters; this phenomenon is common among decapod crustaceans on oceanic islands, and also includes colonization of fresh waters in Ireland by the normally estuarine amphipod shrimp *Gammarus duebeni*, in the absence of *G. pulex* which dominates streams in the rest of north-west Europe. In some cases, these crustaceans retain larval stages that require salt water, resulting in annual migrations to the sea to reproduce. In contrast, few organisms move from the sea into rivers to breed, but the salmonid fish that do so (Section 4.3) are important in transfer of marine-derived nutrients into fresh waters.

3.7 Succession and change

3.7.1 Colonization and succession

Primary succession

Longitudinal distribution patterns illustrate spatial succession in running waters in relation to changing physicochemical conditions, but primary succession over time is more difficult to study, because formation of new streams is rare. Among the few new streams that have been studied are those formed by retreating glacial ice in Glacier Bay in south-east Alaska, which show an interesting pattern of development. Newly exposed meltwater streams are quickly colonized by a small but distinctive biota of algae and cold-tolerant chironomid larvae. The insect fauna at this stage is very specific, and will be present only so long as the low temperature and bed instability impede colonization by other species. The second stage of colonization of these streams is closely related to improvement in the physical conditions, and involves the gradual colonization of species which will then persist in the community.

Rate of stabilization is a key determinant of rate of succession. Unstable streams, characterized by rapid fluctuations in discharge and consequent redistribution of sediment, remain low in species diversity and overall productivity. The presence of a lake in a catchment adds an element of stability, buffering discharge fluctuations and acting as a sediment trap, which allows a more diverse, productive community to develop. Milner (1987) suggests that a long-term factor of importance in stream development in Alaska is riparian vegetation, which will structurally modify the river channel and banks. Development of such vegetation depends upon terrestrial successional processes, but will also require a dampening of discharge fluctuations, as frequent spates will retard growth of riparian plants.

Colonization abilities

Once the physical conditions are acceptable, rates of colonization by aquatic organisms will be the limiting factor in stream community development. Despite often very severe seasonal fluctuations in discharge, almost all river systems have at least some stretches of perennially flowing water and can persist for many millions of years, providing evolutionary stability. As a consequence, river specialists rarely require a capacity to move between river systems and their powers of dispersal between catchments are limited. Some fish species, including many salmonids, are tolerant of salt water (see Section 4.3) and can disperse through the sea to different river systems, but most river taxa are confined to the freshwater habitat. Insects have the ability to fly, but river specialists—including caddis flies, stoneflies and mayflies—have adult stages which are typically short-lived or weak flying. Therefore, their aerial dispersal abilities, apart from accidental displacement, are limited to flights which rarely exceed several hundred metres and are normally, therefore, confined to the catchment from they emerged.

Wolf Point Creek in Glacier Bay is a rare example of a young stream whose age is precisely known, as it was created by a receding glacier. Furthermore, as the glacier originally discharged into the sea, the stream now runs directly into salt water, with no connection to other streams. Thus, 25 years after formation, it supported a fauna containing only a handful of insect species, which flew as adults from an adjacent catchment, and salmonid fish, which entered from the sea. In contrast, a stream formed in an area containing plenty of established streams can experience rapid colonization. Flugströmmen, an artificial stream adjacent to other water courses in southern Sweden, supported a faunal assemblage similar to that of adjacent natural streams after one year. Colonization of Flugströmmen continued throughout the 18 months of the study, with new species being recorded on every sampling occasion, whereas in established streams that cease flowing for a short time, most of the expected species return within a few weeks of flow returning. In the former case, it appears that colonization was determined by the rate at which new species could reach the site by drifting from upstream or in the flying stage, whereas in the latter, recolonization is probably precipitated by the presence of eggs or other dormant stages within the river channel, which hatch or become active when conditions improved.

Drift and upstream colonization

Active drift and passive dislodgement both result in downstream movement of organisms (Section 3.3), and one of the most enduring conundrums in river ecology is why this does not result in depletion of upstream populations. Müller (1954) proposed that this could be explained for aquatic insects by a colonization cycle: larval movement is essentially downstream, so adults will fly upstream to compensate for this. Unfortunately, the evidence available provides little support for this hypothesis. Some studies have found no evidence for a tendency to fly upstream and even those which do find support for the idea, such as Flecker and Allan's (1988) study of mayflies in Colorado, are not as clear as may be expected. These authors showed a significant tendency for *Baetis* spp. to fly upstream, although in several of their samples, more were flying downstream, whereas *Rhithrogena hageni* tended to fly downstream. A further problem with the colonization cycle hypothesis is that it does not explain how species without a flying stage, such as crustaceans, can persist in upstream stretches. Upstream movements by benthos moving over or within sediment (positive rheotaxis) have been recorded but, again, are inadequate to compensate for downstream drift. Upstream movement may not be necessary to replace losses, as drift from headwaters is loss of surplus individuals and enough will remain to replenish the population. The evidence is,

however, equivocal and, for the moment, the phenomenon remains unexplained.

3.7.2 Secondary changes

Temporal change of physical conditions in established river systems is mainly at a local scale—redistribution of the features present, such as erosion and redeposition of gravel bars or severance of meanders to form oxbow lakes. From the perspective of organisms, these processes simply shift the microhabitats that they occupy spatially. Recolonization after disturbance events such as spates can be very rapid, but although a predictable sequence of species colonizing a disturbed patch may occur, this is caused by different rates of dispersal among species, and is not a true succession. Continual spates, moreover, act as a reset mechanism, ensuring that the pattern is continually repeated, rather than reaching an end point. Only major spates, such as that illustrated in Figure 3.6, will have long-term effects upon the entire channel. Changes in water quality are, however, more likely to affect the entire community. A pulse of organic pollution into an otherwise clean river will eliminate sensitive species, and its effects may be measurable for a long period of time afterwards, even if the pollution event itself was only of a few hours' duration. This will be the case particularly if the pollution event occurs close to the source of a river, so that there is no opportunity for recovery through drift, and restoration of the eliminated species requires recruitment from the next generation (Section 3.3). One of the advantages of using living organisms as indicators of aquatic pollution is that in addition to giving an indication of water quality at the time of sampling, they may provide evidence for any previous short-lived pollution events.

Recolonization by organisms drifting from upstream can take a matter of days or weeks once the perturbation has ceased, but if the entire river is affected, so that there is no upstream reservoir from which to drift, then recovery can take rather longer. Even insects

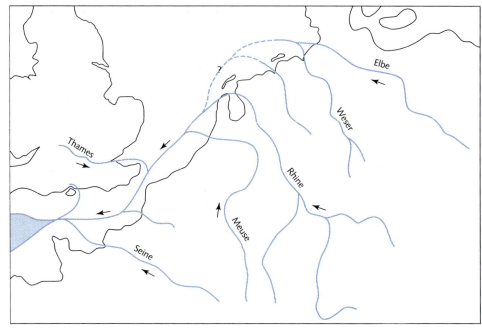

(a)

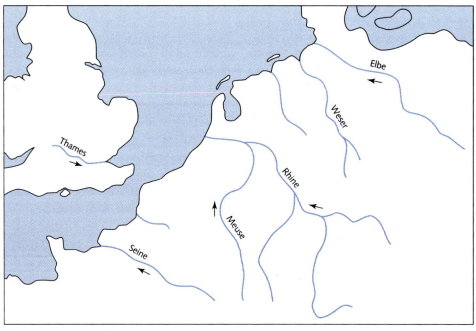

(b)

Figure 3.15 The relative positions of major rivers of north-west Europe (a) during the last glacial, around 20,000 BP; (b) today. Arrows indicate the direction of flow. After Gibbard (1988).

3

with flying adult stages will disperse across catchment boundaries very slowly, and natural restoration may take years or even decades. Species confined to upstream reaches and with no terrestrial stage in their life cycle may never recolonize the upper reaches of a river impacted from its source. At a geological time scale, poor dispersal between catchments can be countered by events which cause separate river systems to merge. In eastern England, for example, the River Thames is separated from rivers elsewhere in Europe by sea, but when sea levels fell during glacial periods it became part of the River Rhine system which drained much of the area now covered by the North Sea (**Figure 3.15**), allowing an exchange of flora and fauna. For this reason, the rivers in eastern England are in many ways faunistically more similar to those in central Europe than to those elsewhere in the British Isles (Wheeler 1977).

3.7.3 Effect of human activity

Long-term or permanent changes to the physical or chemical attributes of a river will occur if it erodes through different geological strata, or even in response to climate changes, particularly changes in amount or seasonality of precipitation. Human modifications, too, are often of long duration or permanent, and can give insights into the ecological response to perturbations. Structural modifications to the river channel, such as the construction of dams or barrages, have clear and effectively permanent effects upon a river system. Construction of barrages along the River Danube in Austria illustrates well the effects of flow reduction and altered patterns of sedimentation. The Austrian stretch of the Danube is 351 km in length, along which it descends in altitude by 156 m. Until the early 1950s it was free-flowing, but since then ten barrages have been constructed, converting the river into a series of impoundments, with only two free-flowing stretches remaining (**Figure 3.16a**). The effect of the barrages has been to reduce flow and increase sedimentation in the impounded stretches, so

that a bed dominated by rocks and gravel in free-flowing stretches changes into one of fine silt immediately upstream of a barrage (**Figure 3.16b**). Sediment transport has been reduced: in one stretch of the river, mean annual suspended load has declined from 6.95 million ta^{-1} before impoundment to 2.75 million ta^{-1} now, while bedload transport, formerly 490,000 m^3a^{-1}, has now dropped to zero. As a consequence, many stretches that were formerly depositional areas are being starved of annual sediment inputs but are still being eroded, so the channel is now cutting into its bed, lowering the water table and threatening lateral wetlands.

In addition to barrages disrupting longitudinal movements, many large rivers have been channelized to improve navigation or reduce flood risk. Again, the Danube provides a good example of this. Comparison of a 10.25 km reach of the current river channel with records dating back to 1812 shows that straightening reduced the area covered by the main channel from 588 ha in 1812 to 288 ha in 1991, and total water bodies, including side channels, were reduced from 739 ha to 413 ha. In addition to straightening, construction of levees reduced the hydrological connectivity of the river with its flood plain (Hohensinner *et al.* 2004).

Biological changes are equally profound. Free-flowing stretches of the Danube, such as that above the Altenwörth dam, have a low density but high taxonomic diversity of benthic macroinvertebrates, whereas the fine silt in the impoundment immediately above the dam supports very high densities of a few silt-tolerant species—98% of individuals are oligochaetes, the remaining 2% are almost entirely chironomids and the freshwater polychaete *Hypania invalida*. Towards the upstream end of the impoundment, physical conditions and therefore the invertebrate fauna more closely resemble those of the free-flowing stretch, though total numbers of individuals are much greater, mainly due to the presence of numerous small Oligochaeta and Diptera (**Figure 3.16b**).

The impoundments have more subtle effects upon fish assemblages. Free-flowing stretches

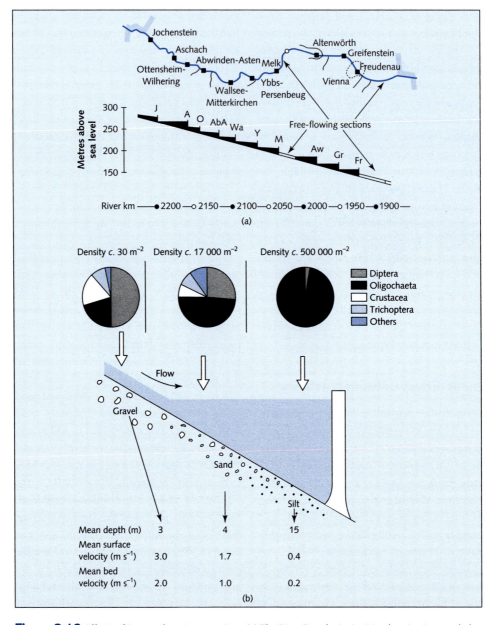

Figure 3.16 Effects of impoundment upon a river. (a) The River Danube in Austria, showing impounded and free-flowing sections. (b) Diagrammatic representation of physical and biological changes that occur moving from a free-flowing to an impounded stretch of the River Danube above the Altenwörth dam. Pie charts show changes in relative numbers of different macroinvertebrate groups at each of three points; note the major increase in total density. Depth and mean flow at each of the three points are also given. After Humpesch (1992).

are dominated by running water species, particularly nase (*Chondrostoma nasus*), whereas the impoundments offer conditions more suitable for species more typical of lakes, such as roach (*Rutilus rutilus*). Impoundments have not developed a true lake fish fauna, however, because of the lack of shoreline diversity and low densities of plankton. The Altenwörth impoundment is still dominated by the original running water fish assemblage, capitalizing on

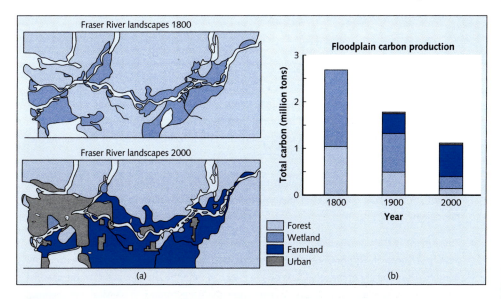

Figure 3.17 Changes in land use and organic carbon production in the Lower Fraser River valley in western Canada over the past 200 years. (a) Land use in 1800, dominated by forest and wetland, and in 2000, dominated by urban development and farmland, with forest confined to the edges of the valley. (b) Estimated total organic carbon production on the floodplain, showing the decline associated with forest and wetland loss. Total organic production has declined, and flood control strategies ensure that a lower proportion of this enters the river. From Healey and Richardson (1996).

the rich macroinvertebrate food resource, but these species do not breed in the impoundment because suitable spawning grounds are lacking, and rely upon juvenile recruitment from the free-flowing stretch upstream. As most stretches of the Austrian Danube are bordered by impoundments at their upper and lower ends, with no free-flowing river, their native fish have little chance for recruitment and will probably become locally extinct without major human intervention.

Reduction in organic matter inputs may be one of the most profound effects of anthropogenic catchment modification. Transformation of a formerly wooded catchment to agriculture or urbanization will inevitably reduce the inputs of organic matter as numbers of trees or area of wetland are reduced. Healey and Richardson (1996) estimate that the total annual carbon production in the lower Fraser River valley in western Canada is now only 20% of that 200 years ago, as productive forests and wetlands have been replaced by cultivated land and urbanization (Figure 3.17). Furthermore, dyking to prevent flooding has cut the river off from some of the remaining sources of carbon inputs. Clearly, a land use modification on this scale will have more than a local effect upon the river, but will modify its productivity throughout the downstream reach and out into the estuary.

Bulleted summary

- Rivers are characterized by unidirectional flow of water. The conditions at any one point on a river will therefore be determined not only by the surrounding catchment but also by inputs from upstream. Thus any process occurring upstream, in the river channel itself or in the catchment from which run-off originates, can potentially affect the ecology of the river further downstream.

- Rivers are intimately coupled with their catchments, with which they interact through direct run-off, groundwater transfer and flooding. Rivers which flood regularly have a clearly defined floodplain.

- The current is the primary determinant of community structure. It ensures that pelagic communities, if present, are poorly developed and that most species, particularly in eroding rivers, are benthic. The diversity of microhabitats on the river bed or in the lateral margins of the channel determines species diversity. Species assemblages change according to physical and chemical parameters. Longitudinal zonation is often present, mirroring abiotic changes that occur along a river. Water chemistry is a key influence on most river species and is used as the basis for biotic indices of water quality.

- A benthic organism entering the water column will drift downstream. This may be initiated as an active strategy if rapid movement is required, but will only be in a downstream direction. There is a variety of refugia employed by river benthos to avoid accidental dislodgement during periods of high flow, or to facilitate recolonization following major disturbance. In many parts of the world, adaptation to drought is also essential for river organisms.

- Rates of primary production are generally low in rivers, particularly outside their middle reaches. The major source of energy is therefore detritus, from the catchment or from upstream. Detritus processing is a key ecological function in rivers.

- Biotic interactions have a role to play in structuring the majority of river communities, but are continually tempered by disturbance. Unpredictable disturbance ensures that the most competitive species do not dominate to the exclusion of all others and, in this way, moderate disturbance can enhance the patchy nature of river beds and therefore their species diversity.

- Colonization of new streams can be slow, unless there is a direct fresh water connection to a potential source of colonists. Disturbance ensures a continuous turnover of species at a small spatial scale. Modification of physical conditions will profoundly affect community structure.

3

Further reading

Allan J.D. (1995) *Stream ecology. Structure and function of running waters*. London, Chapman and Hall.

Calow P. and Petts G.E. (eds) (1992) *The rivers handbook, volume 1. Hydrological and ecological principles*. Oxford, Blackwell Science.

Dudgeon D. (ed.) (2008) *Tropical stream ecology*. London, Elsevier.

Giller P.S. and Malmqvist B. (1998) *The biology of streams and rivers*. Oxford, Oxford University Press.

Hynes H.B.N. (1970) *The ecology of running waters*. Liverpool, Liverpool University Press.

Gordon N.D., McMahon T.A., Finlayson B.L., Gippel C.J. and Nathan R.J. (2004) *Stream hydrology. An introduction for ecologists*, 2nd Edn. Chichester, John Wiley and Sons Ltd.

Petts G. and Foster I. (1985) *Rivers and landscape*. London, Edward Arnold.

Sigee D.C. (2005) *Freshwater microbiology*. Chichester, John Wiley and Sons Ltd.

3

Assessment questions

1 Compare and contrast ecological impacts of floods in low-order streams and in high-order rivers.

2 What are the main causes of drift in running waters?

3 How does detritus processing illustrate longitudinal links in a river system?

4 How may a river influence the ecology of an estuary at its mouth?

5 Why is river channel straightening and channelization detrimental to ecological processes?

Estuaries

■■■■■■■■■■■■■■■■■■■■■■■■■■■■■■■■■■■■

4.1 Introduction

An estuary may be defined as 'a semi-enclosed body of water having a free connection with the open sea and within which the sea water is measurably diluted with fresh water deriving from land drainage' (Cameron and Pritchard 1963). Estuaries are not, however, simply zones where freshwater and marine ecosystems merge with each other. There is a gradation of physical and chemical characteristics of estuarine conditions, particularly salinity, from the fresh water to the marine end but, of equal importance, estuaries are in a constant state of flux. River imports may vary seasonally or in response to local weather conditions, while marine inputs are determined by tidal movements and wind forcing. Zones of mixing, and the degree to which mixing occurs, can therefore vary on a daily or even more frequent basis.

In addition to water chemistry variations, estuaries are generally zones of deposition of sediment and are therefore structurally dynamic. The complex physical and chemical gradients in the estuary and the way they fluctuate impose physiological limits on the organisms able to exploit them. Estuaries therefore have an ecology that is very different from the adjacent freshwater or coastal systems.

4.2 The abiotic environment

4.2.1 Tides in rivers and estuaries

Tides are the periodic, normally twice daily, change in depth of the sea, resulting in an inflow (flood or rising tide) and outflow (ebb or falling tide) in coastal areas; they are described in detail in Chapter 5.

Estuaries are generally wide at the mouth and narrow at the head. As the tide moves into the estuary, it becomes constrained by the topography of the basin, being forced into a narrowing channel. Initially, this raises the height of the tidal wave, as a smaller estuary volume tries to accommodate the same volume of water, so that tides in the lower estuary are higher than those along adjacent open coasts. Indeed, some of the highest tidal ranges in the world, such as the Bay of Fundy in Nova Scotia, Severn Estuary in southern England and the Baie du Mont St Michel in northern France (Figure 4.1), are in rapidly narrowing estuaries.

As one goes further up the estuary, the tidal range begins to decline as the bottom and the sides of the basin both exert a frictional drag on the advancing tidal wave, removing energy as it moves up the estuary. Waves of lower energy have lower amplitude for a given wave length, so the tidal range is progressively reduced up the estuary, until at the tidal limit it is zero. The frictional force of the bed declines, however, with increasing depth. Therefore, as the tide rises and water depth increases, the incoming water experiences less friction and moves more rapidly. In contrast, as the tide falls, depth is also reduced and friction slows the ebb of water. As a consequence, tides in estuaries are asymmetrical, with a rapid flood tide and a slow ebb tide (Figure 4.2).

In some estuaries, either as the result of sudden narrowing or a steepening of the river bed, a wall of water, or tidal bore, is produced when the rising tidal front forces the water to move

Figure 4.1 At low tide, a drowned river valley type estuary typically has a low residual volume of water and extensive areas of intertidal sediment (Baie du Mont St Michel, northern France). Photo M. Dobson.

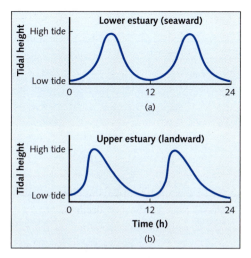

Figure 4.2 The effect of estuary topography on the tidal cycle. (a) In the lower estuary, as on the open coast, the flood tide and ebb tide occur at the same rate and a symmetric tidal wave is produced. (b) In the upper estuary, where the channel is narrower and shallower, the flood is faster than the ebb and an asymmetrical pattern is produced (see text for explanation).

more rapidly than a shallow water wave can freely propagate in water of that depth. It is analogous to the sonic boom, in which a pressure disturbance is forced to travel faster than the speed of sound. Most tidal bores are small, less than 0.5 m high, but the Severn bore in southern England can be 1–2 m high, while the Amazon bore (the pororoca) can reach 5 m and move upstream at over 20 m s^{-1}.

4.2.2 Estuarine salinity

Estuaries may be classified according to their origin (**Box 4.1**). A more meaningful classification from an ecological perspective, however, is related to their salinity (**Box 4.2**). An estuary will contain a gradient of salinity from fully saline (35–37) at the seaward end to fresh water (less than 3) at the landward end. This gradient is not a simple one, being influenced by the topography of the estuary, the tidal regime

4

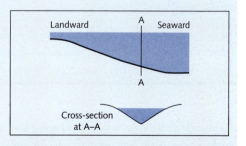

Box 4.1   Topographic classification of estuaries

Most of the world's river systems are in part tidal in their lower reaches, as a result of the rise in sea level during the last glacial recession, which ended 5000 years ago. During this period, sea level rose 100 m in 10,000 years, flooding low-lying areas of what had been land during the glaciation, including river valleys and glacial channels.

Drowned river valley

This is the most common type of estuary in Europe, including, for example, all of the larger estuaries in the British Isles. Such estuaries are often V-shaped in section but are normally relatively wide and shallow, with a very high width:depth ratio (100:1 or more). At low water, extensive mud flats are revealed. They have low sedimentation rates and a low river flow.

Fjord

Fjords are drowned glacial valleys and occur at high latitudes, including those in Norway (from which the name derives) and the sea lochs of western Scotland. They tend to be U-shaped in section and are

Box 4.1   Continued

much deeper than drowned river valleys, with a width:depth ratio as small as 10:1.

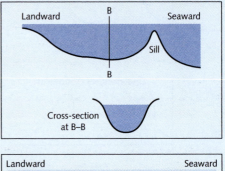

Fjords often have a sill at the mouth, a shallow lip created by the barrier-forming processes acting at the foot of the glacier which scoured out the depression (Box 7.1). Therefore, the mouth of the fjord can be very shallow relative to the rest, restricting water circulation in the body of the fjord and, in some cases, leading to development of anoxia in the bottom waters. Some fjords have several sills, dividing the estuary into a number of distinct basins which vary in the level of exchange of water, so that while some may always contain oxygenated water, others may be permanently anoxic. Intermediate basins may be periodically

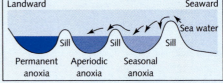

recharged with oxygenated water by storm surges pushing water over the sill. As the biota use the oxygen, and as organic matter builds up leading to more oxygen being used up by decomposition processes, so anoxia develops again. This persists until the next storm surge recharges the basin with oxygenated water.

Fjords typically have very low freshwater inputs, which may be highly seasonal, for example most occurring in spring following the melting of winter snows. Sedimentation rates are low.

Bar-built estuary

Bar-built estuaries are predominantly tropical in distribution. They are drowned river valleys with high sedimentation rates, the sediment being deposited across the estuary mouth to form a partial barrier. The barrier, or bar, is formed of sediments which are not necessarily of river origin, but river sediments will be deposited behind it to generate extensive mud flats. Seasonal pulses of water such as floods associated with monsoons may periodically flush out the estuary, destroying all or part of the bar. This then gradually reforms after the monsoon when normal conditions prevail once more.

Tectonically formed estuary

A limited number of estuaries have been formed by the inundation of the land through land subsidence as a result of tectonic processes, usually at a geological fault. San Francisco Bay, in California, lying in the junction of two plates, is an example of an estuary formed by this mechanism.

and the nature of the residual circulation. The position of the gradient also moves up and down the estuary with the tidal cycle.

Four fundamental types of estuaries can be differentiated on the basis of the relative distribution of salt and fresh water and the degree to which they mix (Figure 4.3).

Salt wedge estuaries

Salt wedge estuaries develop where a river discharges into a virtually tideless sea. Fresh water is less dense than salt water and the river water spreads out over the surface of the seawater, which, if there is no tidal movement, can be

Box 4.2   Salinity

Salinity refers to the amount of salt dissolved in a set volume of water. This seems like a very straight-forward thing to measure, by simply evaporating 1 kg of seawater and weighing the salts left behind. However, because of the complex nature of the salts present in seawater it is actually very difficult. As the salts become concentrated they undergo chemical reactions, some of them dissociate and volatile substances are lost. The traditional approach was to estimate the salt content by chemical analysis of the concentration of chloride ions and to then apply a mathematical scaling to give the total salt. This relies on the constancy of composition of seawater. Using this method, salinity was expressed as grams of salt per kg of seawater, or parts per thousand (‰). In practice by the mid twentieth century most determinations were made by measuring the electrical conductivity of a water sample and scaling this to give a value of salinity (Section 1.3).

Given the technical difficulty of doing the chemical analyses routinely and the development of robust electronic measures for measuring the conductivity, a new definition of salinity was introduced in 1978. Salinity is now expressed as the ratio of the conductivity of the sample to that of a standard solution of KCl. The standard constructed and the ratio are scaled in such away that the salinity of a water sample of oceanic water of 35‰ give a ratio of 35. This is known as the Practical Salinity Scale. As ratios have no units, salinity has also become a dimensionless parameter. However, as many people find it difficult to construct sentences or refer to a physical condition without units, salinity is often referred to as having 'practical salinity units' (psu): for example the ocean's salinity is around 35 psu.

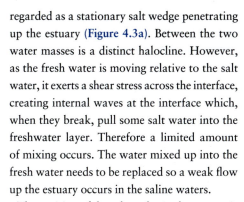

regarded as a stationary salt wedge penetrating up the estuary (Figure 4.3a). Between the two water masses is a distinct halocline. However, as the fresh water is moving relative to the salt water, it exerts a shear stress across the interface, creating internal waves at the interface which, when they break, pull some salt water into the freshwater layer. Therefore a limited amount of mixing occurs. The water mixed up into the fresh water needs to be replaced so a weak flow up the estuary occurs in the saline waters.

The position of the salt wedge in the estuary is dependent on the freshwater flow. Under low flow conditions, the volume of fresh water input is small and salt water can penetrate much further inland than under higher flows. Open salt wedge estuaries only occur if sedimentation is low; high sedimentation produces a delta at the mouth of the estuary, such as those of the Nile and Mississippi.

Partially mixed estuaries

A partially mixed estuary is formed where a river discharges into a sea which has a moderate tidal range. Tidal currents are significant, so that the whole water body moves up and down the estuary with the tide. Therefore, in addition to the shear stress between the saline and fresh waters, there is a shear stress at the bed. This generates turbulence which is effective at mixing the two water masses so that not only is salt mixed upwards, but fresh water is mixed downwards (Figure 4.3b). This two-way mixing leads to a less pronounced halocline and, as more salt water is now flowing seawards in the fresh water layer, the flow up the estuary in the bottom layer is also much larger. The total outflow at the mouth of the estuary is large and equals the inflow of fresh water plus the large inflow of saline water.

4

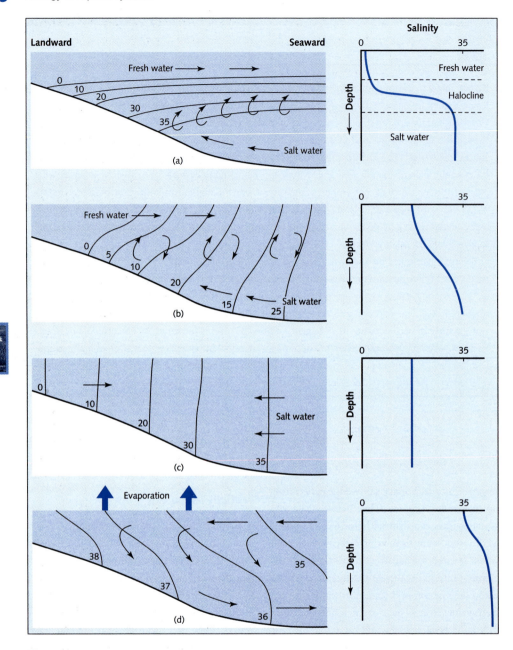

Figure 4.3 The idealized pattern of tidally averaged water circulation and salinity distribution in (a) a salt wedge estuary; (b) a partially mixed estuary; (c) a well-mixed estuary; (d) a negative circulation estuary, showing salinity contour lines (see text for explanation). The graphs to the right show a salinity–depth profile from each estuary, illustrating how salinity changes along a vertical line from a single point on the surface.

The currents produced by the mixing process are termed the residual currents, and are typically only 10% of the magnitude of the tidal currents which are superimposed on them. Residual currents are, however, important when considering transport processes in estuaries, of both sediments and pollutants.

Well-mixed estuaries

In broad shallow estuaries where the tidal range is high and tidal currents are strong relative to the river flow, the water column becomes completely mixed. Such estuaries include the lower Thames and the Severn in southern England. In well-mixed estuaries, salinity hardly changes with depth and there is no halocline (**Figure 4.3c**). As a result, although the estuary may have a large volume, the total outflow essentially equals the inflow of fresh water.

Although there is little vertical structure to a well-mixed estuary, there may be horizontal variations in salinity. Most estuaries are funnel-shaped, wide at the mouth and tapering inland. Coriolis force (**Box 1.1**) tends to generate circular movement of water within the relatively confined estuary basin, swinging both the inflowing seawater and the outflowing river water to the right in the northern hemisphere and to the left in the southern hemisphere. This means that, looking seaward from the river mouth, the seawater flows up estuary on the left-hand side while the river flow tends to flow on the right-hand side in the northern hemisphere, the reverse occurring in the southern hemisphere. Some mixing occurs laterally so that a horizontal residual circulation is produced (**Figure 4.4**).

Negative estuarine circulation

In areas such as the Arabian Gulf, high evaporation rates at the head of the estuary produce hypersaline waters which, being denser than normal seawater, sink and flow seaward, drawing in a surface flow of seawater. This results in a distortion of salinity gradients, or isohalines,

to seaward in the bottom waters and is referred to as a negative estuarine circulation (**Figure 4.3d**).

A classification by type of tidal exchange

In contrast to the temperate northern hemisphere, catchments in temperate Australia are characterized by irregular flood and fire, which strongly influence estuary hydrology and nutrient inputs (following bush fires, ash and nutrients enter the water as a distinct pulse). Based on this, Roy *et al.* (2001) identify three main types of estuary—tide-dominated, wave-dominated and intermittently closed—determined by their geological nature and the nature of the seaward entrance and so the degree and type of tidal exchange. Within the estuaries, four zones are described and these can be recognized in each type of estuary. From seaward the four zones are (i) marine flood-tidal delta, (ii) central mud basin, (iii) fluvial delta and (iv) riverine channel/alluvial plain. The ecology of a zone is influenced by estuary type, which determines the salinity regime, stage of sediment filling (geological maturity), which controls the spatial distribution/size of the zones, and human impacts from exploitation and development.

4.2.3 Sedimentation in estuaries

Estuaries are zones of sedimentation. River water emptying into the relatively large volume of an estuary and seawater entering its relative shelter both lose their momentum and are unable to retain the larger sediment particles that they are carrying. Aggregation of fine particles, described below, adds to sedimentation, so estuaries are typically dominated by soft sediment substrata, particularly fine muds.

Deposition of muds is encouraged by two processes which cause aggregation of very fine particles, increasing their weight and their tendency to be precipitated: biological aggregation and flocculation. Biological aggregation results from the ingestion of fine particles by filter-feeding animals and their subsequent egestion

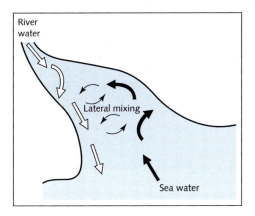

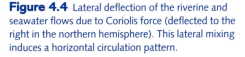

Figure 4.4 Lateral deflection of the riverine and seawater flows due to Coriolis force (deflected to the right in the northern hemisphere). This lateral mixing induces a horizontal circulation pattern.

Box 4.3   Estuarine flushing and the conservation of salt

Flushing time

It is possible to calculate an estuary's flushing time, the time it takes for the average particle of water to move through the estuary. To do this we use the mixing of salt water as a marker. The flushing time T_F is given by:

$$T_F = Q/R \qquad (1)$$

where Q is the volume of river water in the estuary and R is the river inflow rate.

In reality this is calculated by measuring the fractional fresh water concentration (f) in a number of subdivisions of the estuary,

$$f = \frac{S_s - S}{S} \qquad (2)$$

where S_s is the salinity of seawater and S is the salinity (tidal average) at a particular point in the estuary, then

$$Q = \int f \, dv \qquad (3)$$

(i.e. integrated over entire volume of estuary).

This is a very laborious method, requiring a large amount of data to be collected.

Estuarine tank model

An alternative is to use an analogue approach, such as the tank model. Here the estuary is visualized as a basin or water tank with a tap allowing in fresh water, another allowing in saline water and an overflow pipe through which the outflow water leaves the estuary. Such models assume that the sea is well mixed so that the salinity of the inflowing seawater is not influenced by the salinity of the outflowing estuary water. If we assume a steady state, then the amount of salt entering the estuary will equal the amount of salt leaving. If the salinity of river water, $S_R = 0$ and the river inflow is R per tide, and the salinity of seawater is given by S_s and the volume moving in per tide is P, the volume of the tidal prism, then the incoming volume of salt is given by:

$$(R \times 0) + (P \times S_s) \qquad (4)$$

The amount leaving the estuary is the total outflowing volume ($R + P$) multiplied by the salinity of the outflow (S), i.e.

$$(R + P) \times S \qquad (5)$$

These two must balance, so that:

$$(R \times 0) + (P \times S_s) = (R + P) \times S \qquad (6)$$

Rearranging gives us:

$$S/S_s = P/(R + P) \qquad (7)$$

Box 4.3   Continued

From this we can estimate the fractional fresh water content of the estuary from:

$$\hat{f} = \frac{S_s - S}{S} \tag{8}$$

$$\therefore \hat{f} = 1 - \frac{S}{S_s} \tag{9}$$

$$= 1 - \frac{P}{R + P} \tag{10}$$

$$= \frac{R}{R + P} \tag{11}$$

Therefore, based on the assumption that our estuarine tank is a good model,

$$\hat{Q} = \hat{f}v = \frac{RV}{R + P} \tag{12}$$

where V is the mean (over a tidal cycle) volume of the estuary.

$$\hat{T}_f = \frac{\hat{Q}}{R} - \frac{V}{R + P} \tag{13}$$

It is, therefore, possible to estimate the flushing capacity of an estuary from the river input (e.g. data from river gauging stations), the tidal prism (calculated from published tide tables and nautical charts) and the volume of the estuary (estimated from nautical charts). This approach assumes that the entire estuary is mixed within a tidal cycle. This is not true for most estuaries, but is acceptable for smaller basins.

as components of faecal pellets, much larger than the original particles and therefore with higher settling velocities. Flocculation is aggregation of small clay particles into larger particles, resulting from molecular attractive charges. In fresh water, clay minerals tend to carry a negative charge and repel each other, whereas in saline waters these charges are swamped by free cations, allowing the natural molecular attractive forces to dominate. This is further enhanced by the absorption of clay particles on to the surface of organic molecules, which brings them into close proximity and allows the molecular forces to take effect. This change in particle behaviour occurs around a salinity of 10 and so this part of the estuary is often very turbid.

Muds are only deposited in areas of low flow, but as mud particles are cohesive once deposited, they are resistant to resuspension by flows much stronger than those under which they were deposited. Muds are rich in organic matter, but the small particle size leads to small interstices between the particles. This restricts the available habitat for meiofauna and reduced oxygen penetration as the flow of oxygenated water is reduced. The latter, combined with the high bacterial oxygen demand associated with decomposition of the organic matter, leads to anoxic conditions away from the surface. There are, therefore, two distinct ecological zones in the sediment, oxic and anoxic, separated by the redox discontinuity layer (RDL), a zone of rapid change in oxygen

Figure 4.5 Exposing a piece of sediment clearly shows the change from oxic (pale) to anoxic (dark) sediments at the RDL. It is also possible to see the local oxidation of sediment along the sides of macro-faunal burrows, in this case the amphipod *Corophium volutator* (Alnmouth, northern England). Photo C. Frid.

levels and hence redox conditions (**Figures 4.5** and **4.9**).

Sedimentation in salt wedge estuaries

Because the dominant flow in salt wedge estuaries is a seaward movement of fresh water at the surface, with only a weak inflow in the saline layer, virtually all suspended material in the estuary will be of river origin. Some of the coarser material settles through the halocline, but most is carried to sea, where flocculation and the reduced flow rates allow deposition. If flow rates are low and suspended load high, then deposition may occur immediately to seaward, forming a delta. Larger particles are carried by rivers as bed load, rolling along the river bed rather than carried in the water column. At the head of the estuary, where the fresh water meets the salt wedge and flows over the saline water, the currents pushing this bed load along rise above the bed, and any material being transported as bed load will be deposited here, forming a shallow bar of coarse sediments immediately upstream of the salt wedge.

Sedimentation in partially mixed estuaries

In partially mixed estuaries, the potential to move sediment seaward by the freshwater flow is balanced by the landward flow of the saline waters, ensuring that export of sediment from such estuaries is generally low. The landward flows adjacent to the seabed are, however, sufficiently strong to move material. A turbidity maximum is produced where the net landward transport ceases: this is essentially an area of extremely high concentration of suspended sediment, containing small particles up to 10 μm grain size, and is produced by the residual currents acting as a sediment trap.

In estuaries with low tidal ranges, suspended loads in the turbidity maximum are of the order 100–200 mg L^{-1}, whereas in estuaries with high tidal ranges, such as the Severn, they may reach up to one hundred times this concentration. The turbidity maximum does not remain stationary, but moves back and forth with each tide and with the spring–neap cycle (Box 5.1), while unusually high fresh water discharges may push it down and even out of the estuary.

Sedimentation in well-mixed estuaries

The horizontal flow asymmetry in well-mixed estuaries leads to the deposition, in the northern hemisphere, of marine sediments on the left bank (looking seaward) and river sediments on the right bank, the converse occurring in the southern hemisphere.

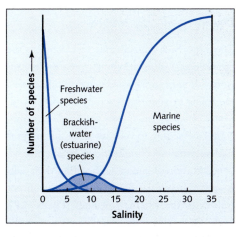

Figure 4.6 The pattern of species richness and the proportion of the biota comprising freshwater, brackish water and marine species along the estuarine salinity gradient.

4.3 Ecological effects of the abiotic environment

The estuarine environment is rigorous and requires that its inhabitants possess adaptations to cope with a wide range of salinity concentrations, salinity fluctuations over short time periods and continual movement and deposition of sediment.

4.3.1 Salinity

Estuarine organisms are divisible into three groups: those of marine origin, those of freshwater origin and true estuarine species. The majority of organisms in an estuary are marine species, adapted to a fully saline environment, which are however euryhaline: able to tolerate the lowered and variable salinities of estuaries. Each species has its own salinity tolerance, but most can exist in salinities as low as 18, and some can cope with salinities as low as 5. Marine organisms which lack the ability to cope with the variable salinity regime in estuaries (stenohaline species) will occur at the mouth of estuaries but rarely penetrate into areas where the salinity falls below 25. Freshwater

organisms (oligohaline species) are generally unable to tolerate salinities of more than 5 and so they, too, will not penetrate into the estuary proper. There is a range of salinity at around 7–10 in which there is a break in faunal distributions (Figure 4.6). While freshwater species dominate in waters of lower salinity than this critical level and marine species become diverse at higher salinities, the number of species occurring within this range is very low. It would appear that it is difficult for marine organisms to osmoregulate down to such low concentrations, while freshwater organisms are unable to cope with the cell volume changes that these salinities induce (see Section 2.1).

The majority of truly estuarine species are able to cope with fully saline conditions, having evolved from marine lineages, but they are restricted to the estuaries by competition with stenohaline species or by a requirement for fine sediments on the bed. They are most common in the 5–18 salinity range where the number of potential competitors is limited (Figure 4.6), and require physiological adaptations to tolerate salinity fluctuations, often supplemented by behavioural responses which reduce the exposure to salinity stress.

Physiological adaptations

There has been considerable research into the physiological adaptations of estuarine organisms. For most soft-bodied invertebrate species, the ability to osmoregulate is poorly developed. Many estuarine organisms are however osmoconformers, able to tolerate wide fluctuations in their internal concentrations, by altering the concentration of amino acids within their cells to maintain cellular fluids at a similar concentration to the external medium (Figure 4.7). Such mechanisms are usually only possible down to salinities of 10–12, but some species are able to osmoregulate to maintain a constant concentration of ions in their body fluids, rather than allowing it to fluctuate according to external conditions (Figure 4.7). This ability occurs in polychaetes such as the lugworm (*Arenicola* sp.) and molluscs including mussels (*Mytilus edulis*), but is most developed in the crustaceans, where the presence of a shell

reduces the permeability of the body, and there is evidence of the active uptake of ions and in some species the production of very dilute (hypotonic) urine. Intermediate between the osmoconformers and the regulators are those species which show partial regulation. This usually involves conforming to changes in environmental salinity over a certain range, but initiating active osmoregulation when environmental concentrations move outside this range.

Behavioural mechanisms

Most resident estuarine species employ behavioural mechanisms to reduce the exposure to salinity stress and hence obviate either physiological stress or the need to expend energy osmoregulating. Among molluscs, the most common behaviour is the closing of the shells, either by an operculum in the case of gastropods or by closing the valves in the case of bivalves, in which state the organism is unable to feed or respire. As intertidal species (Section 5.3) have adaptations to withstand aerial exposure, including sealing the valves and respiring either at low levels or anaerobically, these provide pre-adaptations to survival in estuaries.

Anadromy, catadromy and amphidromy

The most dramatic changes in salinity are experienced by fish species which at some stage in their life cycle pass through the estuary on migration between the sea and fresh water.

Anadromy is the term given to life history patterns in which the adults migrate from the sea into fresh water to breed, salmon (Salmonidae) and sturgeon (*Accipiter* spp.) being the best-known examples. All anadromous fish have well developed osmoregulatory capacity, but the pattern of movement from saline to fresh water varies between species, with some moving rapidly through the estuary while others, such as Atlantic salmon (*Salmo salar*), spend a period of several days resting and acclimatizing in the estuary. Anadromy may be a mechanism for reducing competition between the adults

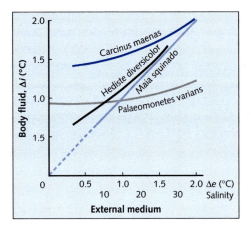

Figure 4.7 The varying concentration of body fluids of various animals in response to changing salt concentrations in the surrounding media (measured as depression of freezing point [°C]; approximate equivalent salinities are also shown). The spider crab (*Maia*) is unable to regulate its fluids. The ragworm (*Hediste*) is an osmoconformer: it is a poor regulator but has a high tolerance of reduced salinity. The shore crab (*Carcinus*), mitten crab (*Eriocheir*) and prawn (*Palaeomonetes*) will tolerate some dilution of their body fluids, but are good osmoregulators, holding their body fluid composition despite further changes in external concentrations. From Nelson-Smith 1977.

and young by placing them in different environments, thus allowing exploitation of a different set of resources, or to avoid high levels of predation when the fish are young. It is much more common at high latitudes than in the tropics, possibly due to the greater diversity of the freshwater fish fauna in tropical regions filling all the available niches, so that the high levels of competition or the higher predation risk in these communities negates any advantage of moving out of the adult environment to breed. The importance of competitive exclusion is suggested by the observation that, in the absence of a typical freshwater fish fauna, marine species have invaded.

Catadromy is the term given to the life history pattern of species such as the eel (*Anguilla* spp.) in which the adults live in fresh water but migrate to sea to spawn. It is much less common than anadromy among fish, but occurs commonly in crustaceans (**Sections 2.3; 3.6**). For these species, the need to return to the sea is probably related to an incomplete adaptation to the fresh water environment rather than the need to exploit a different habitat to the adults.

On the islands of the Caribbean and Indo-Pacific the river systems are insular, oligotrophic and subject to extreme seasonal variations. These systems have been colonized by various species of gobies (Gobiidae) that exhibit a life history pattern that is termed amphidromy. The adults spawn in fresh waters and the developing embryos drift downstream, through the estuary and into the open sea where they undergo a planktonic phase. They then return to the rivers to grow and reproduce (Keith 2003).

Seasonal changes in salinity

The strength of the tide varies over the tidal cycle, the spring–neap cycle (see **Box 5.1**) and over longer time periods, including an annual component. The amount of fresh water entering an estuary also follows a seasonal pattern, with certain times of year having a greater fresh water inflow than at other times. As the distribution of salt in the estuary and the mixing

regime are determined, in part, by the balance between the inflow of fresh water and the mixing derived from tidal flows, so these patterns also vary seasonally. It is possible for a single estuary to display characteristics of a well-mixed, partially mixed and a salt wedge estuary at different times of the year, and mobile elements of the biota will move in response to these changes. Marine organisms often penetrate further up estuaries during drier seasons, when fresh water flows are lower than in wet seasons, and this is enhanced in north temperate regions by the fact that many species of marine zooplankton are able to tolerate lower salinities at higher temperatures, corresponding to the drier summer season. In the St Lawrence Estuary (Canada) the zooplankton can be divided into three groups, whose spatial distribution varies seasonally in response to the changing distribution of salinity, temperature, turbidity and circulation. It is, however, often difficult to separate a response to a seasonal change in the salinity distribution from a true seasonal behaviour. For example, movement of marine species into an estuary during the summer may be the result of dispersal as a response to increased population densities following reproduction, rather than simply following increases in estuarine salinity. Periodic events such as droughts can provide opportunities to separate these responses. During the period 1989–93, fresh water flows into the

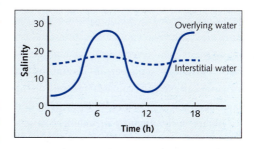

Figure 4.8 Comparison of salinity variation in interstitial water with the overlying water column through a tidal cycle. Salinity in the sedimentary environment varies much less than above the sediment surface, and is intermediate between the high and low tide values.

Thames estuary were very low as a result of low rainfall and heavy extraction of water from the river upstream. During this period, salinity in the estuary remained high throughout the year and certain marine species, which had previously entered the estuary only during the summer months, persisted all year. Flow may also be an important component; the Chinese mitten crab (*Eriocheir sinensis*) successfully colonized the Thames estuary during a drought period after 60 years of occasional records, probably because winter flows, previously too rapid to allow establishment of young crabs, were significantly reduced (Attrill and Thomas 1996).

4.3.2 Sedimentation

Estuarine organisms must be adapted to sedimentation and, indeed, most have adopted a burrowing lifestyle, which has the advantage that it offers a refuge from harsh abiotic conditions and from predation, particularly for those individuals which occupy the tidal flats exposed at low tide. Burrowing species need to remain near the sediment–water interface in order to feed and respire, but remain buried whenever possible to reduce the risk of predation. A rapid burrowing ability allows any exposed individuals quickly to regain the safety of the sediment and, conversely, to return to the surface of the sediment following burial by sediment deposition.

While estuaries are regions of high sedimentation when assessed on geological timescales, the rates are rarely high enough to cause smothering of the fauna. Nevertheless, the high quantities of mobile sediment in certain turbid estuaries can exert an ecological pressure on the fauna, a problem compounded by the temporal variability in the quantity of sediment in suspension, which varies with both the ebb–flood and the spring–neap tidal cycles. Few species are able to tolerate the extreme conditions in the turbidity maximum. Those that can occupy the more turbid regions of estuaries are either mobile epifauna such as the brown shrimp

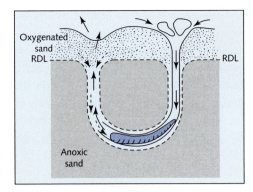

Figure 4.9 Effect of a lugworm (*Arenicola*) burrow on the position of the RDL, and hence the extent of oxygenated and de-oxygenated sediments. The lugworm creates a respiratory current (solid arrows), drawing oxygenated surface water into its burrow, so that it may respire without surfacing; sediment upon which it feeds moves in the opposite direction (dashed arrows). Lugworms grow up to 15 cm in length.

(*Crangon crangon*), or rapid burrowing infauna with heavy shells such as the cockle (*Cerastoderma edule*). The heavy shell lowers the chances of them being transported in the flows and reduces predation pressure on individuals exposed by the scouring away of sediments.

In tropical estuaries the roots of mangroves, and in temperate areas vegetation such as sea grasses and saltmarshes act as sediment traps, reducing flow and hence promoting deposition (Section 8.7). This can lead to the creation of marked changes in the sediments at the junction between vegetated and unvegetated areas.

4.3.3 Residual flows and dispersal

Irrespective of the type of estuary, all estuaries have a net seaward residual flow. This is not a problem for large mobile organisms, but organisms with limited mobility risk being flushed out of the estuary. Estuarine plankton overcome this by the use of behavioural mechanisms which utilize the residual circulation. Many estuarine zooplankters use a tidally synchronized vertical migration to maintain their position in the estuary, travelling up estuary in the near bed flow during the ebb, rising to the

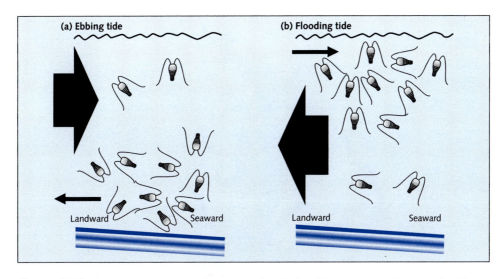

Figure 4.10 Vertical migration as a mechanism to reduce flushing from an estuary. (a) During the ebb tide, individuals migrate into the bottom waters, or adopt a benthic habit; (b) on the flood tide, individuals migrate into the upper part of the water column.

surface waters to feed during the flood when the distance carried seaward will be least, before migrating down to be carried back up estuary on the next ebb (Figure 4.10). Such cycles may also occur in response to seasonal changes. For example, the red-tide forming dinoflagellate *Prorocentrum mariae-lebouriae* is carried around Chesapeake Bay (Figure 4.11) in an annual migration cycle that covers 240 km. The centre of the *Prorocentrum* population is carried in outflowing surface waters to the mouth of the bay by late winter, where it is mixed into the more dense coastal waters. Fresh water flow into the bay is low over winter because most of the precipitation into the catchment is retained as snow and ice. In early spring, however, meltwater generates a large volume of fresh water run-off, establishing a strong density difference, or pycnocline, in the estuary, with salt water forced towards the bed as a salt wedge. The coastal water, containing *Prorocentrum*, flows up estuary as a near bed flow in the salt wedge. *Prorocentrum* accumulates just below the pycnocline and, in late spring, reaches the northern part of the bay, where the rapid shelving of the bottom forces the pycnocline upwards and increases turbulent mixing. This mixes nutrient-rich deep waters containing the previously light limited *Prorocentrum* to the surface, allowing a red-tide bloom to occur in the upper part of the bay. The population is then carried seaward in the surface waters. Precipitation in summer is low and the seaward flow is therefore weak, so the centre of the *Prorocentrum* population is once again in surface waters at the mouth of the bay late the following winter.

4.4 Energy inputs

4.4.1 Primary production

The variety of primary producers is high in most estuarine systems, including phytoplankton, saltmarsh plants or mangroves growing at the margins, macroalgae and sea grasses growing in the intertidal and microphytobenthos—dinoflagellates and diatoms—inhabiting the spaces between the sediment particles. Estuarine environments therefore rank among the most productive habitats on Earth, but primary production is unevenly divided among these groups, and is dominated by higher plants in adjacent wetlands, rather than within the water

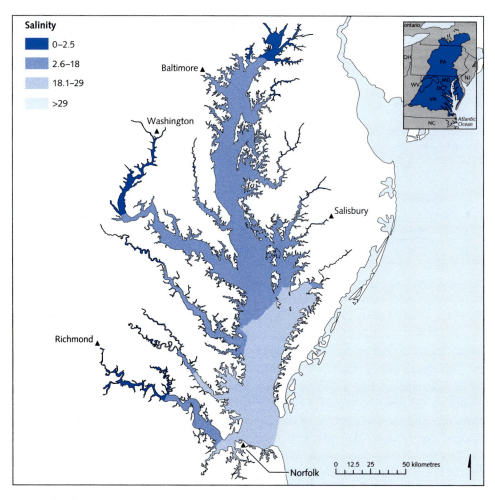

Figure 4.11 Chesapeake Bay, eastern USA, one of the world's largest and most complex estuaries. Isohalines are marked where mean salinity is approximately 1, 18 and 30, delimiting the three sections of the estuary.

itself. Primary production by phytoplankton, although high relative to that in the open sea, is therefore relatively low (Table 4.1). Processes occurring in estuarine wetlands—saltmarshes and mangals (mangrove swamps)—are considered further in Chapter 8; their contribution to energy budgets within the open water estuarine system is discussed below.

The intertidal and subtidal zones (Section 5.2) support few macroalgae, as these require solid substrate on which to grow. Rooted flowering plants may, however, be present, including sea grasses, such as *Zostera* spp. and *Thalassia* spp., which can grow over extensive areas, creating what are known as sea grass meadows. The

intertidal zone of estuaries along the Atlantic coast of North America is often dominated by *Spartina alterniflora*, a saltmarsh grass. Other *Spartina* species are, however, generally confined to the upper part of the shore, which is less frequently flooded, and therefore European saltmarshes, from which *S. alterniflora* is absent, do not extend far into the intertidal zone (but see Section 8.7).

The rate of primary production within the estuary itself is determined particularly by the availability of light. Nutrients are rarely limiting in estuaries as they are carried in from adjacent coastal waters and with the freshwater inflow as well as being produced by decomposition

Table 4.1	Comparison of primary production by various components of the estuarine ecosystem and equivalent marine systems		
	Locality	gC m^{-2} yr^{-1}	dry wt g m^{-2} yr^{-1}
Estuarine			
Phytoplankton	Lynher, UK	82	(164)
	Grevelingen, Netherlands	130	(260)
	Barataria Bay, USA	210	(420)
Microphytobenthos	Lynher, UK	143	(286)
	Grevelingen, Netherlands	25–37	(50–74)
	Barataria Bay, USA	240	(480)
Sea grass (*Zostera*)	Range:		116–856
Saltmarsh	*Limonium*, UK		1050
	Salicornia, UK		867
	Spartina, UK		970
	Spartina, east coast USA		445–3990
	Spartina, west coast USA		1360–1900
Mangrove	*Rhizophora*, SE Asia		340–1580
	Sonneratia, SE Asia		790–1700
Marine	Kelp (*Laminaria*)	1200–1800	(2400–3600)
	Coastal plankton	100	(200)
	Open sea plankton	50	(100)

From data summarized in Chapman (1977), McLusky (1989) and Wafar *et al.* (1997). Algal production is generally quoted as g C m^{-2} yr^{-1} (see Box 7.3), while macrophyte production is quoted as dry wt g m^{-2} yr^{-1}. Approximate conversions for algal production are given (in parentheses) to allow comparison: dry weight production is approximately twice that of C.

and recycling processes within the estuary itself. The principal restriction on primary production in most of the estuary is the high turbidity—this limits phytoplankton growth in the water column and restricts microphytobenthos, macroalgae and sea grasses to a narrow littoral zone. Phytoplankton production may be further limited due to the flushing of the population out of the estuary.

4.4.2 Detritus

The principal source of organic matter in estuaries is detritus, which forms the basis of the estuarine food web (Figure 4.12). Most of the permanent fauna of the estuary are supported by detritus, either directly as detritivores, or indirectly as predators of detritivores. Estuaries are, therefore, similar to rivers in the importance of detritus inputs (see Section 3.4). Indeed, much of their primary production is not utilized directly by grazing organisms, but incorporated into estuarine food webs as detritus. The sources of detrital material vary between estuaries but include autochthonous material—the remains of saltmarsh plants and mangroves for example—and allochthonous inputs, washed in by the river and also brought in from the open sea. Allochthonous material includes the carcasses of freshwater or marine organisms washed into the estuary and killed by osmotic stress, as well as marine macrophyte remains, while detrital inputs from the river, at the lower end of the river continuum (Section 3.6), are dominated by FPOM, and the physical structure of adjacent wetlands, acting as traps for large particulate detritus, ensures that most export into the estuary is of FPOM or DOC broken down by processing within the wetland. Most estuarine detritivores are therefore adapted to consume FPOM, and the shredding functional group (Section 2.3) is rare or absent.

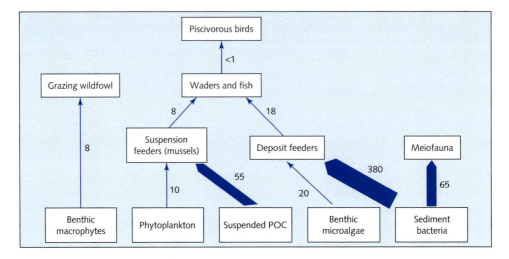

Figure 4.12 The estuarine food web, showing major flows of energy (g C m^{-2} yr^{-1}), in the Ythan estuary, Scotland. Sediment bacteria derive their energy primarily from POM. From Raffaelli and Hawkins (1996).

Estuarine detritivores, in common with their fresh water equivalents, are unable to utilize much of the detrital carbon directly, as it is too refractory. They derive little nourishment from plant material itself, but rely instead upon it being conditioned by microbial action. In a process little different to that in rivers (Section 3.4), estuarine detritus is utilized by a range of specialist decomposers—bacteria and fungi—which contain the necessary enzymes to break down and assimilate the refractory material and, in some cases, are able to synthesize amino acids from inorganic nitrogen. Predatory protists living on the particles consume these primary decomposers, and each other, creating a microscopic food chain on the particles. Most estuarine detritivores digest off the entire microbial community which grows on the detrital surface, including microbial predators; they also digest any labile organic matter present in the sediment including, in the case of intertidal or shallow water populations, microphytobenthos.

The rate at which detritus is conditioned is determined by its surface area. Microorganisms are restricted to the surface of detrital particles by their requirements for oxygen and dissolved nutrients, so the smaller the particle, the greater the microbial activity per unit volume it will support. Small particles, as a consequence, are more nutritious to detritivores than are large ones. The smaller detrital particles are poorer in carbohydrate but richer in protein than larger, fresher material, as a result of degradation of complex structural carbohydrates and the synthesis into microbial biomass (**Figure 4.13**). Physical damage in the gut of the macrofauna means that the particles egested tend to be smaller than those ingested, but they are often bound up into faecal pellets which are large and heavy and so sink relatively rapidly. Once on the seafloor, the mucus binding them together breaks down and the particles are colonized again by microbes, the increased surface area to volume ratio producing a greater relative area over which aerobic microbial breakdown can occur. This allows a higher relative biomass of microbes to be supported and so gives smaller particles a higher food value for the benthos. Thus, while progressive reingestion leads to the digestion of the more labile portion of the detritus, this may be compensated for by the development of richer microbial assemblages. Successive passages through macrofaunal guts leads to progressively smaller particles until they are too small to be handled by macrofauna and the remaining degradation proceeds by microbial activity.

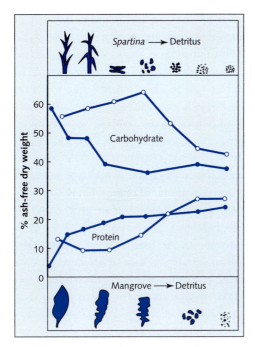

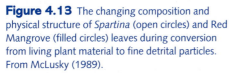

Figure 4.13 The changing composition and physical structure of *Spartina* (open circles) and Red Mangrove (filled circles) leaves during conversion from living plant material to fine detrital particles. From McLusky (1989).

Filter feeders are uncommon in most estuaries, despite the high density of detritus particles in suspension, probably because the concentration of suspended inorganic sediments means that they would catch large numbers of useless mineral particles, severely reducing the efficiency of their filtering apparatus. Filter feeders are more common at the seaward end of estuaries and some species, such as the small bivalve mollusc *Macoma baltica*, can switch between filter feeding and detrital gathering. The majority of macroinvertebrate detritivores rely instead upon collector-gathering as a feeding mechanism. Most are infaunal species, feeding on detritus which aggregates on the bed of the estuary whilst at the same time remaining in their burrows. They feed using appendages, with which they typically collect the entire surface layer, removing organic matter and egesting mineral particles. The European lugworm (*Arenicola marina*) is a good example of an

estuarine detritivore (Figure 4.9). It lives in a U-shaped burrow, spending most time in the bottom chamber, but periodically reversing up the tail shaft to defecate at the sediment surface. The tail shaft is kept clear of sediment and water is drawn down it to provide a respiratory current over the gills, after which it is pumped up the head shaft. The head shaft is filled with sediment, kept semi-fluid and oxygenated by the respiratory flow, which is ingested by the worm. The depression at the top of the head shaft preferentially traps small organic particles and the oxygenation of the head shaft sediments may stimulate bacterial conditioning of this material and production of bacterial biomass, hence the term 'gardening behaviour' for this activity (Hylleberg 1975).

Decomposition and deoxygenation

High concentrations of detritus in estuaries mean that microorganisms can become very abundant, their large density leading to a heavy oxygen demand and, beneath the sediment surface, almost inevitable anoxia. Only the surface 2 or 3 mm is oxidized by oxygen from the overlying water. Below the RDL, sediments are usually black, due to the build up of iron sulphide, and decomposition proceeds by anaerobic processes which are much slower than the aerobic ones. There is increasing evidence, however, that some species of nematodes utilize the special conditions around the RDL for chemoautotrophic production by migrating across the RDL, carrying chemoautotrophic symbiotic bacteria into alternatively reducing and oxidizing concentrations. When beneath the RDL, they load up with reduced organic material, which is then oxidized using oxygen when above the RDL, yielding metabolic energy in the form of carbohydrates (Ott and Novak 1989). This is analogous to the chemoautotrophy occurring in hydrothermal vents in the deep sea (Sections 1.4 and 6.7), but is, however, still ultimately based on solar radiation as the source of reduced compounds is the anaerobic breakdown of organic matter.

4

4.4.3 Energy budgets

Energy flow in estuaries is more complex than that in rivers because, in addition to longitudinal and lateral interactions based upon the single direction of flow of the river, there is the complication of tides alternating the direction of flow of seawater. Rivers entering an estuary act as a source of energy and nutrients. Estuarine tides, by pushing back inflows from the river as the tide rises, can create tidal conditions in fully freshwater wetlands upstream, but the effect of this is generally to reduce export of wetland production into the river, rather than supplying inputs from the estuary itself. Saline wetlands on the periphery of an estuary may act as a net source or a sink, depending upon the structure of the wetland. On the Atlantic coast of the USA, immature saltmarshes act as a sink for dissolved and particulate materials, whereas mature saltmarsh is a source, exporting materials into the estuary. European saltmarshes, in contrast, are generally net importers of energy, probably because they have little or no vegetation in the mid-

intertidal zone, an area occupied by *Spartina alterniflora* in North America.

This difference is illustrated by the carbon energy budgets for two estuaries shown in Table 4.2. In the Dollard estuary in the Netherlands, the main sources of carbon are from outside the estuary and phytoplankton production is very low, being limited by turbidity. In contrast, in Barataria Bay, Louisiana, the main source of carbon is the large stands of *Spartina* marsh in the estuary, but phytoplanktonic and microphytobenthic production are also important. These two estuaries represent extremes on the continuum of estuary type. There are clearly large differences between them, but both act as traps for nutrients and POM, which are consumed and recycled within the estuary, although much is eventually exported into the coastal sea.

Detritus derived from the saltmarsh grass *Spartina* and microalgae are important for a wide range of invertebrates, both benthic and pelagic, and their fish and bird consumers in US east coast estuaries. However, the exact contribution of the various components varies

Table 4.2   Organic carbon budgets, expressed as g C m^{-2} yr^{-1} (with percentage), for the Dollard estuary, The Netherlands, and Barataria Bay, Louisiana

Source		Sink	
Dollard estuary			
Particulate C from North Sea and River Ems	37,100 (46%)	Consumed in estuary	35,560 (44.1%)
From potato flour mill	33,000 (41%)	Utilized in water column	7200 (8.9%)
Phytoplankton primary production	700 (0.9%)	Utilized in sediment	18,200 (22.6%)
Microphytobenthos primary production	9300 (11.5%)	Buried in sediment	9900 (12.3%)
From saltmarshes	500 (0.6%)	Bird feeding	260 (0.3%)
		Dissolved C to North Sea	45,040? (55.9%)
Total	80,600	Total	80,600
Barataria Bay			
From saltmarshes	297 (39.6%)	Consumed in estuary	432 (57.6%)
Phytoplankton primary production	209 (27.9%)	Exported to sea	318 (42.4%)
Microphytobenthos primary production	244 (32.5%)		
Total	750	Total	750

Adapted from van Es (1977) and Wolff (1977) respectively.

between regions of the estuary. For example, in the upper reaches of the Plum Island Sound estuary (Massachusetts) material derived from freshwater marshes and local (upper estuary) phytoplankton dominate, while in the mid and lower reaches the dominant sources are saltmarsh vegetation, benthic microalgae and marine phytoplankton. Interestingly even in the upper estuary, carbon derived from riverine inputs of terrestrial material was a minor source of organic matter. Sobczak *et al.* (2005) developed a budget for the estuarine delta systems feeding San Francisco Bay. They demonstrated that detritus dominated the organic matter supply to the system but that phytoplankton primary production was ecologically more important, despite being a relatively small proportion of the total.

The movement of organic matter around the system is complex and not unidirectional. For example, the principal source of carbon for a number of invertebrates in an Australian estuary is sub-tidal seagrass meadows rather than *in situ* microalgae or fringing vegetation. There is also some evidence that particulate material tends to be retained in the estuary and so used by the estuarine food web, whereas terrestrial and freshwater-produced DOM is rapidly flushed out to sea and is therefore not available for uptake (McCallister *et al.* 2006). In the tropics, mangrove forests act as a net exporter of particulate organic carbon and they show tight ecological coupling with nearby seagrass beds (Section 5.4), but there is less evidence for the often hypothesized link between mangrove forests and coral reefs (Hemminga *et al.* 1994).

4.5 Community structure

■■■■■■■■■■■■■■■■■■■■■■■■■

Estuarine food webs are relatively simple (Figure 4.12), as the range of species capable of tolerating the fluctuating environment is small and the majority have ecologically similar roles.

The harsh physical conditions and fluctuating salinity of estuaries may suggest that their community structure should be dominated by disturbance. Secondary production can, however, be very high, those species which can tolerate the conditions reaching very high densities. This, in turn, gives scope for competition or predation to be important structuring processes.

4.5.1 Competition and predation

Given that detritus is such an important food source for estuarine species, one would expect competition for it to occur. There is, however, little evidence for competitive interactions. Some species apparently utilize different parts of the detrital resource spectrum, particularly particles of different sizes: Fenchel (1975), for example, showed that the range of particle sizes ingested by two species of the mud snail *Hydrobia*, which correlates with snail size, differs where the species occur together from the sizes ingested by each species when they occupy a site alone. While this represents one of the most widely quoted examples of character displacement, there are a number of serious problems with interpretation of data from such 'natural field experiments' due to, for example, the difficulty of getting truly matched sites (see Cherrill and James 1987).

Subtle differences in salinity tolerance allow closely related species to replace each other along the estuarine gradient (Figure 4.14). Here one can envisage the equivalent niche being occupied along the estuary by a series of species which differ in their salinity tolerance, so that their competitive ability for resources varies depending upon the environmental conditions.

Most of the secondary production in estuaries is in the macrofauna of the intertidal flats, whose populations are subject to intense predation pressure. Some predators, such as ragworms (*Hediste*), predatory gastropod molluscs and nemertean worms, occur as permanent residents of the estuarine benthic habitat, but the greatest predation pressure is applied by

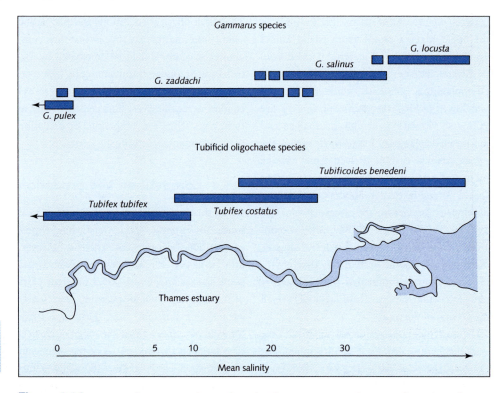

Figure 4.14 Zonation of macroinvertebrates along the Thames estuary in relation to salinity. Several closely related species may occur within the same estuary, partitioning the habitat according to the salinity tolerance of each.

temporary inhabitants of the estuary, which exploit both the infaunal predators and the detritivores directly. At high tide, fish move on the intertidal flats to feed, while at low tide these areas can support vast densities of wading birds (**Figure 4.15**). Temporary predators are seasonal in their occurrence, wading birds using estuaries either on migration in the spring and autumn or as winter feeding grounds, while fish predation tends to be greatest in summer when adults are building up reserves for the winter and the following breeding season and when young fish are present. The production consumed by migratory predators is exported from the estuary when they migrate away and, in the case of birds, feeding is at low tide on mud flats while much defecation will occur in terrestrial roosting sites, with energy consequently being lost from the estuarine system (this may be compared with concentration of energy around roosts in fresh water wetlands; see Section 8.5).

The role of predation in structuring estuarine benthic communities is equivocal. Many studies have concluded that it is important, in that epifaunal predators reduce deposit feeder densities, preventing competitive exclusion and helping to promote a diverse guild of deposit feeders. The estimated consumption of prey biomass by wintering birds in Europe, for example, varies from 6% in the Grevelingen estuary in the Netherlands to 44% in the Tees estuary in northern England. However, other studies have shown no effect on deposit feeder communities of excluding epifauna (Frid and James 1988). This may be because infaunal predators make use of the extra production, a form of competitive release. Alternatively, it may be that the epifauna selectively predate infaunal predators in preference to deposit feeders, as they tend to be larger, and therefore represent more valuable prey. Moreira (1997) estimated that wintering birds were consuming

(a)

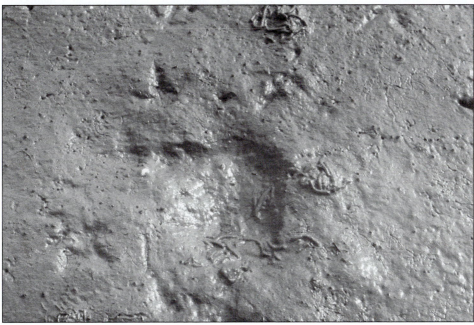

(b)

Figure 4.15 Predation in estuaries. (a) At low tide, large flocks of birds feed on the exposed benthos of the mudflats. Here, as the tide rises, oystercatchers (*Haematopus ostralegus*), formerly dispersed throughout the mudflat, aggregate on exposed rocks (Dee estuary, northern England). Photo S. Dobson. (b) Bite marks in the sediment provide evidence for fish feeding activity at high tide. Photo C. Frid.

12% of benthic biomass in the Tagus estuary in Portugal, mainly in the form of a single bivalve species, *Scrobicularia plana*. He suggested, however, that much of this consumption was of the feeding siphons of the bivalves, rather than the entire organism. Bivalves are able to regenerate lost siphons, providing a continuously renewed food supply for birds. It would therefore appear that epibenthic predators, especially birds but also crabs and fish, are ecologically important. This can however be via direct predation that reduces biomass and alters species composition of the basal species, or indirect involving grazing on siphons (and so exerting a cost on the grazed species and hence altering competitive interactions) or selective predation of intermediate predators in the food web.

4.5.2 Species diversity

Estuaries often contain a high diversity of habitats within a small area, including rocky shores, sand flats, mud flats, saltmarshes, sand dunes, lagoons and sub-tidal sediments. However, with the exception of the more terrestrial habitats, diversity within each habitat is usually highest at the seaward end and decreases up estuary, largely as a result of the progressive loss of species intolerant of low or fluctuating salinity and the absence of a specialist estuarine fauna to replace them (Section 4.3). The dominance of soft sediments is also a determinant of diversity, however, as it limits the fauna to those species able to tolerate or exploit this single habitat; solid substrates, such as sunken ships, will produce zones of dramatically increased diversity. The shipping channel dredged through the sediments in the Thames estuary is an area of particularly high diversity, because dredging has increased habitat heterogeneity in a location sheltered from physical disturbance by the steep sides of the channel.

4.5.3 Ecological functioning

In estuaries the ecological functions that have been most discussed are the role of organisms in elemental cycling (in particular the role in recycling detritus), productivity and the supply of food (see Section 4.4) and the provision of nursery habitat for coastal fish species. Estuaries do however provide to a varying extent all the main ecological functions (see Section 2.5) but as each estuary is unique so too will be the range of ecological goods and services produced.

Estuaries might be regarded as unusual as high biogeochemical rates are produced by systems with relatively low taxonomic diversity and low numbers of species. In low-diversity ecosystems, there is potentially a greater chance that certain species or groups of species may be critical to the maintenance of function (i.e. there is a lower level of redundancy).

In the case of estuarine sediments, a considerable amount of biological activity is associated with a thin layer of microscopic algae that inhabit the sediment surface (Cahoon 1999). The algal layer is important in a number of ecosystem processes. Microphytobenthos are an important source of new organic carbon, while the microphytobenthic layer can also control the rate and direction of inorganic nutrient exchange between benthic and pelagic compartments. The physical properties of the layer, in particular the smoothness and the presence of the mucus matrix, alter the erodability of the sediment. Forster *et al.* (2006) have shown that rates of primary production, a key ecosystem function, on the tidal flats of the Westerschelde estuary varied between sites but in a simple manner correlated with either biomass or diversity of the biofilm. Similarly, the diversity of the biofilm was unrelated to the biomass, the latter controlling the habitat modification effects of the assemblage. So for these two functions there was no simple relationship whereby high species diversity delivered high levels of functionality. This contrasts with the dominant paradigm for terrestrial and some other aquatic systems, where high diversity systems deliver more in the way of ecological functions (see Section 3.4), thus the highly stressful nature of the estuarine environment has resulted in a low diversity system that still delivers high levels of production and nutrient remineralization.

Given the vulnerability of estuaries to invasion by non-native species (see Section 4.7) there is scope for marked alterations in an estuary's ecological functioning from the addition of 'new' species with novel ecological functions. For example, in Willapa Bay on the US west coast there are a documented 45 non-native species established. Two of the non-native species of bivalve molluscs are harvested, and levels of secondary production harvested from these species is over 250% of the maximum level of production ever taken from the native oyster beds of the estuary. Furthermore, two species of emergent saltmarsh plant that have invaded the estuary have increased total primary production in the intertidal by more than 50%, most of this extra material entering the detrital pathway. Thus, the impacts of four non-native invasive species have resulted in the estuarine system undergoing marked changes in primary production, secondary production, levels of filtration/water clearance, detrital production and export and the availability of biogenic habitat (Ruesink *et al.* 2006).

4.6 Succession

Geologically, estuaries are short-lived. Most of those we see today were formed when sea levels rose at the end of the last glaciation and many will have disappeared given a similar length of time into the future. The reason for this is that estuaries act as zones of accumulation of sediment, by which process they are gradually infilled. Accretion of sediment allows colonization by plants which stabilize mud flats and enhance further accretion and vegetation development. In the absence of strong currents or storms, coastal wetland vegetation will gradually encroach deeper into the estuary. Alluvial sediments raise the estuary above sea level, enhancing terrestrialization or development of fresh water wetlands, while longshore drift may

form a bar across the mouth of small estuaries, eventually excluding inflow of seawater and creating a lagoon which gradually becomes fresh water.

Intermittent exclusion of seawater leads to widely fluctuating conditions. Lake St Lucia, on the Indian Ocean coast of South Africa, is a lagoon connected to the sea by a channel, the Narrows, 10 km in length. Maximum fresh water flow into the lagoon is during the summer, when excess water readily discharges to the sea through the Narrows and salinity in the lagoon ranges from negligible levels at river mouths to 35 towards the estuary mouth. During drought periods, however, water ceases to flow out through the Narrows, whereas occasional tidal incursions flow in, followed by evaporation which produces salinities typically approaching 50 towards river mouths. Hypersalinity is compounded by strong littoral drift, which can temporarily close the mouth of the estuary, stopping further marine incursions which would dilute the lagoon water; under these circumstances, salinities of up to 100 have been recorded. The biota of Lake St Lucia is a fluctuating mixture of marine and freshwater forms, the salinity fluxes creating instability and, occasionally, extensive mortality. In due course, sedimentation will dam the Narrows and the lake will become fresh water, but until such time the conditions, and therefore community structure, will remain unstable.

4.7 Human impacts on estuaries

4.7.1 Development on estuaries

Estuaries have been utilized by humans for many thousands of years. Initially they were good sites for the collection of shellfish and hunting fish and birds. Later they provided important transport functions with trading vessels using the relative shelter of estuaries to unload or as a route inland. As a result, trading

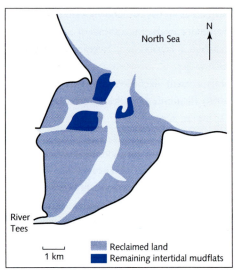

Figure 4.16 The extent of former tidal mudflats and salt marshes reclaimed from the Tees estuary, north-east England, since the early nineteenth century. Most is now used for industrial development.

centres developed on estuaries and these later became large towns and cities. Seven of the world's ten largest capital cities are built on estuaries and it is estimated that, in the UK alone, around one-third of the population lives in towns and cities associated with estuaries.

The extensive tidal flats of estuaries have long been seen as easy sites of land claim, to form grazing meadows or other agricultural land. Large areas of land around the Wash in eastern England were claimed from the sea as early as Roman times. In other estuaries the land is utilized for industrial development (Figure 4.16), with the advantage that the estuarine waterway provides a good route for transport of raw material and products. The development of urban and industrial centres on estuaries has frequently been facilitated by the availability of good building land on the wide level floodplain and by land claim of intertidal flats. Land claim removes the most productive part of the estuarine system—the tidal flats—fundamentally affecting its overall productivity. It also tends to increase sedimentation in the main channel, which then raises the costs of maintenance dredging for port operators and may affect the ability of the estuary to assimilate waste prod-

ucts. Most estuaries have lost at least some of their natural intertidal areas to land claim and many are now constrained to a channel flowing between artificial banks. Thus the current state of most estuaries is not a sustainable one, and many would consider the system to be ecologically poorer as a result.

4.7.2 Pollutants in estuaries

As the population living adjacent to an estuary grows, so does the need to dispose of waste, and the proximity of the estuary means it appears to be a convenient waste disposal route. Initially the dominant waste inputs were organic material, particularly sewage, but as industry developed and chose estuarine locations for ease of transport, so industrial waste disposal in estuaries increased. It is estimated that during the 1980s the combined discharges from the estuaries of the Elbe, Humber, Rhine, Scheldt, Tees, Tyne and Wear were second only to atmospheric deposition as a source of contaminants into the North Sea. The use of estuaries for human waste disposal has led to many environmental problems, including the build up of heavy metals from industrial wastes in the sediments and the food chain (including their presence in fish for human consumption) and the production of long stretches of anoxic conditions in estuaries receiving large quantities of organic pollution. Recent decades have seen major international and national initiatives aimed at reducing these inputs and in many estuaries water quality is now the best it has been for several hundred years (see Frid and Dobson 2002).

One of the most comprehensively studied estuaries is the Ythan estuary in north-east Scotland. This is a small estuary and has not been subject to large development pressures, but over the last 40–50 years it has seen increased inputs of nutrients from agricultural run-off in the catchment and discharges from sewage treatment works. Consideration of the changes in the estuary over the period 1960s–2000 show that the physical environment has changed little, numbers of shore birds (waders) increased and then declined but the most dramatic change has been the increase in mats of algae blanketing the surface of the mudflats (Raffaelli *et al.* 1999). Where the mats occur numbers of infauna and hence food for waders are decreased: this might be the reason for the recent decline in bird numbers. The exact cause of the increase in algal mats is contentious but the increases in nutrient inputs may play a key role.

4.7.3 Non-indigenous species invasions

The establishment of populations of non-indigenous species can have major impacts on the ecology of a system. In aquatic environments the biggest threat of accidental introduction of an 'alien' species occurs in estuaries. For example in the Elbe estuary in Europe a total of 31 species of non-native benthic macrofauna have been identified. Of these 21 occur mainly, or entirely, within the brackish middle reaches of the estuary. A number of (not mutually exclusive) hypotheses have been advanced to explain this pattern of invasion (Nehring 2006). The three mechanisms are:

1 As estuaries often have intensive international shipping there is a higher potential infection rate than other aquatic habitats;

2 Brackish water species have, due to their specific physiological characteristics, a better chance of being transported alive than stenohaline or freshwater species and they also probably have a higher establishment potential after release;

3 Brackish waters are naturally regions of low species richness and so potentially hold vacant niches that an alien species might invade into.

Many estuarine species are pre-adapted to being good invaders. For example polychaete worms of the genus *Marenzelleria* are native to the estuaries of the east coast of North America. They were first recorded in the North Sea region in 1979 when a population was dis-

covered in the Forth estuary (Scotland). Over the next decade they spread to virtually every North Sea estuary and invaded the Baltic in 1985. They subsequently spread north to reach as far as the Gulf of Finland (Bastrop and Blank 2006). *Marenzelleria* has a planktonic larva and it is thought that larval transport in ships' ballast water secured the initial invasion. Subsequently, the high fecundity of the worm and its dispersive larvae have allowed it to spread and colonize other estuaries.

4.7.4 Barrages

The fundamental importance of the dynamic physical regime to the functioning of the estuarine ecosystem is clearly revealed when physical barriers are erected across the estuary, interrupting these natural processes. Barrages or dams may be erected for a number of purposes, including providing permanent water bodies for amenity reasons (e.g. Cardiff Bay barrage in South Wales), preventing up estuary penetration of water carrying high concentrations of pollutants, storm surge barriers, or to generate electricity from the power of the tides. Storm surge barriers, such as the Thames Barrier in London, are designed to allow free propagation of normal tides into the estuary, and are closed only to prevent unusually high storm surges penetrating the estuary and flooding low-lying areas. If current predictions of sea level rise are accurate, considerably more storm surge barriers will be built in the near future but, as they are only closed relatively infrequently, they have low impact on estuarine ecology.

Barrages constructed to impound water for amenity use or to maintain water quality are generally closed for most of the time. They therefore act as a barrier to the propagation of tides up estuary, prevent sediment movement and interfere with migration patterns, although the latter are frequently addressed for fish by the construction of fish passes or 'ladders'. If the barrage is rarely opened, the impounded water body will become increasingly fresh in character, initially taking on the characteristics of a saline lagoon and, if isolation is prolonged,

4

a freshwater lake. The impounding of Cardiff Bay in Wales has led to the development of anoxic conditions as the organic load in the sediment decomposes. In response to this the impounded lagoon has air bubbled through it to stimulate water movement and promote oxygenation.

The large tidal range in some estuaries, caused in part by the channelling effect of the estuary on the tidal wave, has led to a number of proposals to build tidal barrages to generate electricity. The basic premise is that the tide flows in through open sluices, which are closed at high tide, impounding the water. After the tide has fallen to a certain level, sluices are opened to allow the impounded water to run out through turbines, which generate electricity. At La Rance in northern France, a tidal barrage has been operating since 1966, producing 544 million kilowatt-hours of electricity each year from the 13.4 m tidal range. Pilot scale plants operate in the Bay of Fundy, eastern Canada, and in Russia and China. Unfortunately, no ecological studies were carried out at La Rance before construction, whereas the proposed Severn Tidal Barrage in the UK has seen a large amount of work over many years directed towards predicting the ecological consequences of the development.

Theoretically, the principal problems associated with tidal barrages are that the reduced flow rates will alter patterns of sedimentation, salinity and flushing of pollutants, thereby affecting water quality. Pollutants discharged above the barrage will be impounded, giving them a longer residence time and potentially a greater exposure to the biota. Riverine sediments will build up above the barrage, but if, as is the case in the Severn, marine sediments normally dominate, the area above barrage would be starved of sediment. The decreased turbidity could lead to algal blooms, especially if nutrients were also being impounded behind the barrage. Following an algal bloom, or as a result of detrital matter accumulating above the barrier, problems of anoxia could be encountered. Sub-tidal communities would be directly impacted by the changes in sedimentation and indirectly by any changes in algal production. Intertidal communities would also be affected by the changed tidal regime. Below the barrage, the tide ebbs more slowly than normal, and may never fall as low as previously—reducing both the extent and duration of exposure of the intertidal. Above the barrage, the regime imposed is a rapid flood tide, and a slow ebb with the tidal range much reduced—again significantly reducing the extent and exposure of the intertidal. The most profound effect of this is on the bird fauna, the shortened feeding times and smaller areas available having potentially dramatic consequences for both resident and migrating wading birds. However, these predictions remain untested.

Bulleted summary

- Estuaries are zones of transition between freshwater and marine environments. They are complex systems containing many environmental gradients, but are characterized particularly by salinity gradients. The physicochemical conditions in an estuary are determined mainly by relative volumes of salt and fresh water. Tidal influences ensure that conditions are continually fluctuating.

- Estuaries are zones of deposition of sediment from both rivers and the sea. Soft sediments, therefore, dominate.

- The salinity distribution in the estuary is the primary determinant of the distribution of organisms. Few freshwater organisms can tolerate salinities in excess of 5, while few marine organisms can cope with salinities below 18, although most can survive in salinities between this and full seawater. True estuarine species, which are therefore most common in the salinity range 5–18, are few in number. There appears to be a critical salinity range at around 7–10 which causes a break in faunal distributions.

- Estuarine species adapt to sedimentation mainly by adopting a burrowing lifestyle. Planktonic species withstand the net seaward flow by taking advantage of differential surface and near bed flows, and follow daily or seasonal cycles of movement.

- Estuaries are amongst the most productive of habitats, yet much of this production is based upon detritus derived from external sources and is exported to coastal systems. Primary production is concentrated into peripheral wetlands, that within the water being light-limited due to turbidity. Most estuarine invertebrates are, therefore, generalist deposit feeders.

- Most predators are temporary members of the estuarine fauna: wading birds at low tide and marine fish at high tide. Estuaries provide important feeding grounds for large populations of wading birds and waterfowl and are utilized by coastal fish populations as nursery areas.

- In contrast to productivity, diversity is low, a combination of fluctuating salinity, lack of habitat heterogeneity and geological transience of estuaries. Of these, habitat heterogeneity is probably the primary determinant.

- Urban and industrial development has taken place on many estuaries, often on land claimed from the sea. Waste produced from these developments has frequently been discharged into the estuary and many estuaries show signs of human impacts, including high levels of heavy metals and sewage contamination.

Further reading

Hemminga M. and Duarte C.M. (2000) *Seagrass ecology*. Cambridge, Cambridge University Press.

Kennish M.J. (1986) *Ecology of estuaries: Vol. 1: Physical and chemical aspects*. Boca Raton, FL, CRC Press.

Kennish M.J. (1990) *Ecology of estuaries: Vol. 2: Biological aspects*. Boca Raton, FL, CRC Press.

Little C. (2000) *The biology of soft shores and estuaries*. Oxford, Oxford University Press.

McLusky D.S. (1989) *The estuarine ecosystem*, 2nd Edn. Glasgow, Blackie.

McLusky D.S. and Elliott M. (2004) *The estuarine ecosystem: Ecology, threats and management*, 3rd Edn. Oxford, Oxford University Press.

Mitsch W.J. (1994) *Global wetlands: Old world and new*. Amsterdam, Elsevier Science B.V.

Assessment questions

1 If salinity is the amount of salt dissolved in seawater, why does it have no units?

2 What are the two main determinants of the type of water circulation an estuary has?

3 What are the main ecological factors operating in estuaries?

4 What are the main types of waste that cause contamination and pollution in estuaries?

5 What are the main challenges faced by plankton in estuaries?

4

5

Coastal Seas

■■■■■■■■■■■■■■■■■■■■■■■■■■■■■■

5.1 Introduction

Coastal seas may be defined as those lying between the shore and the edge of the continental shelf, generally at around 200 m depth. They differ from the open ocean not just in their depth, but in the greater significance of physical processes such as tidal mixing and wave action. Inputs from land can influence water chemistry and mixing processes, particularly around the mouths of large rivers, and large discharges of fresh water can extend estuarine conditions (see Chapter 4) out into the open sea, as occurs along the southern coast of Alaska and the mouth of the Amazon. Conversely, a heavily indented coastline in an arid subtropical zone, such as the Gulf of California or the Red Sea, will create areas of high salinity, as evaporation exceeds the rate of mixing from the open ocean. In most of the world, the most important determinant of community structure along the coast itself is the action of tides —the vertical displacement of water which effectively moves the margin of open water up and down the shore at regular, frequent intervals.

5.2 The abiotic environment

■■■■■■■■■■■■■■■■■■■■■■■■

5.2.1 Tides

Tides are the periodic movement of the sea, generated by the gravitational attraction of the Sun and Moon on the hydrosphere (Box 5.1). The effect of tidal activity is the periodic immersion and emersion of a strip of the coastal zone, as the sea level rises (flood tide) and falls (ebb tide) in sequence, normally twice per day. Additionally, over a period of about 14 days the tidal range follows a spring–neap cycle (Figure 5.1). Spring tides, with the largest tidal range, occur at approximately fortnightly

Box 5.1   Principal tide-generating forces

Tides are the response of the hydrosphere to the gravitational attraction of the Moon and the Sun. This gravitational force also raises tides in the Earth's crust, but it is so dense and viscous that they are not visible.

The lunar tide

The Earth and Moon form a single system, revolving around a common centre of mass with a period of 27.3 days. As the Earth is much larger and heavier than the Moon, the common centre of mass lies within the Earth, although offset from its centre. Tidal movements are caused by a combination of centrifugal force, which acts away from the Moon as the Earth moves around the Earth–Moon system, and the Moon's own gravitational pull, which acts towards the Moon.

Ignoring the spin of the Earth on its own axis for now, the Earth–Moon system rotates, eccentrically, about the common centre of mass. The total centrifugal force within the Earth–Moon system exactly balances the gravitational attraction between the Earth and the Moon. It is this balance which keeps the Moon in orbit, so that we neither lose the Moon nor collide with it. The centrifugal forces operate in parallel to a line joining the centre of the Earth to the centre of the Moon.

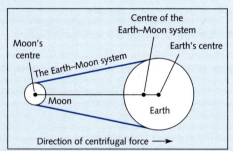

Every point on Earth experiences the same centrifugal force. The gravitation attraction of the Moon, in contrast, varies with distance from the Moon, as the inverse square of the distance, and is not the same everywhere. Its direction is always towards the centre of the Moon and therefore, except on the centre line, is never parallel to the direction of the centrifugal forces. The result of the centrifugal and gravitational forces at any point on the Earth is the tide-producing force.

The tide-producing force has two components, one vertical and the other horizontal. The vertical force acts against the pull of the Earth's gravity,

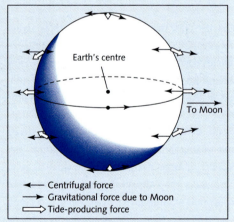

whereas the horizontal component, the tractive force, is essentially unopposed, except by seabed friction, and it is the primary mover of water. The tractive force acts to move water over the surface of the Earth to produce two bulges of water directly towards and away from the Moon. As the Earth rotates upon its axis, a point on its surface will pass under these two bulges of water and experience a semidiurnal pattern of two high tides with periods of low water in between.

Box 5.1   Continued

Variations in the lunar tide

The Moon's declination

The Moon's position is not fixed above the Equator, but moves 28° either side of the equatorial plane every 27.2 days. When the Moon is at a large angle of declination the plane of the two tidal bulges will be offset, and their effect at a given latitude will be unequal, particularly at mid latitudes. This produces a diurnal inequality in the semidiurnal tides, a large high tide being followed by a small high tide.

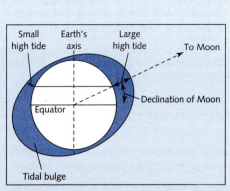

The Moon's elliptical orbit

The Moon's orbit is an ellipse, hence it moves closer to and further away during its orbit. At its closest approach, perigee, the tide-producing force is 20% above average, while at apogee, its greatest distance away, the force is 20% below average. The interval between successive perigees is 27.5 days.

The solar tide

The Earth–Sun system behaves in exactly the same way as the Earth–Moon system, producing two bulges of water on a line joining the centre of the Earth and the centre of the Sun. However, although the Sun is more massive than the Moon, it is considerably further away, and as a result of the inverse square law, its influence is smaller. The solar tide-producing forces are about 0.46 of the lunar ones.

The solar tide is affected by the Sun's declination, which varies by 23° each side of the equatorial plane over a yearly cycle. The Earth's orbit is elliptical, thus at perihelion, closest approach, the solar tidal forces are increased, while at aphelion (most distant) they are decreased, but this effect only causes a variation of about 4%.

Interaction of lunar and solar tides

At new Moon, when the Moon and Sun are in line (i.e. the Moon is in syzygy), such that a single line will pass through all three centres of mass, the two tidal bulges are additive and the difference between high and low water is greatest, creating a spring tide. At full Moon, the Moon is once again in syzygy, and spring tides again occur.

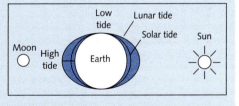

These events are (27.3)/2 days apart, i.e. approximately two weeks.

Between the new and full Moons the Moon is in quadrature, either the first quarter, or third quarter. Now the solar tide does not act to increase the height of the high waters, but does increase the height of the low tide, creating a neap tide.

5

Box 5.1   Continued

The highest astronomical tide

The highest astronomical tide is produced by the maximum tide rising force, that is with the Earth at perihelion, the Moon at perigee, the Sun and Moon in conjunction (new Moon), with the Sun and Moon both at zero declination and with the planets in optimum alignment. This combination would produce a tide of greatest height—HAT— and lowest low—lowest astronomical tide (LAT).

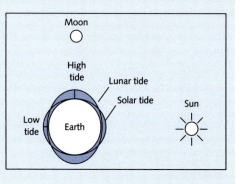

At Newlyn in Cornwall, used as the standard for sea level on British Ordnance Survey maps, the normal tidal range is about 3.5 m, the spring tide range is about 5 m and the range, HAT to LAT is 6 m. The next HAT is due in about 6580 AD!

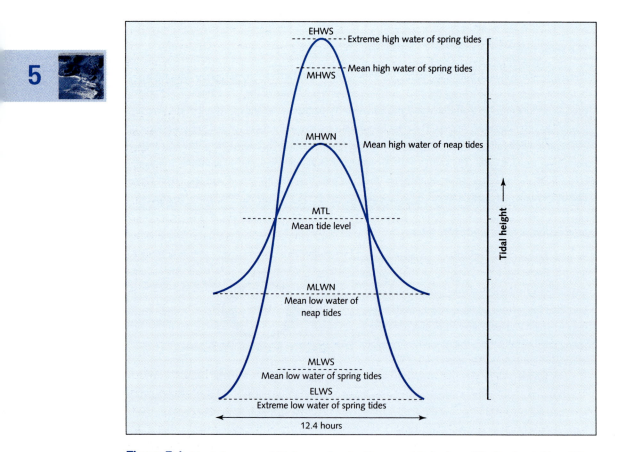

Figure 5.1 A typical open coast tidal curve, showing the symmetrical nature of the flood and ebb and the slow rise from low water followed by a rapid period of flood and a slowing just before high water. The various abbreviations used in describing tidal heights are defined. Note the difference in tidal range between spring and neap tides.

intervals, following which the tidal range decreases to the neap tides (approximately 7 days after the spring tides) and then increases again to the next set of spring tides. Local effects can modify these patterns, especially in bays which may have their own harmonic frequencies. In enclosed seas, such as the Mediterranean, the topography of the basin and the local tide-generating forces produce a tidal range so small that the sea appears tideless, whereas the funnelling of estuaries produces a tidal range much higher than that experienced along the adjacent open coast (Section 4.2).

Tides create a series of zones at the land–water interface, defined by the degree to which they are affected by the variation in water level. The area periodically covered and uncovered by the sea is referred to as the intertidal or littoral zone, the area permanently covered by the sea being the subtidal or sublittoral zone. Immediately above the littoral is the splash zone, or supralittoral, which, while never covered by the sea, is influenced by the splash and spray of saline waters. To these three zones may be added the supralittoral fringe, which is only covered by splash from storm waves, and the sublittoral fringe, the lower part of the shore only uncovered by particularly low spring tides. The supralittoral fringe marks the lowest extent of terrestrial vegetation, while the sublittoral fringe is a zone of much higher diversity than the littoral zone as a result of the presence of a number of species unable to withstand prolonged desiccation. Note that the term 'littoral' is used in a different context when applied to lakes (Section 7.3).

Tidal variations within the littoral zone create a range of conditions. Only the zone within the range of neap tides is covered and exposed with every tidal cycle, while the highest and lowest parts of the littoral are covered and exposed, respectively, only by spring tides. During neap tides, the very top of the shore is never submerged, while the lowest part is never exposed to the air. A result of this variation in the pattern of inundation is a vertical zonation of physical conditions, which is most clear on

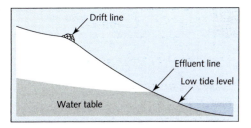

Figure 5.2 Zonation of physical conditions on sandy shores. The water table within the sediment may be higher than the surface water, extending the sublittoral up the shore, although this varies with beach type. The drift line marks the position of high tide, while the effluent line marks the elevation to which the water table extends. Adapted from McLachlan and Jarmillo (1995).

rocky shores. Sandy shores retain water within their sediments, which can effectively raise the sublittoral to a higher elevation than the low water mark (Figure 5.2a), particularly in dissipative beaches, which are low gradient and experience reduced wave action relative to higher gradient reflective beaches (Figure 5.2b).

The intertidal is a harsh environment which varies between extremes of submergence and aerial exposure over very short time periods. A temperate rocky shore in summer, with low water at midday, will be exposed to high light levels, temperatures, desiccation rates and salinity (due to evaporation); when the sea returns, coastal seawater conditions are restored within seconds. In winter, low water just before dawn will expose the shore biota to extremely low temperatures, while rain at any time of year may reduce salinity in pools almost to fresh water in a few hours.

5.2.2 Waves and currents

Waves are the result of energy being transferred between two fluids moving at different rates. They are a familiar feature at the sea surface, caused by the differential motion of the air and sea, but internal waves are also produced within the ocean along surfaces of different density, such as the thermocline (Box 1.3) in the open sea and haloclines (Section 3.2b)

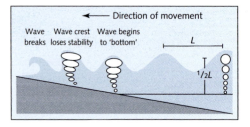

Figure 5.3 Water movement and wave breaking on the shore. L, 1 wavelength. See text for more details.

in estuaries. Waves are transient phenomena which transfer energy but not mass. Water below a wave in the open ocean undertakes a circular path as the wave passes. At the surface, the diameter of this path equals the wave height but, as one goes deeper in the water column, the diameter of the path decreases until, below a depth of approximately half the wavelength, no motion is felt (Figure 5.3). As a wave enters shallow water, these orbits interact with the sea floor, which exerts a drag, initially deforming the orbits into ellipses as the top of the wave travels more rapidly than the base. As the water further shallows, the drag slowing the base of the wave causes it to become unstable and the wave breaks (Figure 5.3), releasing its energy content. Table 5.1 shows the magnitude of the flows associated with breaking waves; velocities up to 10 m s^{-1} have been recorded on rocky shores, but the flow of a breaking wave is turbulent and velocities above and below these averages will be experienced. Furthermore, the level of wave action varies continuously, the large waves and strong drag forces which occur under storm conditions being interspersed with more moderate conditions and occasional periods of calm.

While the passage of waves introduces oscillatory motions with periods of seconds, tidal forces produce further oscillatory motions with periods of hours. These forces are strongest where sudden shallows or narrow straits constrain the flow. In the entrance to Lough Ine, a sea lough in south-west Ireland, the ebbing tidal flows reach 3 m s^{-1}, but when the tide turns, velocities up to 5 m s^{-1} in the opposite direction can be recorded only 5 minutes later. Tidal currents are comparable in magnitude to flows in stream and river systems, while the flows associated with large ocean waves may exceed anything experienced even in spate conditions in a river.

5.2.3 The nature of the substratum

The nature of the sea floor has a profound effect on the communities that develop there. Hard substrates are restricted to areas where water flows are sufficiently strong to keep them clear of sediments or where vertical surfaces have been produced by geological processes. In progressively less dynamic environments, the size of particles comprising the substrate decreases. There is a clear distinction between 'hard'—essentially non-mobile substratum— and 'soft'—mobile particles of any size. Soft sediment is the most extensive type of environment in coastal seas, covering much of the shore, including the extensive tidal flats of many estuaries (Section 4.7), and most of the sea floor.

Table 5.1 The velocity of water at the substratum at 3 m depth and under the breaking wave for waves of various heights		
Wave height (m)	Water velocity at 3 m deep	Substrata (m s^{-1}) breaking
1	0.9	4.4
2	1.8	6.3
3	2.7	7.7

Adapted from Denny (1987). The maximum water velocity associated with a wave occurs at breaking and varies with height of the wave.

Chapter 5 Coastal Seas **121**

Table 5.2	The effect of particle size on the physicochemical properties of soft littoral sediments	
	Fine-grained sediments	**Coarse-grained sediments**
Slope	Shallow	Steep
Wave action	Low	High
Stability	More stable	Unstable
Water table	At surface	Sinks at low water
Capillarity	High	Low
Permeability	Low	High
Oxygen penetration	Low	High
Organic matter content	High	Low
Bacteria	Abundant	Few
Redox discontinuity	Present—boundary near surface	Absent, or if present, boundary is deep

Particle size is an important determinant of conditions within soft sediments, because a number of environmental properties scale with the size of the particles (Table 5.2). This is partly the result of the processes which cause particles of a particular size to accumulate at a location, but also as a result of the environment produced by the particles themselves. A sediment in which the particles are of a similar size is referred to as well sorted, in contrast to a sediment of many different particle sizes, which is described as poorly sorted.

5.3 Ecological responses to abiotic conditions

5.3.1 The tidal cycle

Only a limited number of species are able to withstand the stressful conditions in the intertidal zone. They are almost all marine in origin, and the shore therefore represents a gradient of conditions in which the lower shore, being immersed for longer periods, is more benign than progressively higher levels. On a rocky shore, the upper distribution limits of most species correspond to some physiological bar-

rier, normally the ability to tolerate desiccation, and therefore many species do not manage to extend above the midshore region. Typically the upper shore contains relatively few highly adapted species such as lichens, isopod crustaceans and gastropods. Sandy shores, too, show some zonation in relation to physical conditions, although it is not so clear as on rocky shores. Normally, up to three zones can be distinguished (Figure 5.4) and the boundaries of these biological zones coincide with physical boundaries. As the fauna are mobile, the location of the zones varies each day as the tidal range changes. Reflective beaches support fewer species and fewer zones than dissipative beaches (Figure 5.4).

5

5.3.2 Wave action

Community structure in coastal seas is strongly influenced by wave action. Organisms entrained in the water column experience the passage of a wave but, if neutrally buoyant, merely follow a circular trajectory, whereas breaking waves exert a drag on benthic biota. In general, a large body form experiences a greater force, creating a selective pressure for small body size in wave-swept environments.

The variability and unpredictability of wave-induced flows has reduced the scope for adaptation by intertidal organisms (Denny 2006).

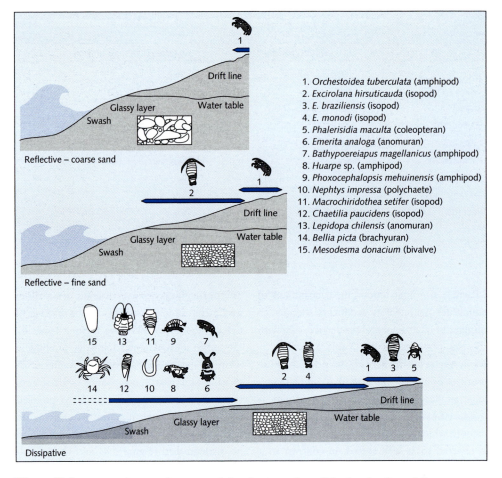

Figure 5.4 Zonation of macroinfauna in sandy beaches in southern Chile, showing the variation across a range of beach types. The insets illustrate the relative particle size on each of these three beaches. Note the absence of the lower zones on the most reflective beaches. From McLachlan and Jaramillo (1995).

1. *Orchestoidea tuberculata* (amphipod)
2. *Excirolana hirsuticauda* (isopod)
3. *E. braziliensis* (isopod)
4. *E. monodi* (isopod)
5. *Phalerisidia maculta* (coleopteran)
6. *Emerita analoga* (anomuran)
7. *Bathypoereiapus magellanicus* (amphipod)
8. *Huarpe* sp. (amphipod)
9. *Phoxocephalopsis mehuinensis* (amphipod)
10. *Nephtys impressa* (polychaete)
11. *Macrochiridothea setifer* (isopod)
12. *Chaetilia paucidens* (isopod)
13. *Lepidopa chilensis* (anomuran)
14. *Bellia picta* (brachyuran)
15. *Mesodesma donacium* (bivalve)

Storm waves, in particular, can profoundly affect shore communities. The size of storms tends to follow a log-normal distribution, such that rarely, at intervals of years or even decades, a very large storm occurs. Organisms may, therefore, recruit and grow under calm conditions only to be removed by an unpredictable storm event. Every organism has a certain ability to resist wave action, honed by evolutionary processes to meet its needs. Whether evolution has responded to a rare storm event that occurs at intervals of many generations depends on the other selection pressures operating but, as there is variation in the population, some individuals may survive and some will be lost following the disturbance. Others will survive through being 'sheltered' by topographic features, such as crevices, or other species, including clumps of mussels, which act as refugia (compare with Section 3.3).

At the shore scale, there are clear differences between shores with contrasting levels of exposure to wave action. Very wave-exposed temperate rocky shores in much of the world support a near monoculture of barnacles, with occasional mussels (*Mytilus*) (**Figure 5.5a**). As

(a)

(b)

5

Figure 5.5 (a) Exposed rocky shores are relatively species-poor; in this case, there is a zone of black lichen (*Verrucaria* sp.) above a zone of barnacles (*Chthamalus*), with the mid and lower shore covered in mussels (*Mytilus edulis*): St Ives, southern England. (b) On sheltered shores, the shore is dominated by algae, while the extreme shelter allows terrestrial vegetation to extend to the very top of the shore, where a lichen zone containing a high number of species merges into the intertidal, with *Verrucaria* extending into the upper shore. The upper shore has a zone of *Chthamalus* and channel wrack (*Pelvetia canaliculata*), the midshore has a zone of spiral wrack (*Fucus spiralis*) and the lower-mid and lower shore are covered in a dense canopy of egg wrack (*Ascophyllum nodosum*): Lough Ine, south-west Ireland. Photos C. Frid.

conditions become calmer, the cover of *Mytilus* initially increases, but then decreases as fucoid algae appear. In the North Atlantic, the fucoid cover varies with exposure: the midshore of wave-exposed shores is covered by *Fucus vesiculosis* var. *linearis*, which lacks gas bladders, but as the level of wave action decreases, this is replaced by the bladdered form of *F. vesiculosis* and, on sheltered shores, by *Ascophyllum nodosum* (**Figure 5.5b**). At high latitudes, where ice rafting tends to denude the shore each winter, each spring sees a succession from early colonizers through to the summer flush of rapidly growing macroalgae before autumn storms and the return of the ice closes the shore down for winter. This tends to reduce the degree of difference between sheltered and wave-exposed shores.

5.3.3 The nature of the substratum

Non-mobile substrata

Hard substrates clear of sediment provide secure anchorage points, allowing sessile filter feeders and macroalgae access to the water column. They are, however, a limited resource in the sea and competition for attachment space is intense, newly exposed hard surfaces—for example as a result of a rock fall, or the legs of an oil rig—being quickly colonized (see Section

5.5). Boulders are occupied in the same way, although any movement will lead to partial or total mortality of the attached fauna by abrasion or crushing. Overturning of a particular boulder is a probabilistic event—the larger a storm the more likely it is to be moved. Below a certain mass, boulders are moved so frequently that we can regard them as mobile substrates (see below). As boulder size increases, the probability of being moved within an organism's lifetime decreases, so that large boulders tend to have established communities while smaller boulders have communities typical of early succession stages—they are overturned so frequently that any succession is continually being reset to the beginning (Sousa 1979).

Mobile particles

Benthic organisms live in intimate contact with the substratum, usually within it **(Figure 5.6)**. Macrofauna will move particles aside as they burrow through the matrix or construct chambers, whilst meiofauna live in interstitial spaces or on the surface of particles. They are therefore responsive to changes in the physical and chemical nature of the sediment: species occupying a coarse sand beach (well drained, low organic matter, high oxygen) differ from those in a sub-tidal mud (poorly drained, high organic matter, low oxygen).

Well-sorted gravels are highly mobile, so construction of tubes or burrows is difficult. Offshore shingle deposits may, however, be relics of former dynamic environments which are now stable enough to incorporate fine particles deposited more recently; these support opportunistic sessile species attached to the larger stones, epibenthic forms and an infauna in the fine material in the interstices between the gravels. Sands accumulate in sub-tidal areas of moderate tidal flows, and in areas with moderate wave action. On the upper shore, the coarse particle size leads to rapid draining at low tide. This, and the mobility of the sediment, serve to restrict macrofauna to crustaceans **(Figure 5.4)**, which are well adapted to withstand desiccation and able to swim in the water column when they are washed out. The higher water content of lower shore sands allows a rich fauna to develop and, in areas of relative stability, extensive beds of biogenic structures may be produced, modifying the environment further. Dense beds of polychaete tubes, for example, act as baffles to reduce near-bed flows.

Muds are only deposited in areas of low flow, such as estuaries and the lee of fringing islands. Their physical nature is described in Section 4.2. Coasts with embayments or many offshore islands, reefs and skerries form macroscale gradients that are incredibly variable at smaller scales of <1 km with considerable local variation in sediment sorting.

Figure 5.6 The majority of the bed of coastal seas is covered in sediment, most of whose inhabitants live hidden within the sediment matrix. The seven-armed sea star (*Luidia ciliaris*) is a roving predator within this system. Photo C. Frid.

5.4 Energy inputs

Photosynthesis in coastal waters is restricted to the intertidal, shallow areas of the sea floor and the surface layer of the sea. Phytoplankton are responsible for autochthonous production in the water column while, on shores and in shallow water, macrophytes or microscopic algae may be important. On coral reefs, a large proportion of the primary production is by symbiotic flagellates—zooxanthellae—living within coral tissues.

5.4.1 Detritus

In many coastal systems, as in other aquatic systems, primary production is generally exploited only after it becomes detritus. Birds, particularly geese and sea ducks, manatees and sea cows and some turtles are the principal large grazers of marine plants. Microbial primary producers are exploited by filter feeders and deposit feeders along with the detritus. Dead plant material, fronds ripped off by storms, mangrove leaves shed by the trees and dead phytoplankton cells all contribute to the pool of POM, as do faecal material, crustacean moults, shed gametes and carcasses from animals. Production and utilization of detritus in the oceans is considered further in Section 6.4, while its exploitation by benthos is described in Section 4.4.

5.4.2 Phytoplankton production

The physical conditions of light and mixing ultimately control the rate at which phytoplankton growth can occur, but other factors, particularly the availability of inorganic nutrients, may prevent production being expressed at this rate.

Nutrient control of phytoplankton

The rate at which phytoplankton can uptake nutrients can be estimated as a half-saturation parameter, usually referred to as the K_s value. These values differ for each nutrient and between species and in some cases different clones of the same species have different K_s values.

Nitrogen is the principal limiting nutrient in marine waters (Elser and Hasset 1994). Given a thermally stratified water column, no horizontal input of nutrient-rich water and no significant nitrogen fixation by Cyanobacteria, primary production will consist of two components. The first is 'new' production, which utilizes nitrate crossing the thermocline by turbulent diffusion. Zooplankton consume some of the production and, in due course, excrete nitrogenous compounds, which are utilized for the second form of production—regenerated production. This model of production, referred to as the Dugdale and Goering (1967) model, provides for a basic rate of primary production, set by the rate of turbulent diffusion across the thermocline, which may be enhanced up to 10 times by nitrogen recycling within the photic zone. This model has been shown to work well in coastal seas and explains the frequently observed phenomenon of a sub-surface chlorophyll maximum. Below the thermocline, light is limiting, but nutrients are available in relatively high concentrations as a result of regeneration by heterotrophs and the lack of primary production. Above the thermocline, a zone exists in which nutrients are present as a result of turbulent diffusion from below and in which light levels allow net primary production, giving a high standing stock of phytoplankton. Some of the production is grazed, liberating nitrogenous compounds which support further, regenerated, production. In the surface layer, nutrient concentrations are insufficient to support net primary production, due to their removal from the water column by the organisms in the region just above the thermocline.

Eutrophication

Background levels of nitrates in the North Atlantic in winter are typically 12 μmol dm^{-3} while concentrations in rivers flowing into the North Sea are in the range 350–600 μmol dm^{-3}, most of which is anthropogenic input. These waters are diluted by mixing with the saline waters, but concentrations in the coastal waters of the North Sea in winter, before spring bloom utilization (see below), are now many times the expected background levels. The spatial distribution of these excess nutrients follows the major sources of inputs, and the interannual variability in river flows and concentrations influences the distribution further. There is, therefore, clear evidence for anthropogenic eutrophication (Box 7.3) in the North Sea, but little corresponding evidence for a biological response. Phytoplankton abundance in

5

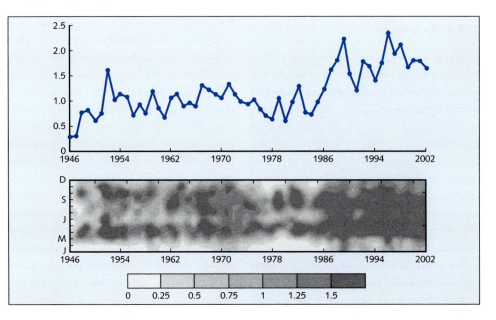

Figure 5.7 A contoured plot of monthly means of phytoplankton colour averaged for the North Sea (1946–2002) with above a graph of mean annual data. Updated from Reid *et al.* (1998) and DEFRA (2005).

5

the North Sea away from the nearshore region declined in the period 1960–1985, and it has subsequently increased so that for most of the period from 1987 it has been above the long-term mean **(Figure 5.7)**. Retrospective analysis of plankton colour data and satellite imagery suggest that while nutrient levels have increased it is climate change, notably sea surface warming and increasing transparency, that are primarily responsible for the shift in phytoplankton abundance (McQuatters-Gollop *et al.* 2007). Analyses of phytoplankton time series data for the island of Helgoland in the North Sea suggest that, while marked changes in the abundance occurred in the mid to late 1980s, there was a delay of up to ten years before major changes in the functioning of the ecosystem followed (Wirtz and Wiltshire 2005).

While the total abundance of phytoplankton has varied, there is evidence for a shift in species composition and an altered annual cycle. Some authors claim that recent 'unusual' algal blooms—producing discoloration (red tides) (e.g. *Noctiluca*), slicks of foam (e.g. *Phaeocystis*), mortality of marine life (e.g. *Chrysochro-*

mulina, *Gyrodinium*) or toxicity in humans (e.g. *Alexandrium*)—are manifestations of such altered species compositions (Reid *et al.* 1991). Such blooms can occur naturally, so it is unclear whether anthropogenic inputs are increasing the frequency, intensity or duration of such blooms. In the western part of the Wadden Sea (south-east North Sea) there is evidence of an increase in dominance of *Phaeocystis* (cell numbers and duration of blooms) between the mid 1970s and the late 1980s. Standing stock of phytoplankton and primary production both also increased during this period, and eutrophication has been cited as the reason for increased biomass of infauna on the intertidal areas of the Wadden Sea. Since the levels of nutrients have declined following regulation, the composition of the phytoplankton, macrofauna and bird communities have altered significantly, presumably in response to changes in nutrient dynamics (Philippart *et al.* 2007). The pattern of annual production by the phytoplankton in the North Sea in the 1960s and 1970s clearly showed a spring and autumn bloom with a period of reduced abundance in summer (Figure 5.7b). However, since the mid 1980s this has disappeared with the phytoplankton showing a single, prolonged, bloom starting in spring and extending well into the autumn. This has resulted in a marked increase in annual productivity.

5.4.3 Grazing control of phytoplankton

The uptake of light and nutrients by single-celled algae is enhanced by small size, yet relatively large phytoplankton, which should therefore be competitively inferior, are numerically dominant. The probable explanation is grazing pressure (Kiørboe 1993): larger phytoplankters are only accessible to large grazers, whose long reproductive times ensure that their numerical response to increased phytoplankton abundance only occurs after a significant lag. Large phytoplankton show marked temporal and spatial variability in abundance and their grazers cannot rapidly respond to the increased food which the bloom represents. Only after the bloom develops will grazers become an important check, although once present, zooplankton can significantly reduce phytoplankton concentrations. Nanophytoplankton are also kept in check by their predators, but these tend to be of the same size order to the algae, with similar generation times, so are able to match any tendency to bloom by increasing the predation pressure and holding population levels more constant.

5.4.4 Phytoplankton seasonality

Seasonal variation in the determinants of phytoplankton production results in seasonal patterns of abundance, of which a common feature is a rapid increase in biomass—the phytoplankton bloom. The idealized sequence for the initiation of a phytoplankton bloom consists of four phases, illustrated in **Figure 5.8**.

1 Photoinhibition phase, in which high light levels at the surface initially restrict the

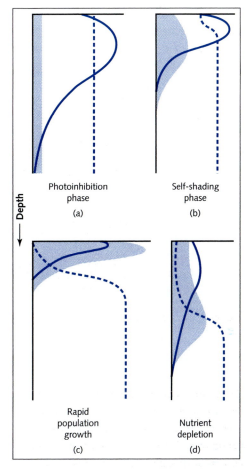

Figure 5.8 The process of bloom development in phytoplankton, illustrating phytoplankton biomass (shaded), daily net photosynthetic rate (solid line) and nutrient concentration (dashed line). See text for more details.

photosynthetic maximum to an intermediate depth (**Figure 5.8a**). This production results in an increase in biomass.

2 Self-shading phase, caused by the increase in biomass, which reduces light beneath the upper layers by self-shading, so maximum photosynthesis occurs towards the surface (**Figure 5.8b**).

3 Rapid population growth, exploiting the available nutrients (**Figure 5.8c**).

4 Nutrient depletion, causing the population to sink until it is concentrated above thermocline, both due to the physical barrier associated with the change in

density and slightly elevated nutrient availability (Figure 5.8d). Photosynthetic production is, however, limited by the low light levels at depth.

There are four classic patterns of phytoplankton seasonality (Figure 5.9), based upon the pattern of bloom which occurs.

Temperate North Atlantic pattern

In the coastal seas bordering the North Atlantic and in the open ocean, phytoplankton abundance shows a distinct, bimodal, annual cycle (Figure 5.9a). It is at a minimum over winter—low light and a deep mixing depth prevent net production despite the availability of nutrients. In spring, increasing light levels and stability of the water column allow a rapid increase in phytoplankton standing stock—the spring bloom. During summer, phytoplankton biomass falls as grazing zooplankton crop the phytoplankton and production rates become limited by the increasing depletion of nutrients above the thermocline. In autumn, storm waves erode the thermocline and mix nutrient-rich waters up into the surface layer, allowing a second bloom of phytoplankton before light levels and mixing restrict primary production in the winter.

Increasing primary production in the spring provides an increasing food source for zooplankton, whose numbers increase in response, but at a slower rate due to their lower doubling times. This increase is short-lived, as the summer decline in phytoplankton numbers leads to a decrease in zooplankton numbers later in the summer. Zooplankton numbers increase once more following the autumn bloom, but do not become high enough to check phytoplankton growth before winter mixing, low light and low temperatures reduce growth (Figure 5.9a).

This pattern is altered in coastal seas, where mixing by tides and nutrients from terrestrial sources alters the balance. In the north-western North Sea, the autumn phytoplankton bloom tends to be larger and longer than the spring

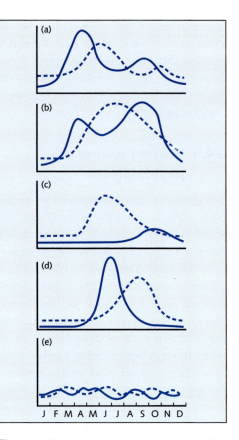

Figure 5.9 Annual cycle of phytoplankton (solid line) and zooplankton (dashed line) abundance in (a) the temperate North Atlantic; (b) the North Sea off the north-east coast of England; (c) the temperate North Pacific; (d) at high latitudes; and (e) in the tropics.

outbreak and zooplankton show a very broad single peak of summer abundance (Figure 5.9b).

Temperate North Pacific pattern

In the temperate North Pacific, phytoplankton standing stock remains low for most of the year, with a period of higher abundance in autumn (Figure 5.9c). Zooplankton biomass shows a distinct annual cycle with high levels from spring through the early summer. Therefore, in contrast to the North Atlantic, neither zooplankton breeding nor standing stock is dependent on phytoplankton growth in spring.

5

This is the result of the different life history strategies of the dominant grazers in the two regions. In the Pacific, the nauplii larvae of the copepods *Calanus plumchrus* and *C. cristatus* are hatched from eggs produced by adults at depth during the winter. They migrate up through the water column, as non-feeding larvae, and arrive at the surface in spring (Section 6.8). These young stages are able to make immediate use of the spring increase in phytoplankton productivity and their grazing pressure limits the standing stock of phytoplankton. It is only in the autumn, when zooplankton numbers decrease, that any appreciable increase in phytoplankton can be observed.

High latitude pattern

Light is limiting for most of the year. As it becomes available in late spring, productivity is extremely high as a result of the abundant nutrients, allowing a rapid and intense phytoplankton bloom to develop (**Figure 5.9d**). Zooplankton respond to the availability of phytoplankton with their own bloom, which may or may not limit the phytoplankton bloom before the onset of winter brings decreasing light levels and hence decreasing productivity.

Tropical pattern

There is little scope for broad seasonal patterns in the tropics. Rather a succession of increases and decreases in phytoplankton and zooplankton occurs (**Figure 5.9e**) in response to local weather conditions and the movement of water masses, which bring in nutrients and populations of plankton.

Seasonal changes in species composition

While the broad patterns described above relate to overall biomass, individual species distributions tend to occur as a series of pulses rather than a smooth transition. The early bloom of phytoplankton in the North Sea, for example, is dominated by diatoms but, as silica becomes limiting, they are replaced by dinoflagellates. A similar turnover in response to silicon depletion occurs in many freshwater phytoplankton assemblages (Section 7.4).

5.4.5 Benthic primary production

Rocky shores and the rocky sublittoral

Rocky shores and sublittoral rocky reefs in sheltered, well-illuminated waters generally support dense growths of macroalgae, whose primary production in the well-illuminated sublittoral kelp forests of California reaches 1000 g C m^{-2} yr^{-1} and, when sheltered from wave action, may exceed 1250 g C m^{-2} yr^{-1} on rocky shores. Physically, these algal forests are structured in the same way as a terrestrial forest. The large macroalgae form a canopy, below which shade-adapted, usually red, macroalgae grow, while below this shrub layer there may be a turf or encrusting film of photosynthetic algae or microbes. The productivity of the lower layers is set by the degree of penetration of light through the canopy and the extent to which nutrients are inputted. It is also affected by direct physical abrasion: in wave-swept areas, the fronds of the macroalgae can keep the surrounding area clear of microalgae or macroalgal spores by sweeping over the substrate.

Coral reefs

Primary production in coral reefs occurs by two mechanisms, macroalgae growing on the reef and symbiotic zooxanthellae, living within the coral tissues. It was originally thought that the low levels of nutrients in the water column controlled algal growth and thus provided a distinct advantage to the zooxanthellae, which recycle nutrients from the waste products of their hosts. However, removal of the fish fauna

5

from a reef, for example by dynamite or cyanide 'fishing', is often followed by a rapid bloom of algae, which can smother coral polyps and bring about a profound change in the reef system (Hughes 1994). This implies that it is grazing which controls the macroalgal community.

At the base of the reef, corals usually give way to a sandy sediment plain. If sufficient light reaches this plain, algal growth can occur, although it is limited by the availability of suitable attachment sites (coral fragments or shells) and frequency of storm wave dislodgement. In sheltered lagoonal areas behind reefs, and in depressions on the reef flat where sediment accumulates, it is often colonized by sea grasses (see below).

Soft sediments

In most places sedimentary systems show no obvious signs of primary production, other than where beds of sea grass develop. Seagrasses are higher plants with a true root system and so most species need a sedimentary environment to establish. Seagrass productivity is much higher (up to three times), per unit area, that macroalgae and so seagrass plays an important role in the productivity of areas where it occurs. Seagrass is grazed by some invertebrate grazers, but in general these are feeding more on the epiphytic microalgae than the seagrass itself and most of the primary production is exported as detritus (Duarte and Chiscano 1999). The principal specialist grazers of seagrass are gastropod molluscs and large vertebrate grazers such as turtles and the dugong (*Dugong dugon*), a situation analogous to that in freshwater wetlands, where most direct grazing is also by vertebrates (Section 8.4). Seagrasses occur in coastal regions of all the oceans except the Antarctic but are limited to locations where the physical environment (i.e. type of sediment, degree of water movement, lack of wave action and ice scour) allows them to colonize. The seagrass acts as a major habitat engineer, altering flow regimes and patterns of sedimentation, often enhancing

fluxes of organic matter into the sediment. The seagrass beds are a biogenic habitat providing attachment space for epiphytic plants and filter-feeding animals and providing predator refugia for mobile species such as shrimps and small fish. Seagrasses are harvested by humans for use as a textile.

In areas of poorly mixed and coarse sediments that are not exposed to high levels of wave or ice scour, large particles such as pebbles and shells provide attachment for a diverse range of mainly ephemeral macroalgae. The abundance, and hence productivity, of these is limited by the availability of attachment sites, but like the seagrasses they both contribute to productivity and provide physical habitat structure.

The vast majority of marine sedimentary habitats lack any macro primary producers but wherever levels of light are sufficient photosynthetic primary production does occur. Microalgae, particularly diatoms, cyanobacteria and photoautotrophic protists and eukaryotes contribute to this production. These organisms often migrate vertically in the interstices between sediment particles in response to the light regime and can, through the production of extracellular organic material, play a critical role in stabilizing the sediment.

In addition to photoautotrophic production, it is now increasingly recognized that chemoautotrophy is widespread in marine sediments. While first coming to light in the dramatic plumes emerging from hydrothermal vents (Section 6.7), chemoautotrophic bacteria also occur at the interface of reduced and oxygenated conditions in sediments. The bacteria appear to migrate between the reduced conditions, where they 'pick up' reduced compounds, and the oxygenated layer where they harness the energy released from the oxidation of these substances to drive cellular processes. In addition to these free-living bacteria there are a number of multicellular organisms, nematodes and Pogonophora, which have developed symbiotic relationships with these chemoautotrophs in a manner exactly analogous to that seen in vent organisms (Cavanaugh 1994).

5.5 Community structure on hard substrata

Rocky shores, sub-tidal reefs, pier legs and other artificial structures develop assemblages including species attached to the substratum, the basal species and mobile forms. On shores and in shallow water the basal species include photosynthetic primary producers including Cyanobacteria, microalgae and macroalgae. Below the photic zone the basal species are dominated by filter-feeding animals, bryozoans, hydroids, sponges, bivalve molluscs, soft corals, barnacles, sea squirts (tunicates) and tube worms. Filter feeders will also establish in shallow water if space for colonization is made available. In areas of strong flow, large growth forms are impeded as they exhibit too much drag and low-growing/encrusting forms dominate. Mobile grazing gastropods, shrimp, crabs and fish utilize the system created by these attached species, feeding and/or sheltering in the matrix of the basal organisms.

5.5.1 Disturbance

The overriding determinant of community structure on rocky shores is disturbance, particularly through wave action. Benthos are unable to burrow into the substrate but must remain exposed to the full force of waves. Physical disturbance inevitably leads, therefore, to some losses, particularly during storms of unusual strength (Section 5.3). Those individuals that are lost release resources, such as attachment space, allowing colonists into the community. In the absence of further disturbance, the most competitive species will eventually dominate, but another disturbance event may reset the succession, allowing less competitive species to coexist. The patchiness of disturbance events will, in this way, determine the patchiness of the community (**Figure 5.10**), with frequent

Figure 5.10 Patchiness in a mussel-dominated system. Disturbance events, such as the impact of wave-borne debris or the failure of mussels to remain attached, creates a mosaic of patches. Regenerated resources—attachment space in this case—are colonized and begin to follow a succession towards the 'climax' state. In the meantime, further patches are created by additional disturbance events—so a high-diversity system is produced as a result of the mosaic of patches at different stages of the successional continuum. Photo C. Frid.

disturbance at a small spatial scale maintaining a series of patches at different successional stages (see Section 3.5 for an example of a similar process in rivers).

The biomass of fucales and laminarians on the shores and in the shallow sublittoral of temperate regions has attracted commercial harvesting operations in some areas of the world. Small-scale harvesting of seaweeds for iodine and potash extraction has a long history, while today the harvest is mainly concerned with the extraction of alginates and other polysaccharides, that of the kelp *Macrocystis pyrifera* in southern California alone being around 144,000 t yr^{-1}. Harvesting is, in many ways, analogous to natural disturbances that remove the biomass, and as such the effect of harvesting varies with the type of harvest (vegetative fronds, fronds including reproductive structures or whole plants), time of year, areal extent of harvested area and frequency of harvesting. For example winter harvesting of *Ascophyllum*, when it is reproductively active, results in very slow recolonization due to a lack of viable spores. Large scale and repeated harvesting acts like an increase in natural disturbance frequency, causing a shift in the whole community towards one typical of wave-swept areas.

Archaeological evidence suggests that the coastal habitats of Chile have been exploited by people since at least 700 BC. The principal harvest is gastropods, particularly limpets, for human consumption, but some species of red algae are also collected and sold as a cash crop. Exclusion of humans from experimental areas in the late 1970s showed that the entire coast appears to be in its present state, dominated by barnacles and small mussels, as a result of this long history and continuing exploitation. When the human collectors were excluded, the abundance of the red alga *Iridaea boryana* declined dramatically while the abundance and mean size of the limpet *Fissurella picta* increased. The traditional gatherers were effectively managing

the shore, unwittingly, to provide a cash crop not otherwise available to them.

5.5.2 Competition

Interspecific competition contributes to patterns of zonation found amongst sedentary species on rocky shores. Whereas the upper limit of a species' distribution is determined by its ability to withstand desiccation, its lower limit is generally set by competitive interactions. Coping with extreme physiological conditions requires metabolic energy, so species able to cope with the harsh upper shore environment tend to be poorer competitors than similar species with less physiological tolerance. The result is that, as one moves downshore into more benign conditions, the upper shore species find themselves in competition with species which lack their physiological resistance but are better competitors.

One of the classic examples of interspecific competition is the study by Connell (1961) of two species of barnacle, *Chthamalus stellatus* and *Semibalanus balanoides*, on rocky shores at Millport in western Scotland. *Chthamalus* generally occurs higher on the shore than *Semibalanus*, yet small *Chthamalus* are able to settle on the lower shore (Figure 5.11). Connell

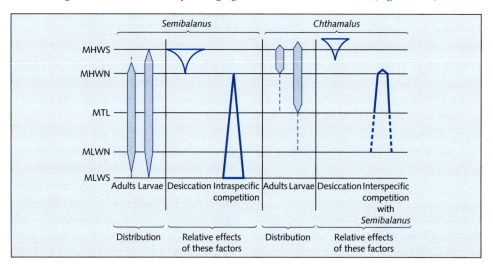

Figure 5.11 The distribution of adults and newly settled larvae of the intertidal barnacles *Semibalanus balanoides* and *Chthamalus stellatus*, with an indication of the relative importance of physical (desiccation) and biological (competition) controls. After Connell (1961).

followed the fate of *Chthamalus* young that settled in the *Semibalanus* zone, removing *Semibalanus* from some plots so that they did not encroach on the *Chthamalus*. He found that young *Chthamalus* barnacles kept free of *Semibalanus* did not suffer mortality resulting from the increased submergence time, while in unmanipulated plots, *Semibalanus* undercut, smothered and crushed *Chthamalus* individuals as they grew. Those *Chthamalus* that survived one year in the *Semibalanus* zone were smaller than higher shore barnacles of the same age. Menge (1976) demonstrated that barnacles, in turn, are competitively inferior to mussels (*Mytilus edulis*). Exclusion experiments carried out on rocky shores in New England showed that, in the absence of competition, barnacles thrive in the lower middle shore zone normally dominated by mussels.

A similar process occurs amongst rocky shore algae, which dominate moderately wave-exposed rocky shores. The midshore in New England is covered by *Fucus* spp. which are replaced on the lower shore by *Chondrus crispus*. Removing *Chondrus* from plots in the lower shore allows colonization by *Fucus*, which grows more rapidly than midshore populations. The lower extent of *Chondrus*, in turn, is defined by grazing pressure from sea urchins (primarily *Strongylocentrotus droebachiensis*), which are confined to the sublittoral.

Thus the rocky shore can be seen as a series of physiologically (upper limit) and ecologically defined (lower limit) zones. The resultant algal zonation is repeated worldwide but is most dramatic when seen on temperate shores with reasonable tidal ranges. Here the dominant algae differ in colour and form so that the shore, when seen from a distance, appears banded (**Figure 5.5**).

5.5.3 Predation

The effect of predation on rocky shores is well established and, indeed, it was work carried out on rocky intertidal systems that allowed Paine (1966) to develop the concept of a 'keystone'

species. The starfish *Pisaster ochraceous*, a common species on coasts of western North America, eats a variety of different sessile species: barnacles, mussels, limpets and chitons. Removal of *Pisaster* results however in the loss of many of its prey species, because a single competitively superior mussel, *Mytilus californianus*, monopolizes space and excludes other species, including algae (Paine 1966). *Pisaster*, therefore, is beneficial to several of the species that it predates, by keeping interspecific competition in check. Similarly, Menge and Sutherland (1976) demonstrated that mussels and barnacles could coexist, albeit in low numbers, in the presence of the dogwhelk *Thais lapillus*, which preyed upon both, but preferentially upon the more competitive mussels.

On most shores, the dominant grazers are gastropod molluscs, while in the sublittoral they are replaced by sea urchins. Grazers tend to be small relative to the macroalgae and only able to exploit the microflora, including newly settled macroalgal spores, which coat the rock surface and may be present as epiphytes on the macroalgae. Consumption of this film may be selective: the periwinkle, *Littorina littorea*, for example, which feeds on seaweed germlings in the biofilm, consumes some species but will avoid others.

While studies on the marine intertidal have led to advances in ecological understanding, these now 'classic' studies often contain methodological flaws and it was only the result of the overriding size of the effect that led to such a clear outcome (Raffaelli and Moller 2000). More recent studies of keystone predators suggest that they are not universally such influential controllers of community dynamics. In fact the term 'keystone predator' places the emphasis in the wrong place, as it is the dynamics of the prey species and not the predator that determines the outcome, and any perturbation, such as a storm, can influence the dynamics of the interactions between competing species. Only in a system where there is competition for a limited resource and a strong competitive dominant do you get the classic 'Paine-type'

5

response. In many systems the competitive interactions between species are finely balanced and no strong dominant emerges, or the system may be constrained not by the resources available but by the supply of colonists to the system, a process referred to as supply-side ecology.

5.5.4 Supply side ecology

On many shores, the crucial factor determining which species are present and their abundance is the availability of settling larvae. The importance of a ready supply of new recruits has long been recognized in marine ecology but the experimental studies of the late 1960s and 1970s emphasized the role of post settlement processes: competition for space and the modifying influence of predators. It is now clear that, by chance, the shores used in these experiments were ones which received a good supply of larvae every year. Roughgarden *et al.* (1987) showed that other shores suffer variability in the supply of larvae, due to variation in currents or predation, and it is this which controls the type of community that develops. This emphasis on the availability of recruits as a determinant of community dynamics has been termed supply-side ecology. The factors controlling the supply of larvae include hydrography, meteorology and food availability, as well as biotic interactions with other planktonic organisms (larval and the permanent plankton). This represents a series of challenges to understanding the dynamics of rocky shore communities that involve factors outside the shore itself.

In areas where there is a ready supply of larvae, other factors will influence the settlement pattern. Jenkins (2005) used meticulous examination of newly settled barnacles and the identification to species of larvae in the plankton to examine the role of supply-side ecology in barnacle dynamics around south-west Britain. Two species of chthamalid barnacle co-occur: *Chthamalus montagui* and *C. stellatus*. He was able to show that, while there were spatial differences in the abundance of early stage

larvae, *C. stellatus* larvae being more abundant near wave-exposed shores and *C. montagui* near sheltered shores, there was no difference in abundance of late stage larvae. This implies that sufficient mixing occurs during the larval period to wipe out these initial differences. Examination of the shore showed however that the abundance of newly settled individuals matched the pattern of adult abundance. While post-settlement mortality reinforced this pattern, it was the initial choice of settlement site by the larvae that set the pattern. In this case any pattern in larval supply is negated by larval behaviour, adding a further challenge to our understanding of the causes of the patterns observed in adult populations.

5.5.5 Ecological functioning

It has long been recognized that most of the primary production occurring in the rocky intertidal and on shallow sub-tidal reefs is not utilized by grazers in the habitat but is rather exported as detritus (particulate organic matter) and dissolved organic matter (Section 5.4). Thus the ecological functioning of many coastal ecosystems is dependent on the functioning of these systems. Similarly the structural complexity of the epibiota on hard substrates provide both habitat for associated fauna/flora, refugia from predation, currents and wave drag and nursery habitats/egg attachment sites. This emphasizes the strong functional connections that occur in many marine systems and contrasts with the situation in many terrestrial systems.

Given the long history of manipulative field experiments on ecology in the rocky intertidal, it is surprising that experimental tests on the control of ecological functioning are few (Benedetti-Cecchi 2006). Allison (2004) examined resistance to perturbation and the resilience of intertidal assemblages. He concluded that while more diverse plots were more resilient (they recovered more quickly following a perturbation) they were not, as predicted by the 'insurance hypothesis', more resistant to change. This was interpreted as being due to

the fact that the more diverse plots were dominated by large structural seaweeds and as these tended to be keystone species in terms of providing shelter/habitat for other organisms, their loss was catastrophic for the system.

Bruno *et al.* (2005) used manipulative field experiments and culture in laboratory mesocosms to examine the effect of species identities and assemblage richness on ecological functioning (photosynthetic rate and biomass accumulation) in sublittoral rocky macroalgae assemblages. They found a strong effect, often explaining over 75% of the variation, of species' identities and a consistent but weaker effect of assemblage diversity. These, albeit limited, results for marine intertidal systems are broadly consistent with the emerging patterns described for terrestrial systems.

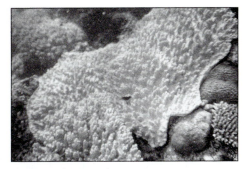

Figure 5.12 Space is limiting on hard substrate communities. Among corals, the competition for space—and hence access to light and nutrients for the symbiotic zooxanthellae and the water column for the polyps—manifests itself as overgrowth, often mediated by the use of biologically active chemicals: Namalatu, Ambon Islands, Indonesia. Photo C. Frid.

Coral reefs

The physical structure of a coral reef is a function of the interaction between the growing corals, which secrete an exoskeleton of calcium carbonate ($CaCO_3$), and the physical environment, in particular wave action and water movement. As much as 50% of the reef may be covered by living coral colonies. For the reef-building corals, access to light to support the symbiotic zooxanthellae is important, so illuminated space is usually the primary limiting resource (Figure 5.12). Space is also limiting for the filter-feeding sponges and soft corals that require access to the water column to feed. Competition occurs through overgrowth, shading and chemical or physical attack, with many sessile forms practising antibiosis, the chemical defence of space. Some soft corals, for example, exude chemicals which cause necrosis of nearby hard corals, and the extension of digestive filaments between colonies, often over distances in excess of 1 m, is used in interspecific encounters, with the attacker literally digesting the tissue of the victim. Coral reef fish, too, are apparently space limited (Sale 1984), hence the very rapid colonization of newly deployed artificial reefs, the fact that specialist microhab-

itats such as anemone tentacles have a fairly constant density of fish and evidence for inter-specific defence of living sites. However, long term monitoring of reefs shows considerable temporal variation in fish numbers and, while some of this variability may be attributable to seasonal patterns, the majority appears to derive from variations in the arrival of recruits to an area of reef, a further example of supply-side ecology.

The high diversity of organisms within coral reef systems is maintained by the fact the community is in a non-equilibrium state, determined by a combination of the rate of competitive displacement and a range of disturbances including heat shock, wave action (including the extremes associated with tropical storms), sedimentation, predation (including grazing) and disease. Diversity on a reef generally increases from the surface (frequently disturbed by wave action and heat stress) to a maximum at about 20 m and then declines in the more stable, but increasingly light limited, conditions. In addition to small- or medium-scale disturbance which promotes local diversity, coral reefs are subject to several large-scale disturbances, including large tropical storms (hurricanes and cyclones), tsunamis, climatic fluctuations, such as the El Niño southern oscillation (Section 6.8) in the atmospheric and

5

ocean currents over the Pacific, and aperiodic outbreaks of predator populations or disease. These all cause widespread destruction of reef communities, allowing colonization by dispersing propagules which introduce a further source of variability into the system.

It has long been recognized that broadcast spawners would either need to breed all year round, giving shed gametes a chance to fertilize gametes shed by another individual, or if spawning in a limited period mechanisms to synchronize spawning to maximize the chances of fertilization (Section 2.3). The use of lunar phases (and hence tidal cycles) and changes in day length as cues has long been recognized. However, in the early 1990s a new phenomenon was observed, that of synchronous mass spawning of many species. At least 60% of the 184 species of scleractinian corals found on reefs in western Australia engage in a late summer mass spawning (Babcock *et al.* 1994). Similar mass spawning has now been observed on other reefs considerable distances from those first observed and furthermore they all occur in the same month, implying that it is not a simple temperature cue or annual cycle but a mechanism to ensure synchronous spawning in populations over a wide area. This also implies that it is not simply an adaptation to ensure high fertilization for each species but more a mechanism to swamp potential predators over a large area and so promote larval survival.

5.6 Community structure in soft sediment systems

At low tide, sediment shores, sandy beaches and mudflats often appear devoid of life. This is far from the case. Intertidal sediments contain a flora of unicellar algae and microphytobenthos, including Cyanobacteria and other photosynthetic bacteria. These live in the film of water coating/between the particles and will migrate through the sediment to keep in the optimum light regime. This *in situ* production, along with the settled organic matter, is consumed by deposit feeders in the sediment, while filter feeders use living plankton and detrital particles filtered from the overlying water. The dominant groups are various groups of worms, notably polychaetes and nematodes, bivalve molluscs and amphipod crustaceans. Larger crustaceans such as crabs, fish and, in intertidal zones, birds, act as top predators in the system. Sub-tidally the system is dominated by a rich guild of larger epibenthos, including echinoderms, crabs, starfish and predatory gastropods.

5.6.1 Types of communities

The dominant feature of sediment communities is that they are patchy on any scale of observation, reflecting a range of biological and physical processes. These include physical disturbances, foraging behaviour and breeding behaviour leading to distinct metapopulations. For example, studies of the genetic composition of populations of the eel grass *Zostera marina* show a step change in the degree of relatedness at distances of around 150 km, implying that this is the 'natural' range for a metapopulation in this species (Olsen *et al.* 2004). In contrast, using a combination of genetic markers, Hoarau *et al.* (2004) showed that the large population size, mobility and relatively recent colonization of the north-east Atlantic by plaice (*Pleuronectes platessa*) meant that there was no genetic isolation between populations in Iceland, Norway, the North Sea and the Irish Sea.

There are, however, major differences in the species assemblages inhabiting different types of sediments, stimulating attempts to map species distributions by sediment types. The earliest and most extensive of these descriptive approaches was that of Petersen who, in 1914–1922, carried out a wide-ranging survey of the shallower parts of the North Sea using a quantitative grab sampling technique. From

Table 5.3 Features and key organisms in Petersen's communities from the North Sea		
Petersen's community	**Description**	**Key organisms**
Macoma community	Shallow 10–60 m, also estuaries	*Macoma, Mya, Cerastoderma* and *Arenicola*
Tellina community	Shallow, exposed, hard sand, 0–10 m	Suspension feeders. *Tellina, Donax, Astropecten*, and *Dosinia.*
Venus community	Open sea, 7–40 m, sand	*Venus, Spisula, Tellina, Thracia, Natica, Astropecten* and *Echinocardium*
Abra community	Sheltered waters and estuaries, with reduced salinity, mixed to muddy, organically rich sediments. Increasing sand merges into the *Venus* community. Increasing silt merges into the *Amphiura* community	*Abra, Cautellus, Corbula, Pectinaria, Nephtys*, and *Echinocardium*
Amphiura community	15–100 m, high silt content	*Amphiura, Turitella, Thyasira, Nucula, Nephtys* and *Echinocardium*. Increasing sand leads to increasing *Echinocardium* and *Turritella*. Increasing silt leads to increasing *Brisposis, Thyasira* and *Maldane*
Maldane—Ophiura sarsi community	Soft muds, shallow estuaries to 200 m	Deposit feeders. *Maldane* and *Ophiura sarsi*
Amphipod community	Muddy sands generally	Communities dominated by different species of amphipods

5

his results he classified the benthos into seven distinct 'communities', which he named after the dominant species (Table 5.3). He implied no linkage between the species in these communities, merely that they tended to occur together in space.

There is a clear implication that depth and the nature of the substrate are important in determining which community is present at a particular location. Thorson (1957) expanded Petersen's work and implied an ecological linkage between the species, concluding that similar environments worldwide should have functionally similar communities. While species occurring together do interact, the primary reason they occur together is, however, a need for similar physicochemical conditions. More recent studies in the North Sea have confirmed that associations of species do, indeed, correlate with the distribution of sediment types and other physical factors (Künitzer *et al.* 1992).

5.6.2 Disturbance

Several scales of disturbance can be distinguished in marine sedimentary environments (Probert 1984). At a small scale (<1 cm²), burrowing and cast production by macrofauna probably influence meiofauna to a greater extent than macrofauna. Combined effects of disturbances at this scale can be large. In the Wadden Sea, for example, it is estimated to be equivalent to the top 35 cm of the sediment being turned over every year. At intermediate scales (cm²–m²) biological activity by large macrofauna—crabs, shore birds, fish and sea mammals—will cause disturbance, much of which is highly seasonal. Estimates include 26% of mussel beds in a Scottish estuary disturbed per month in winter by eider ducks, and side-scan sonar images revealing 6% of the seabed in the Bering Sea being covered by pits from feeding grey whales. At large scales (dm²–km²), physical processes or human impact are responsible for most disturbance, although biological processes may predispose an area to a physical disturbance, for example by altering sediment stability. Large-scale biological disturbance could include disease outbreaks, population explosions or species invasions and introductions. Storms represent perhaps the most obvious large-scale disturbance, but other causes include deoxygenation events, sea floor smothering (by sediment, dredge spoil, fish eggs etc.), disturbance by fishing gears and ice scour in the intertidal at higher latitudes.

One widely observed phenomenon is the effect of disturbance on the mobile fauna. At all scales, from small pits to trawl tracks, immediately following disturbance there is an influx of mobile scavengers, which exploit the individuals exposed at the sediment surface or carcasses of infauna killed by the disturbance. This response is however short lived, so although commonly observed, experimental data are still scarce (Kaiser and Spencer 1996).

Despite the high frequency of small-scale disturbance, benthic communities can remain stable for long periods of time. A monitoring programme has been running off the coast of Northumberland (north-east England) since 1972 at a station in 52 m of water. During the first 10 years, numbers of macrofauna remained remarkably constant. Each year a spring recruitment led to higher numbers in the autumn, which winter mortality reduced by the following spring. In addition to this annual cycle, there was a biennial cycle of alternating high and low years: a low number in March was followed by a high number in September, which was followed by a high March population and subsequently a low number in September. These patterns provide evidence for density-dependent recruitment and mortality, although after 10 years of monitoring this biennial cycle broke down and population values fluctuated markedly. Over 50% of the interannual variation in the spring abundance of macrofauna was explained by interannual variation in the productivity of phytoplankton two years previously (e.g. spring 1992 macrofauna

numbers were related to phytoplankton production in 1990). During the study period of nearly four decades, the relative abundance of the species of macrofauna present at this site has changed little, and the large interannual changes in abundance have been driven by changing inputs of food.

Disturbance is not necessarily detrimental. Small-scale biological processes, including tube building, pit or burrow digging, generate heterogeneity and provide niches for other, usually smaller, species (Dayton 1984).

5.6.3 Competition

Direct evidence for competition in soft sediments is scarce. One proposed mechanism for competitive interaction is the trophic group amensalism hypothesis of Rhoads and Young (1970). Suspension feeders need a stable substratum in which to build their tubes, and water currents to bring in food. Deposit feeders, however, need sediments containing organic matter, which requires lower current speeds. At sites of intermediate current speed, deposit feeders will alter the environment to make it unsuitable for suspension feeders: their faecal pellets, released at the sediment surface, have a high water content and, upon disintegration, small particle size, whose resuspension smothers the larvae of suspension feeders and clogs their filters. Hence one trophic group can exclude the other. The reciprocal may also occur, because tubes of filter feeders stabilize the sediment, reduce the area available to deposit feeders and may intercept sufficient incoming POM to reduce sediment organic matter content, making it unsuitable for deposit feeders.

The generality of the trophic group amensalism hypothesis has been questioned. Wildish (1977) considered bottom roughness and current flows to be the critical factors and suggested, in his trophic group mutual exclusion hypothesis, that both groups are essentially food limited. When current velocities at the seabed exceed a few centimetres per second, then development of a deposit-feeding community is inhibited because biologically produced particles (being of low density) are resuspended and removed.

Following examination of the impacts of organic pollution, sewage and effluent from paper mills, Pearson and Rosenberg (1978) concluded that the primary factor influencing shallow water benthic communities was the level of organic matter in the sediment. The Pearson and Rosenberg model, as it has become known, initially described the changes in the benthos on a gradient from grossly enriched to normal (or through time after the cessation of a discharge). Adjacent to the input the sea floor is anoxic as bacterial breakdown of the organic waste is proceeding so rapidly that it is utilizing all the available oxygen. This excludes all macrofauna. As one moves from the discharge point oxygen levels increase (both as there is less waste using up the oxygen and there is more scope for oxygen-rich waters to flow into the area). These regions support species-poor communities, as only a few specialist species can cope with the physiological challenges of the low oxygen environment. These opportunistic and extremely tolerant species often occur in very high numbers as they exploit the rich food resource of the organic matter and benefit from the lack of competitors. Further still from the discharge point the system becomes richer as more tolerant species are able to invade, but the decreasing level of organic matter results in competition so that a transition can be observed from weakly competitive but highly tolerant species, through intermediate forms to the background community of low tolerance but more competitive species. This pattern is often summarized as a gradient in space or a succession through time from abiotic, through a very high abundance but species-poor system, to an intermediate high abundance and species-rich phase to the background assemblage.

From this impact model Pearson and Rosenberg (1986) developed a more generic model of benthic dynamics. Food supply was the central structuring factor and its influence was moderated by latitude, depth and water movement. Long-term studies of natural assemblages have tended to support the critical role of food supply as a limiting factor in benthic dynamics. With such a clear, strong, limiting factor one would expect to see clear competitive interactions, but this is not the case. The lack of strong competitive hierarchies in soft sediment systems contrasts with the hard substrate ones where competition is either the dominant factor or is clearly mediated by predation or disturbance. Disturbance and predation clearly account for some of the apparent lack of competitive interactions in the benthos and may act in a complex manner. It may be however that variations in environmental factors (temperature, oxygen), physical and biological disturbances, predation, and variations in larval settlement (both larval supply and post-settlement survival) (Olafsson *et al.* 1994) combine with adult mobility with the result that these systems are kept in a dynamic non-equilibrium state.

Relicts and founders

Pure competition for space can be distinguished from founder control, in which the first species to recolonize following disturbance is able to monopolize the resource (Section 3.5). If, for example, a large tube-building filter feeder colonizes a soft sediment patch, then the combined effect of the tube and the interception of incoming larvae by the feeding tentacles may effectively prevent certain other species from colonizing. However, as competition is much less of a structuring force in soft than hard sediments, so founder control is less widespread. In contrast to hard substrata, where disturbance events may remove all the macrobiota from a patch, both the nature of the disturbance generating forces and the biology of the fauna mean that most disturbances only remove a proportion of the biota originally present in a patch.

The individuals remaining after the disturbance have been referred to as the relict fauna (Woodin 1981). They are able immediately to respond to the reduced density of organisms in the disturbed patch and may affect subsequent colonization. The activities of some species may facilitate re-establishment by others, the behaviour of tube-building polychaetes and crustaceans, for example, modifying sediment structure to the advantage of deposit-feeding bivalves and oligochaetes. In other situations, however, they can inhibit the colonization process. Tube-dwelling predators may impede colonization, while burrowing and feeding activities of deposit feeders can inhibit the recruitment of less mobile infauna. There are also neutral relationships between the two groups. McCann and Levin (1989) demonstrated that mobile subsurface deposit-feeding oligochaetes in a saltmarsh tidal creek had no effect upon the settlement of tubiculous subsurface deposit-feeding polychaetes, nor upon molluscs.

5.6.4 Predation

Predation is another process whose role is unclear in soft bottom sediments. Despite predator densities and biomass well in excess of those found on rocky shores, including high densities of wading birds, experimental manipulations in these communities have met with mixed results, some studies demonstrating a clear impact on prey populations while others show no effect (see also Section 4.5). One potential reason for this is the usually overlooked role of infaunal predators. Ambrose (1984) postulated a three-tier system, in which deposit feeders are consumed by infaunal predators which, in turn, are eaten by epibenthic predators. Frid and James (1988) failed to find a response to epibenthic predator exclusion amongst deposit-feeding species, but did see an increase in infaunal predators. This could have been the result of release from selective predation, competition or a combination of the two.

The algal species occupying the sediment plain at the base of coral reefs (Section 5.4) in the Caribbean are distinct from the species occupying the reef itself. Sand plain species are competitively superior to reef-inhabiting species, but are unable to exploit the reef because of the high levels of grazers, whereas the sand plain provides them with a refuge from grazing. This would also appear to be the case for seagrasses growing in the lagoons associated with the reefs. Here, distinct halos of grazed seagrass meadows delimiting the area exploited by grazers from the reef can be observed. The reef provides the grazing fish with refugia from predators and so the further from the reef they venture the higher the risk of predation, hence the halo effect.

5.6.5 Ecological functioning

The majority of the sea floor consists of soft sediments and these habitats therefore have the potential to play a dominant role in the delivery of key ecological functions. A number of ecological functions delivered by coastal marine systems have been identified (Table 5.4). These include the food value of the fauna to fish and birds in higher trophic levels and the breaking down of organic material to release nutrients. The importance of organic matter as a food resource for benthic communities has long been recognized but during the consumption, metabolizing and subsequent release of the material important biogeochemical transformations occur that result in the release of nutrients into the water column. While the macrofauna contribute to this process, the majority of the remineralization is undertaken by microbial organisms. However, the rates of the processes are highly influenced by the activity of macroorganisms, for example the tunnels excavated by ragworms (*Hediste*) are important for making oxygen available to microbes in the sediment.

Most studies of ecological functioning in marine systems have used some rate process as an index of the functioning. The most commonly followed processes are nutrient regenerations (as measured by increasing concentrations of nutrients), bioturbation and oxygen uptake (as a measure of total respiration).

Table 5.4   Key aspects of functioning identified during two international workshops (Frid *et al.* 2008)
Process, property or activity
1 Energy and elemental cycling (carbon, nitrogen, phosphorus, sulphur)
2 Silicon cycling
3 Calcium carbonate cycling
4 Food supply/export
5 Productivity
6 Habitat/refugia provision
7 Temporal pattern (population variability, community resistance and resilience)
8 Propagule supply/export
9 Adult immigration/emigration
10 Modification of physical processes

Ieno *et al.* (2006) set up laboratory tanks containing combinations of three intertidal mudflat-dwelling species—ragworm (*Hediste diversicolor*), cockle (*Cerastoderma edule*) and the mud snail (*Hydrobia ulvae*)—and followed ecosystem functioning by observing the regeneration of nitrogen. They found that a species-rich assemblage did deliver more ecological functions than a species-poor system. However, the identity of the species and the stocking density were also important. Thus, while the three possible combinations of two species all delivered more functioning than any single species, the three pairs also differed in their rates of regeneration. This serves to emphasize that functional diversity (diversity of groups of ecologically equivalent species) may be equally, if not more, important for the delivery of ecosystem services than simple species diversity.

5.7 Pelagic community structure

The pelagic environment is a dynamic and complex system in which three-dimensional motions are the dominant feature. Most organisms are advected by the medium rather than being fixed in a particular location. There is usually a vertical reference frame, in the form of light intensity, but no horizontal structures

except at boundaries such as the surface, the thermocline and the sea bed (Denman 1994).

5.7.1 Determinants of phytoplankton assemblages

Interspecific differences in response to the temperature and light regime and to nutrient uptake kinetics (K_s values) are important in promoting coexistence of many phytoplankton species in a nutrient-limited environment. If we consider two potentially limiting resources, such as nitrate and phosphate, then each species will require certain minimum levels of both nutrients to survive (**Figure 5.13a**). If two species attempt to coexist and species 2 is a better competitor than species 1 for both resources, then it will outcompete and eliminate species 1. If, however, species 1 is the better competitor for resource a, but species 2 is the better competitor for resource b, the situation is more complex and coexistence is possible (**Figure 5.13b**). At low concentrations of resource a, the superior competitive ability of species 1 leads to the exclusion of species 2 whereas, at low concentration of resource b, species 1 is excluded. When both resources are present in concentrations slightly above these minima, coexistence occurs. If further species are added to the system (**Figure 5.13c**), more possibilities for coexistence are produced.

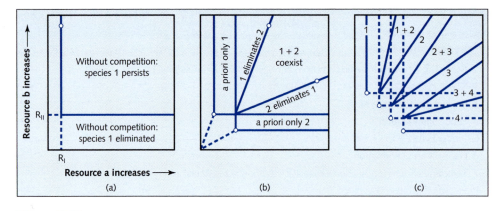

Figure 5.13 Coexistence of phytoplankton species under different concentrations of two limiting resources. (a) A single species (sp. 1) limited only by the two lower concentrations of each resource. (b) Two species (1 and 2) showing the regions of competitive exclusion and coexistence. Unless both resources are abundant, one species will outcompete the other. (c) Four species showing regions of coexistence based on each resource. Species 1 is the best competitor for resource a, 2 is the second best, 3 the third and 4 the least. The competitive order is reversed for resource b. From Lalli and Parsons (1993).

Adding more types of resource or physical factors such as light, temperature or salinity further increases the scope for many species to coexist. Recent studies looking at the possible effect of changing patterns of climate show that the extent and composition of the spring bloom may be altered in response to changes in the timing of the spring increase in temperature relative to the nutrient cycle (Domis *et al.* 2007).

5.7.2 Determinants of zooplankton assemblages

The tracking, with a short time lag, of phytoplankton production by zooplankton (**Figure 5.9**) provides evidence for food limitation. There is, however, no evidence for dominance by competitively superior species or for division of the resource, nor, given the generalist and opportunistic feeding in most species, any niche segregation. Therefore, some process must be operating to keep the community in a disequilibrium state. As zooplankton communities are potentially moved and mixed by physical processes, it could merely be the continual movement—horizontal and vertical—of nutrients, phytoplankton, copepods and their planktonic predators that maintains a disequilibrium.

Copepod assemblages are however able to maintain their integrity in the face of considerable advective flows.

The key role of predation has been emphasized in many littoral systems. The high mobility and low selectivity of marine planktivorous fish reduces the likelihood of them applying a strong regulatory effect, but predatory forms within the plankton are important. Copepod populations in the North Sea are controlled during the winter by their predators—chaetognaths and amphipods. The number of copepods surviving the winter then influences the rate and size of their summer bloom, as planktonic predators are unable to apply sufficient predation pressure to control copepods at this productive time. The autumn decline in copepod numbers may be linked to the presence at that time of year of large numbers of fish larvae, but the role of such temporary members of the plankton in the dynamics of the system remains poorly understood.

5.7.3 Interannual variation in zooplankton abundance

There can be considerable interannual variation in the zooplankton community at a single station. The records of the Plymouth Marine Laboratory show a long term cycle in the English Channel. Until the 1920s, and again in the 1960s, there was a high plankton abundance, the chaetognath *Sagitta elegans* was present and there was a productive herring (*Clupea harengus*) fishery. Between these dates, plankton abundance was low, the herring were replaced by pilchards (*Sardina pilchardus*) and *Sagitta setosa* replaced *S. elegans* (Southward 1974). These alterations seem to have been associated with a change in the volume of Atlantic Ocean water penetrating the English Channel.

The annual productivity of zooplankton in the north-east Atlantic correlates with the position of the north wall of the Gulf Stream where it leaves the American coast (**Figure 5.14**). To the north and west of the British Isles, productivity is highest in years when the Gulf Stream's track is more northerly, but at the same time these conditions depress production in the north-west North Sea and the lakes of the English Lake District. That plankton isolated in lakes respond in the same way demonstrates that it is not a direct interaction between Gulf Stream water and plankton, but rather that the two are responding to a third component. While the detailed mechanisms behind these relationships are, as yet, not fully resolved, it is clear that meteorological- and ocean-scale processes are important in controlling productivity and, in the North Sea at least, the relative abundance of species (Frid and Huliselan 1996). The Gulf Stream, like most major current systems, varies in strength and position in a complex way, and the meteorological and oceanographic factors which influence its path simultaneously affect plankton production. The

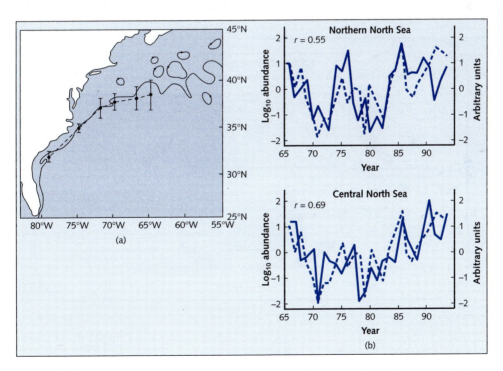

Figure 5.14 The north wall of the Gulf Stream and the relationship between its position and zooplankton abundance around the UK. (a) The position of the north wall off the east coast of the USA during August 1984 (solid line) and its mean position (+95% confidence limits) between 1966 and 1996 (dashed line). (b) The relationship between latitude of the north wall and total copepod abundance in the northern North Sea, off north-east Scotland and the central North Sea, west of Denmark. Figures have been standardized to have zero mean and unit variance; the correlation coefficient (r) between each pair of graphs is given. From Taylor (1995).

most likely mechanism is to do with interannual variations in the climatically determined timing of the onset of stratification relative to the phytoplankton production cycle.

In the North Atlantic the network of plankton recorders operated by the Sir Alistair Hardy Foundation for Ocean Science provide a unique opportunity to observe ecological changes at ocean and multi-decadal scales. Analyses of these data show clear effects of climatic drivers, fisheries and, in some areas, eutrophication (Beaugrand *et al.* 2002).

5.7.4 Ecological functioning

There is a view that the functioning of the global ecosystem is mediated in large part by pelagic marine organisms through their influence on biomass production, elemental cycling, and atmospheric composition (Figure 5.15). The marine pelagic ecosystem is the most extensive on Earth and given, in particular, the role of the oceans in mediating the composition of gases in the atmosphere, the role of pelagic ecological processes will influence global phenomena. While the pelagic zone of the coastal seas is only a fraction of this total area, the higher productivity and rates of ecological processes in the coastal seas, compared to the open ocean, mean that the coastal seas deliver a greater proportion of these ecological functions than their area suggests. In common with other systems studied, the marine pelagic shows a relationship between functional diversity and

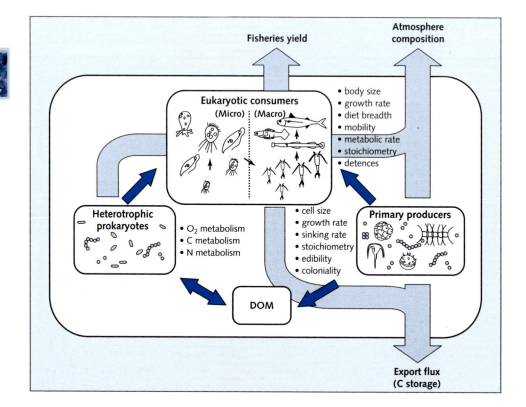

Figure 5.15 Summary of some proposed links between pelagic biodiversity and marine ecosystem processes. A stylized food web, divided into three major components linked by trophic flows (black arrows), is shown. Beside each compartment is a list of traits likely to affect trophic flows, and grey arrows show influences of those compartments on ecosystem services. Compartments are presented as separate enclosed areas for clarity, although in reality ocean microorganisms span a metabolic continuum from pure autotrophs, through facultative autotrophs, mixotrophs, and facultative heterotrophs, to pure heterotrophs. DOM, dissolved organic matter. From Duffy and Stachowicz (2006).

ecosystem functioning, with more diverse systems delivering more functionality and generally being more stable in the face of perturbations.

Understanding of the pelagic zone has often been gained through the use of biogeochemical and ecosystem models. These tend to use the fewest aggregated units to represent the system so that most taxa are not modelled but represented within aggregate classes based on function, such as zooplankton grazers. Recent work on ecosystem functioning in the pelagic highlights the danger of this and identifies how even relatively simple extensions to the models, for example by modelling large and small phytoplankton as distinct groups, can greatly increase their realism and predictive power.

5.8 Succession and change

5.8.1 Colonization on new shore

Hard substrata

Any hard substrate placed into the sea is rapidly colonized by a microbial film, including organic molecules, heterotrophic bacteria, fungi, protists and, in shallow waters, photosynthetic bacteria, microalgae and the early stages of macroalgae. Mobile adults of grazing species may invade this community from adjacent areas at this stage. Mucus from the trails of grazing gastropods adds to the biofilm and may enhance recruitment of seaweed propagules and adhesion of diatoms. This has led to the idea that territorial limpets may use their mucus to 'garden' algae in their territories (Connor and Quinn 1984). There is also evidence that sessile animals recruit preferentially to surfaces after they have developed a microbial coating. Most pelagic larvae start prospecting for a suitable settlement site when they reach the stage when metamorphosis could occur. There is a 'window of opportunity' during which settlement can occur, at the end of

which metamorphosis occurs irrespective of suitability of the site. Thus, early in this period all the settlement cues are required to initiate settlement but, as time goes on, the presence of progressively fewer cues will suffice.

Over time, spores of macroalgae will settle. The microscopic flora is utilized directly by grazers who can also exploit newly settled spores of macroalgae. Once the macroalgae exceed a certain size, they become too large for most of the grazers to exploit and will be kept in check by physical abrasion and competition with other macroalgae. Large numbers of grazers can, however, prevent the domination of macroalgae by grazing the sporelings and effectively interrupting the succession.

Soft sediments

In soft shallow-water sediments, the process of recovery has been best studied in patches in the $10 \, \text{cm}^2$ to $10 \, \text{m}^2$ range. For large-scale disturbances, however, studies have tended to follow the fate of a 'natural event', so these studies lack adequate controls and replication is not usually possible. In general, colonization of small- to medium-sized patches occurs over a few days by the addition of adults of species from the ambient community, but organic matter accumulating in pits excavated by predators encourages initial colonization by a different suite of opportunistic species, not previously present in the area.

5.8.2 Excess siltation and sublittoral communities

The deposition of new particulate material onto the sea floor is a natural process in many areas, the size and chemical nature of the particles and their rate of deposition being determined by local geological and hydrographic processes. Many coastal seas are, however, subjected to deposition of material at much higher rates than those occurring naturally, through direct dumping of material as a waste disposal operation and excess supply of 'natural' sediments due to changes in coastal defences or land use in adjacent catchments. The latter, in particular, has caused widespread degradation of coral reefs. Deforestation and other changes in land use in the catchments often lead to erosion of soil, which is washed into rivers and transported downstream. In coastal seas, this sediment-rich water discharges from the estuary mouth as a discoloured plume and, as it spreads out, the velocity decreases and the sediment settles onto the sea floor. The extra silt causes direct mortality by smothering the corals, while increased turbidity reduces the light reaching the photosynthetic zooxanthellae. With the death or reduced growth of corals, the reef community shifts: if turbidity is not too great, it may become covered in rapidly growing fleshy algae, but the diversity of the system is generally dramatically reduced and large areas become dominated by a few tolerant species.

Sea disposal of wastes is an economically attractive option for a number of industries which produce significant quantities of solid, low toxicity wastes. The three types of waste which have most frequently been disposed of in this way are mine tailings, china clay wastes and power station coal ash. In each case, the waste undergoes very little chemical change in the marine environment and chemically is often very similar to natural sediments of the area. The principal impacts arising out of the disposal occur as a result of smothering of the natural sea bed and its fauna at the disposal site, changes in the nature of the sediments in the surrounding area and possibly an impact of the plume from the discharge on the water column. Dredge spoil is also dumped at sea. This is material removed from the floor of harbours, ports and shipping channels to allow access by large vessels. As such, it is natural sediment of the source area, but, as many ports are sites of polluting discharges, the sediment may be contaminated by pollutants. For economic reasons, disposal of dredge spoil occurs close to shore, and the fine sediments removed from the calm waters of the port may be very different in character to the natural sediments of the dump site.

Therefore, dredge spoil causes a direct impact by smothering the fauna and indirect effects through altering sediment particle size and possibly by introducing toxic pollutants.

5.8.3 Wrecks and artificial reefs

Artificial surfaces in the sea, such as sea walls, piers, jetties, shipwrecks, oil and gas production facilities and similar structures all represent hard surfaces available for colonization. In general, the communities present on such structures are similar to those on natural hard substrata in the same area. Sea walls and coastal defences tend not to be heavily colonized as their location means they are continually exposed to high wave action and associated sand and sediment scour. Harbour walls often provide an interesting contrast, supporting wave-tolerant communities on the outer face and turbidity-tolerant ones on the inner face.

Much of the ecological interest of artificial structures lies in the opportunity they provide to observe succession. When a wreck sinks or a new sea wall is built we can be certain of the date when the substratum first became available for colonization. One of the most striking features of the communities which develop on shipwrecks below the photic zone is the importance of orientation in determining the dominant members of the community. While most species can be found both on vertical faces and horizontal structures, their relative abundance is very different, probably as a result of a combination of relative susceptibility to siltation and current. Large anemones such as the plumose anemone, *Metridium senile*, dominate on the vertical faces while hydroids and bryozoans dominate on the more horizontal areas.

Artificial reefs act as foci for fish. For some species, it is the requirement for hard substrata that attracts them, but many species common on the surrounding sediment plains are also abundant around the wrecks. It is unclear whether the increased abundance of fish is a direct response to the increased productivity associated with the extensive epibiotic

5

communities, a response to the changes in the physical environment produced by the reef, or the potential reduction in predation from having a shelter to hide in. Whatever the cause, this phenomenon has been responsible for a large number of projects worldwide in which artificial reefs, made from materials ranging from old tyres to concrete blocks, have been deployed to restore degraded systems or to enhance local fish catches.

5.9 Exploitation of coastal fisheries

The current global harvest of marine organisms amounts to approximately 86 million tonnes per year and sea food is the source of around 20% of world protein consumption. Over two-thirds of this catch comes from temperate coastal seas, the North Sea alone accounting for around 3.5% of the world fish catch (Box 5.2). Worldwide, the fishing industry supports over 15 million people and Japan, Russia, China, USA and Chile are responsible for nearly 50% of the catch.

It is estimated that the maximum sustainable yield (MSY) from the sea is between 100 and 170 million tonnes. Currently, harvest levels are close the lower part of this range, although many stocks appear to be in an imminent state of collapse. There can, therefore, be no doubt that fishing is a major ecological force in the coastal seas. We might expect three effects of high levels of fishing: an effect on the target population(s), an effect on the ecosystem via a trophic cascade (Section 7.5) and an ecosystem effect of the fishing activity per se, including mortality of non-target species, changes in the

5

Box 5.2   The North Sea fishing industry

The North Sea accounted for 3.5% of the world fish catch in 1983, a total composed of 2.42×10^6 t of fish and 0.16×10^6 t of shellfish. ICES (International Council for the Exploration of the Seas) data show that, in 1913, the North Sea catch was 1.25×10^6 t of fish, increasing to 1.5×10^6 t by 1960. Between 1960 and 1968 it more than doubled, peaking with a catch of 3.44×10^6 t in 1974, since when it has fallen. The north-east Atlantic currently accounts for around 10 m tonnes per year of the global landings of around 80 m tonnes, so this region is of global importance as a fishing region.

Looking at total catch is, however, a rather coarse analysis, particularly because the pattern of exploitation has changed, with increasing quantities used for fishmeal rather than directly for human consumption. By the mid 1980s, fish for human consumption accounted for only 46% of the catch, the remainder being industrial fisheries.

The fish used for human consumption have traditionally been divided, on the basis of the gear used for capture, into demersal species, caught by trawls, and pelagic species, caught by drift nets, seines etc. With the advent of midwater and surface trawls, this distinction has gone, but variations in the life histories of the two groups still make a convenient distinction for fisheries biologists.

Demersal fisheries

A wide range of demersal fish is exploited, with varying effects upon populations. From 1957 to 1984, the North Sea plaice (*Pleuronectes platessa*) catch increased, as a result of increased fishing effort, from 71,000 t to 150,000 t. The more valuable sole (*Solea solea*) catch has remained fairly constant at around 20,000 t. These species are robust and show no sign of overfishing affecting recruitment.

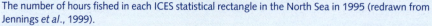

Box 5.2 Continued

The cod (*Gadus morhua*) catch rose during the 1960s and reached 340,000 t in 1972. Since then, it has declined to below the 1960s level and is now under 200,000 t per year. This appears to be, in part, due to a long-term biological trend rather than just the result of overfishing. Haddock (*Melanogrammus aeglefinus*) appear to be subject to wide natural fluctuations, catches varying between 50,000 and 550,000 t. Currently levels are around 100,000 t and the fishery, mainly dominated by UK vessels, is subject to strict quotas.

As a result of concern over the level of whiting (*Merlangius merlangus*) and haddock taken as a by-catch in the industrial fisheries, an exclusion area was established in the northern North Sea, the 'Norway Pout box', from which young individuals of these species cannot be taken.

In addition to effects on fish stocks the dragging over the sea floor of demersal fishing gear such as trawls and dredges has a ploughing effect. This results in mortality of many large, slow-growing or fragile species such as sea pens, corals, and large bivalves. The fauna becomes dominated by small-fast growing species and in particular polychaete worms. Some fishermen argue that this provides better food for fish and so the ploughing effect of fishing is stimulating productivity for fish. Scientific analyses refute such claims and show only negative effects from fishing gear disturbance.

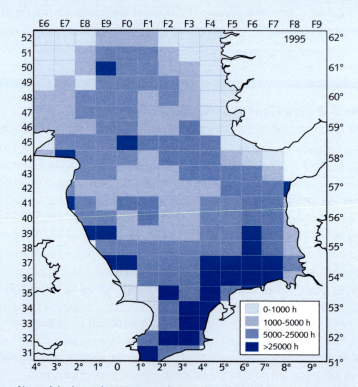

The number of hours fished in each ICES statistical rectangle in the North Sea in 1995 (redrawn from Jennings *et al.*, 1999).

Pelagic fishing

Pelagic fishing for food fish is centred on the herring (*Clupaea harengus*) and the mackerel (*Scomber scombrus*). Their life history characteristics make these species highly susceptible to overfishing,

---

Box 5.2 Continued

---

culminating in the 5-year closure of the North Sea herring fishery between 1978 and 1982. The lesson has, unfortunately, not been learned and, despite repeated calls from ICES for a ban, the mackerel fishery appears to be in a state of imminent collapse.

By-catch

Biological material caught but not the target of the fishery is termed by-catch or incidental catch. Some of this comprises species of economic worth that are landed but much of the material is of no value: species with no commercial value or individuals that are not suitable for landing. This includes individuals of commercial species that are below the legal minimum landing size or of species for which the vessel is not licensed (has no quota) or for which the catch quota has already been caught. This material is returned to the sea as 'discards', but the vast majority is dead as a result of the trauma of being caught and so contributes to the scavenger food chain. Globally over 50% of the global fish catch is discarded while in the North Sea it is around 60%.

It is unlikely that overfishing will ever lead to the extinction of a target species, as it would become uneconomic to fish long before this time. Some by-catch species are, however, susceptible. Skate (*Raja batis*), for example, are taken as part of the catch in trawling operations. Because they are not target species, they are deemed not to be of economic value and no quotas are set on them. Being slow breeders and highly susceptible to fishing, extinction is a possibility.

The industrial fishery

Denmark dominates the industrial fishery. Several small species are taken including sand eels (*Ammodytes* spp.), Norway pout (*Trisopterus esmarkii*), blue whiting (*Micromesistius poutassou*) and sprat (*Sprattus sprattus*). The fishery takes about 350,000 t of Norway pout, and the stock seems robust. Sand eels are fished around the Shetlands, in the north-eastern North Sea and in the southern North Sea. Currently about 600,000 t are taken. There is some evidence of a decline, but this is hard to separate from the wide fluctuations in population which this species undergoes.

Sprats show clear signs of stock depletion. Stock estimates of 700,000 t in the mid 1970s had shrunk to under 100,000 t by the late 1990s.

---

inputs of POM and direct physical disruption by fishing gear.

Effect upon the target population

Fishing selectively removes the larger individuals in a population. At moderate levels, this removes older, generally less fecund, fish and so promotes a healthy population dominated by reproductively active individuals, which are the most efficient at converting food to flesh. At high fishing intensities, the population becomes dominated by young fish. In the North Sea the catch of cod (*Gadus morhua*) in the 1990s was dominated by 3–4-year-old fish. When fisheries scientists first began studying the fish stocks towards the end of the nineteenth century, cod bred for the first time at four years old, whereas now they are breeding at three years old, possibly demonstrating an evolutionary response to the high levels of fishing pressure.

The trophic cascade

Changes in a target population affect species with which it interacts competitively or

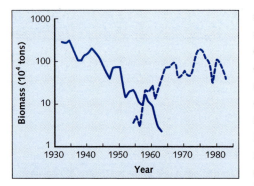

Figure 5.16 Increase in the population of anchovy (dashed line) following the decline of the Pacific sardine (solid line) off the coast of California. From Levinton (1995).

trophically. Until the 1940s, the productive, upwelling waters off California supported a fishery for Pacific sardines (*Sardinops caerulea*) which landed over 100,000 t yr^{-1}. When this fishery collapsed, the sardine was replaced by the anchovy (*Engraulis mordax*) (**Figure 5.16**), and even a relaxation of the fishery did not allow the return of the sardine. While variations in ocean currents and water temperature contributed to this change, these species are competitors and the heavy exploitation pressure on the sardine, along with a decrease in competitive ability due to temperature changes, may have allowed the pilchard to dominate, leading to a new stable community which is resistant to reinvasion by the sardine. A similar explanation has been proposed for the changes brought about by whaling in the Antarctic (see Section 6.8).

Non-target species

The material caught by the fishing process but not required is referred to as the by-catch (**Figure 5.17**). In some fisheries, this amounts to 75% or more, by weight, of the catch and is usually discarded. While some may be alive, much of it is dead or so moribund as to be easy prey to scavenging predators, thus potentially

Figure 5.17 The catch on deck of a trawler in the North Sea. The valuable part of the catch has to be sorted from the unwanted species—the by-catch—which is then thrown back, by which time most of it is dead or so moribund as to fall easy prey to the following flock of birds. The offal from the gutted fish is also thrown into the sea, adding to the flux of organic matter for the scavenging birds and benthos. Photo M. Gill.

Table 5.5	Estimates of the fate of material discarded by North Sea non-coastal fishing vessels	
	Taken by seabirds (t yr^{-1})	Sinking and potentially available to fish and benthic scavengers (t yr^{-1})
Offal	55,000	7800
Roundfish	206,000	56,200
Flatfish	38,000	261,200
Elasmobranchs	2000	13,000
Benthic invertebrates	9000	140,800
Total	**310,000**	**479,000**

Data from Garthe *et al.* (1996).

impacting populations of non-target species. It is estimated that the North Pacific drift net fishery, with an annual harvest of around 300,000 t, sets up to 60,000 km of drift nets each night. In addition to the target catch, these take marine mammals, turtles, birds and non-target fish and probably affect at least some macrozooplankton. Mortality figures are difficult to obtain but estimates amount to over 10,000 bird deaths per year.

The discards and offal thrown overboard when the catch is processed and gutted mean that fisheries are beneficial to scavenging species (**Table 5.5**). Much of the increase in seabird numbers in the North Sea has been attributed to increased food as a result of discarded by-catch and fish offal (Garthe *et al.* 1996). Similarly, benthic scavenging fish, whelks, starfish and hermit crabs appear to respond to the feeding opportunity that the discards represent.

Physical disturbance

Demersal fishing gears directly impact the sea floor, the passage of a dredge, beam trawl or otter trawl being analogous to a plough or rake being dragged over a field. Large, slow-growing and fragile species are most susceptible to damage—studies in the southern North Sea show that beam trawling can lead to decreases of 10–65% in the density of echinoderms, polychaetes and molluscs.

The physical structure and the creviced nature of coral reefs mean that they are not amenable to towed fishing gear, so fishermen have to use relatively inefficient lines and traps. In some areas, poisons and dynamite have been used to kill fish, which subsequently float to the surface for easy collection. These also cause mass mortality of non-target forms and, in the case of explosive fishing, destruction of the physical structure of the reef (**Figure 5.18**). Removal of grazing fish and invertebrates, along with mortality of the corals, often leads to the system becoming dominated by rapidly growing algae which prevent reinvasion by coral, providing another example of a change in stable state.

5

(a)

(b)

Figure 5.18 (a) Unexploited coral reefs are natural high-diversity systems which support a diverse fish fauna. (Copyright Simon Jennings). (b) Exploitation of these systems by poisons or explosives leads to marked changes in the reef. The system shifts to a low diversity one dominated by rapidly growing algae, and following 'dynamite fishing' physical destruction of the reef structure: Pombo, Ambon Islands, Indonesia. Photo C. Frid.

Bulleted summary

- Coastal seas are the most productive region of the oceans, and contain a high diversity of environments. The relative ease of study means that we know more about the ecological functioning of intertidal and shallow water communities than the rest of the marine environment.

- The abiotic environment in coastal regions is dominated by tides, particularly along the shore itself. The ecology of intertidal areas is primarily a function of the degree of water movement and the underlying geology. This determines the nature of the shore—wave-swept rock, sheltered rock, coarse mobile particles or stable muds. Tides produce clear zonation on rocky shores, in which dominant organisms are sessile, but zonation patterns are less clear on soft sediment shores, where most organisms are mobile burrowers.

- Energy inputs in the water column are dominated by phytoplankton production, which generally shows distinct seasonal patterns in response to temperature, light, nutrient ability and grazing pressure. Inputs from rivers can enhance nutrient availability in coastal waters, though this affects species composition of phytoplankton more than overall abundance. Primary production in benthic communities is limited to shallow waters, particularly the rocky intertidal and sublittoral and, in clear tropical waters, coral reefs. Detritus is the primary energy source for most subtidal systems.

- Community structure in hard substratum systems is determined by physical disturbance and predation, which prevent monopolization of resources by dominant species. Competition contributes to patterns of zonation seen on rocky shores. In soft sediment systems, the type of sediment is the primary determinant of community structure. Biological disturbance is important, though its effects may be beneficial to certain species. Benthic community structure is often explained, at least partially, by variability in recruitment (supply-side ecology) and, in soft sediments, by individuals which have survived disturbance (relicts).

- In the water column, there is good evidence for both competitive and predation control of plankton communities, often with the relative importance varying seasonally.

- Human impacts on coastal seas include dumping of spoil, which smothers biota, and construction of hard surfaces, which can be beneficial to taxa adapted to rocky substrates. Harvesting of wild populations of marine organisms has major effects, both upon the target species and upon the environment in which they are caught.

Further reading

Alongi D.M. (1997) *Coastal ecosystem processes.* Boca Raton, FL, CRC Press.

Belgrano A., Scharler U.M., Dunne J. and Ulanowicz R.E. (2005) *Aquatic food webs: An ecosystem approach.* Oxford, Oxford University Press.

Boaden P.J.S. and Seed R. (1985) *An introduction to coastal ecology.* Glasgow, Blackie and Sons Ltd.

Giller P.S., Hildrew A.G. and Raffaelli D.G. (eds) (1994) *Aquatic ecology. Scale, pattern and pro-* cess. *34th Symposium of the British Ecological Society.* Oxford, Blackwell Science.

Longhurst A.R. and Pauly D. (1987). *Ecology of tropical oceans.* London, Academic Press.

Mann K.H. (1982) *Ecology of coastal waters. A systems approach.* Oxford, Blackwell Scientific Publications.

Mann K.H. (2000) *Ecology of coastal waters: With implications for management.* Oxford, Wiley-Blackwell

Assessment questions

1 Describe the conversion of dissolved organic matter to particulate organic matter.

2 What factors and processes affect the supply of larvae to an assemblage?

3 How are tides generated?

4 What causes the pattern of annual productivity in temperate North Atlantic plankton?

5 How do the assemblages found on wrecks differ from those on natural rocky reefs in the same area?

6 Describe the main routes for the cycling of nutrients (nitrogen and phosphorus) in coastal ecosystems.

6

The Open Ocean

■ ■

6.1 Introduction

Beyond the continental shelf, the coastal zone is replaced by the open ocean. In oceanic areas, the sea floor is always below the depth at which sufficient light penetrates to support photosynthesis and, with the exception of vent fields (Section 6.7), is dependent for its energy upon allochthonous production.

The water column can be divided into a series of zones based primarily on the availability of light (Box 1.2). The photic zone, in which there is sufficient light to support photosynthesis, extends to around 100 m depth. Beneath this is the mesopelagic zone, down to about 1000 m, where some light penetrates but not enough to support primary production. Beyond this, and in complete darkness, are the bathypelagic zone (1000–4000 m) and the abyssopelagic zone (4000–6000 m). Only deep ocean trenches, the hadal zone, extend beyond this depth, although the deepest, the Challenger Deep in the Marianas Trench in the western Pacific Ocean, exceeds 11,000 m in depth.

6.2 The abiotic environment

6.2.1 Stratification

The thermocline

In the open ocean, a permanent thermocline is present at depths of 500–800 m (**Figure 6.1a**), effectively isolating surface waters from deep waters. In temperate areas, a second thermocline is produced at a depth of around 30–50 m each summer as intense summer warming at the sea surface, along with a decrease in wave-induced mixing, allows the surface layer to warm up (**Figure 6.1b**). In autumn, increased wave action and decreased solar radiation lead to the breakdown of this seasonal thermocline and the water column becomes well-mixed again. In addition to this seasonal thermocline, daytime heating can produce a daily thermocline, typically at depths of up to 10 m, which is dispersed each night when the input of solar radiation ceases (**Figure 6.1c**).

Water masses

Below the thermocline, a number of layers in the water column, differing particularly in salinity and temperature (and therefore density), can be identified (**Figure 6.2**). Distinct bodies of water with differing characteristics are known as water masses and are usually named after their region of formation. For example, Mediterranean Deep Water is present in the eastern North Atlantic at depths of around 1000 m. This relatively warm and saline water mass is formed in the north-west Mediterranean in winter by cool dry winds blowing over the sea, causing intense evaporation and mixing the water column down to depths of 2000 m or more. This water flows out of the Mediterranean under the inflowing Atlantic Water and then sinks and spreads out as it finds the layer of correct density at about 1000 m.

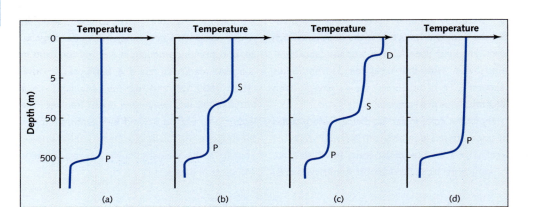

Figure 6.1 Temperature–depth profiles showing seasonal patterns of thermocline development at mid latitudes. (a) Winter. Low inputs of solar radiation and strong wind mixing at the surface produce a deep well-mixed layer. The permanent thermocline (P) is at a depth of 500–800 m. (b) Spring. Decreased wind mixing and increasing solar heating of the surface layer leads to the development of a warm surface layer separated by a seasonal thermocline (S) from deeper waters. As the season progresses the temperature difference across this seasonal thermocline increases and the thermocline depth becomes deeper. (c) Summer. The seasonal thermocline is fully developed at 30–80 m and a daily thermocline (D) may be produced by the strong daytime heating of the surface layer. Other near-surface thermoclines may persist for periods of days when very calm conditions coincide with strong heating. (d) Autumn. Decreasing inputs of solar radiation and increasing wind-driven mixing erode the thermocline, making it weaker and shallower before it breaks down completely to give a well-mixed surface layer (compare with lakes, Figure 7.3).

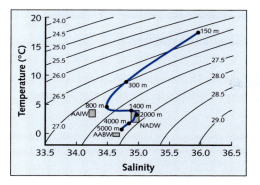

Figure 6.2 A T–S (Temperature–Salinity) diagram is a widely used tool in oceanography to examine the stability of the water column and identify constituent water masses. The density (usually expressed as $\sigma^t = 1000 *$ (specific gravity -1)) of saline waters, illustrated here by the contour lines, is a function of salinity and temperature, but influenced more by temperature, hence the curved σ^t contours. The salinity and temperature of samples at various depths are plotted on the T–S diagram along with the 'core' characteristics of various water masses (boxed regions). In this plot from the tropical South Atlantic, the presence of Antarctic Intermediate Water (AAIW), North Atlantic Deep Water (NADW) and Antarctic Bottom Water (AABW) can all be identified.

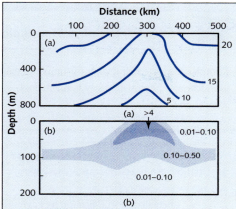

Figure 6.3 Diagrammatic representation of a vertical section through a cyclonic ring in a sub-tropical ocean. (a) Temperature (°C) and (b) Chlorophyll a (µg/l). From Parsons *et al.* (1984).

Denser bodies of water attempt to sink beneath less dense ones, so a stable water column is one in which the most dense water is in contact with the sea floor and the least dense is at the surface. The density boundaries between water masses, marked by thermoclines or haloclines, can be significant impediments to the mixing of nutrients, oxygen and pollutants as well as marking breaks in salinity and temperature.

Upwelling

Surface waters and the deep ocean act as separate water masses, each with its own circulation system, connected only in discrete areas where surface waters downwell and deep waters upwell (Section 1.2). The main areas of upwelling occur where surface ocean currents diverge, forcing deep water to rise (Section 1.2, Figure 1.3a). As they are generated by global air circulation processes (Box 1.1), these are centres of permanent or predictably seasonal

upwelling (but see Section 6.8). In addition to these, however, local conditions, extending over tens to hundreds of kilometres, can cause a breakdown of the thermocline and upwelling of deeper waters. A cold core ring, a cyclonic gyre which spins off a major current system, is an example of this (Figure 6.3).

Movement of surface water into shallows, such as the edge of the continental shelf (the shelf break) and the tops of sea mounts, increases water velocity, while frictional interaction with the bottom causes vertical turbulence. This leads to upwelling (Box 6.1), which can change otherwise low productivity surface water into a region of high production.

Benthic storms and sediment slumps

Over most of the ocean the currents flowing over the sea bed are relatively weak and predictable. However, some of the deep water currents can generate high energy eddies. These eddies, known as 'benthic storms' move over the sea floor with current speeds in the eddy of up to 40 cm s^{-1}. At 4800 m off Cape Cod, for example, the area experiences 8–10 storms per year, each lasting 2–20 days. In addition to producing extra drag and stress on benthic organisms, the currents associated with these storms

6

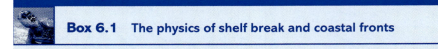

Box 6.1 The physics of shelf break and coastal fronts

Consider the potential energy of the water column and the tidal energy. Let R be the ratio of the production of potential energy in maintaining well-mixed conditions (PE) to the rate of tidal energy dissipation (TED):

$$R = PE/TED$$

We can estimate these two terms in a number of ways and, for a restricted area, many of the terms used in those estimates are constants. However, two terms which vary are the water depth, h, and the average water velocity $|U|$.

We can calculate the stratification parameter S from this by:

$$S = \log_{10}(h/(C_D * |U|))$$

where C_D is the drag coefficient.

Low values of S indicate turbulent mixing, while high values indicate stratification. For coastal waters, S usually has a value in the range -2 to $+2$. At intermediate values of S, around 1.5, the degree of stratification and turbulence are just sufficient to stabilize the water column and hold phytoplankton above the critical depth, while also providing sufficient turbulence to maintain a supply of nutrients to the surface layer (Pingree 1978).

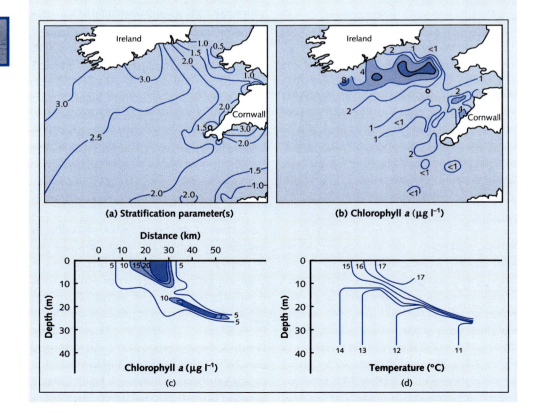

(a) Stratification parameter(s)

(b) Chlorophyll a (μg l^{-1})

(c)

(d)

Box 6.1 Continued

For example, in the Celtic Sea, off south-west England, a summer front develops, to the south-west of which the water is stratified, while to the north-east strong tidal flows and the shallowness of the water lead to a well-mixed water column. The thermocline rises to the surface at the front, producing a surface chlorophyll maximum, in contrast to a deep chlorophyll maximum in the stratified region. The close correlation between the $S = 1.5$ isoline and the surface chlorophyll concentrations is clearly seen in the figure above. This shows (a) the distribution of the stratification parameter (S); (b) the surface chlorophyll (9–12 April 1975) in the Celtic Sea area of the continental shelf; (c) the vertical profile of chlorophyll *a* (upper) and temperature (lower) along a transect through a frontal region.

re-suspend benthic sediments and move them considerable distances. Thus benthic storms may cause smothering of some benthos but for others they may provide additional sources of food. Areas of the sea floor close to the continental slope may be subject to further stress as both internal waves and waves generated by atmospheric storms may impact on the sediments that have accumulated on the slope, triggering slides and slips. The large swell waves associated with high-latitude storms can induce water movements at depths of over 100 m and so can mobilize sediments in the vicinity of the shelf edge. In many areas the slope sediments are unstable and a relatively small disturbance can trigger a slide, which can then trigger further slides. The sediment placed in suspension by these events can 'roll' down slope and spread over the sea floor as a turbidity current; the lighter fractions (including the organic material) may spread out in the adjacent water layer and be moved a considerable distance before they eventually settle hundreds of kilometres from the shelf edge.

NADW is +1.8 to −2°C (Figure 6.2). In deep trenches, the high pressure causes a heating of the water, which is, therefore, slightly warmer than abyssal waters. Deep basins isolated from the rest of the deep ocean by a shallow sill can be deoxygenated, as the shallow barrier prevents oxygen-rich water flowing into the basin to counter oxygen consumption by decomposers. These deep basins, including the Black Sea, may be anoxic and essentially abiotic due to this physical isolation from the deep ocean circulation (see also deep lakes, Section 7.2). AABW and NADW are, however, formed in well-mixed surface regions and maintain a high oxygen concentration through continual circulation of cold water from the surface. Therefore, most deep ocean waters are well oxygenated and capable of supporting a diverse assemblage of aerobic organisms. Deep circulation also maintains constant conditions across large geographical areas, allowing organisms adapted to the conditions to become widely distributed. There is, for example, no climatic differentiation between areas of the deep ocean.

6.2.2 Conditions in the ocean depths

At the sea floor, the physicochemical environment is very constant. Throughout the deep sea, salinity is about 34.8 and water is at high pressure. It is always dark and very cold: AABW temperatures are 0 to −1.8°C, while

6.2.3 The ocean bed

Sediments

The abyssal plain is dominated by soft sediments, many of which derive from phytoplankton. Carbonate sediment is the skeletal remains of coccolithophores and foraminiferans, which

6

have calcium carbonate (CaCO$_3$) skeletons, while siliceous sediment is the remains of diatoms and radiolarians, whose skeletons are formed from silica (SiO$_2$). Red clays, in contrast, have a relatively small biological component; they are composed mainly of tiny particles originating from continental weathering and carried to the open ocean by wind and currents. Their red colour is caused by small quantities of ferric iron oxide (Fe$_2$O$_3$). The thickness of deep ocean sediments can reach several kilometres on the abyssal plain, but their rate of deposition is so slow that they are only a few metres deep at spreading centres of ocean ridges, and often confined to depressions in the underlying topography.

The distribution of the different sediment types is principally controlled by three interrelated processes: climate and current patterns, distribution of nutrients (which determine surface water productivity) and relative solubility of SiO$_2$ and CaCO$_3$ as they sink. Carbonate sediments dominate in shallow water overlying mid-ocean ridges; in areas of high nutrient concentration, these derive from foraminiferans, whereas, in low productivity regions, coccolithophores are the source. Carbonate dissolves most rapidly in cold, deep water, so siliceous sediments dominate in cold waters, where diatoms are most abundant, and in deep equatorial regions, where radiolarian activity is highest. In the least productive parts of the deep ocean, where planktonic activity is very low, red clays are the dominant sediment **(Figure 6.4)**.

Trenches

Trenches are long, narrow features, typically with a flat bottom and convex sides, along whose axes currents can be up to 10 cm s^{-1}. They often contain poorly sorted sediments from a variety of sources, including terrigenous sediments, while the trench sides may include exposed rock, providing a wide diversity of niches.

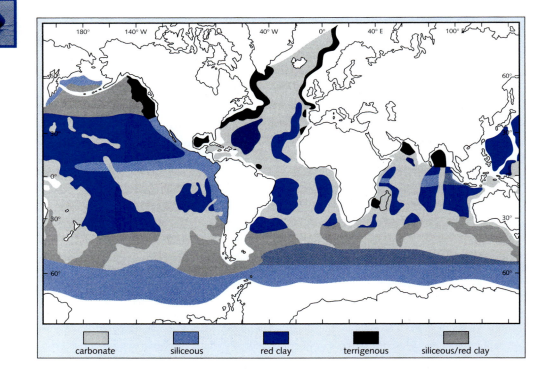

Figure 6.4 The distribution of dominant sediment types in the deep ocean.

Hydrothermal vents

Hydrothermal vents are localized phenomena in seismically active parts of the ocean bed, in which reduced compounds and, often, heat are emitted into the water (Box 6.2). Although accounting for only a tiny proportion of the ocean bed, and very transient in their nature, they are locally extremely important in that they provide optimum conditions for chemosynthesis in chemoautotrophic bacteria (Section 1.4) and in organisms with symbiotic chemoautotrophs such as the vent clams (Section 6.7).

Box 6.2   Hydrothermal circulation in the ocean crust

Hydrothermal vents are areas of the seafloor where hot water is emitted from the Earth's crust. They were first observed in the mid 1970s by submersibles involved in Project FAMOUS (French–American Mid-Ocean Undersea Study), although their existence had been predicted and hinted at by temperature anomalies discovered in the mid 1960s. Vents occur at ocean spreading centres, the most studied ones being the Galapagos Rift system, the East Pacific Rise and the Atlantic Mid-Ocean Ridge at 21°N.

Once the oceanic crust has been formed (Section 6.8), hydrothermal processes start to occur. Sea water, driven by convection, circulates within the newly formed rock. Chemical reactions between the sea water and the usually hot basalt cause major chemical changes to occur. These reactions are the main source of some elements, notably lithium, rubidium and manganese, in ocean waters, and are an important source of barium, calcium and silicon. It is estimated that about one-third of the sea floor contains active water circulation cells, so it is a widespread phenomenon. About 1.7×10^{14} kg of sea water flows through the oceanic crust every year, and the entire ocean probably passes through the crust every few million years.

There are four types of vents. The most spectacular are the black smokers, where hot water, at >350°C, gushes out of the crust in a dense plume carrying particles of metal sulphide—the black smoke. At temperatures in the range 30–330°C, the vents appear as white smokers, carrying particles of barium sulphate. Warm water vents have outflows at 10–20°C while, in cold water seeps, the water has undergone chemical change but no heating.

The nature of hydrothermal circulation

Water enters the crust over a wide area, via cracks and fissures, and percolates down through the crust. Hydrothermal circulation can only occur where the rock is permeable, the high degree of fracturing associated with new crust and the lack of a sediment infill partly explaining why it is so common at sites of new crust formation. The figure shows a cross-section of an idealized hydrothermal vent. Sea water percolates down over a wide area and is drawn down and laterally into the

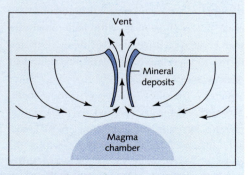

region above the magma chamber by the continual rising of heated water. This rises upwards in a narrow plume carrying with it high concentrations of dissolved minerals, which may be deposited amongst the fractured crust (the stock work) and at the surface.

> **Box 6.2   Continued**
>
> Magma within the crust is at about 900°C, and as the crust is thin and, at ridge centres, not covered in a thick layer of sediment, a strong thermal gradient exists. This drives the convection cells, with the hot water rising through the rocks above the magma chamber in a narrow, strongly flowing band.
>
> While the flows alter the chemical balance of the sea water, they also affect the rocks, being important in the conversion of basalts into greenschists. Magnesium is completely stripped out of sea water during high temperature reactions with rock, so its presence in the outflow from warm water vents indicates that their outflow is composed of some hydrothermal water which has mixed with ordinary sea water before emerging. There is no dilution of hydrothermal water from black smokers. As the solution mixes with sea water, precipitates of sulphides, which make up the 'smoke', are often deposited to form a chimney 10 m high from which the vent emerges.
>
> The transition from one type of vent to another is poorly understood, but can occur at any time. For example, the precipitation of salts within the upper rock layers of a warm water vent may restrict the outflow to a smaller area and so reduce the opportunity for dilution with ordinary sea water, changing the warm water vent into a white or black smoker. Over time, hydrothermal cooling drives the magma chamber deeper, and may be sufficient to solidify it at slow spreading centres. Thus magma chambers and the hydrothermal circulation are relatively short-lived and isolated phenomena, typically 15–25 yr.

6.3 Ecological effects of the abiotic environment

The great depth of the ocean limits photosynthesis to a small proportion of its volume, in the illuminated upper layers. The decrease in light intensity from the surface to depth represents the only major environmental gradient in the open ocean (Box 1.2), with light intensities high enough to drive photosynthesis restricted to the top few hundred metres, even in clear ocean waters.

Water pressure increases with depth, although organisms respond to the external pressure by maintaining the same internal pressure and, so long as they remain within the pressure zone to which they are adapted, suffer no detrimental effects. In the deepest trenches, however, extreme pressures can become an ecological limiting factor. They affect ionic reactions, solubility of gases and the viscosity of the water. The change in the chemical behaviour of some ions, for example the biologically important carbonate ion, and gases such as oxygen restrict the ability of many species to invade the hadal zone. Pressure also affects the behaviour of lipids surrounding nerve cells and so can cause problems for neurophysiology. As locomotion, ventilation of the gills and generation of feeding currents in filter feeders are all dependent on the physical properties of the water this too is a major constraint.

Permanent stratification limits circulation and therefore primary production. The thermocline prevents algal cells from being mixed into deep water where there is insufficient light, but also prevents nutrients produced by the decomposition of material in the deep sea from reaching the phytoplankton. Deep waters are nutrient-rich, as detritus raining down from the photic zone (Section 6.4) is decomposed, but little of this is mixed back into the upper layers, where uptake of nutrients by phytoplankton can severely limit their availability. In contrast to coastal seas (Section 5.4) and shallow lakes (Section 7.2), the permanent thermocline

Box 6.3 Ocean fronts and eddies

Wherever two water masses meet, the horizontal boundary between them will form a front in a similar way to the meeting of stratified and unstratified waters in coastal areas (see Section 5.4). In the oceans these boundaries may be permanent or seasonal but tend to vary in intensity and in exact location from year to year. The fluctuation in the position of the front on the northern edge of the Gulf Stream is influenced by interannual variation in the weather in the North Atlantic and, in turn, predicts some of the interannual patterns in ecology. Similarly variations in the oceanographic conditions off California have major implications for the fisheries of the whole Pacific region.

When two water masses are moving against each other there is a frictional stress along the boundary and this can generate eddies. These initially form as meanders/irregularities in the boundary and end up spinning off as self-contained features. For example, the Gulf Stream spins off, on average, 10 cold-core and 5 warm-core rings a year and each lasts for about a year before it either dissipates through diffusive mixing or is reabsorbed into the current. Warm-core rings are bodies of warm Gulf Stream Water spun off into the Labrador Sea, while cold-core rings are bodies of Labrador Sea water trapped and mixed into the warm subtropical gyre region.

The rings spun off currents such as the Gulf Stream are eddies but the term 'oceanic eddy' is usually used for lower amplitude fluctuations that can occur anywhere in the ocean. While rings can persist (although changing all the time) for a year, eddies usually dissipate much more rapidly. The presence of large numbers of eddies, of varying scales and intensities (current speeds) has led to analogies with a turbulent flow.

caused by the great depth of the oceans means that breakdown of the seasonal thermocline in temperate regions, although resulting in significant nutrient inputs to the photic zone, does not allow nutrients from the deepest part of the water body to recirculate. Only in regions of upwelling are high concentrations of nutrients inputted to the photic zone, making these the most productive parts of the ocean. Permanent upwelling off the edge of continents can extend over many hundreds of kilometres, while cold core rings produce patches of enhanced primary production typically around 200 km in diameter.

The movement of fronts and eddies (**Box 6.3**) can serve to draw water up from depth and so boost productivity. Cold core Gulf Stream rings south of Cape Hatteras (USA) have been estimated to provide about 55,000 tonnes of nitrogen to the outer shelf in the region. Frontal systems and eddies influence the rates of planktonic processes and so produce spatial patterns in productivity and associated ecosystem functions. It is increasingly apparent that many of the top predators in oceans cannot support their biological needs from the background concentration of their prey and must seek out areas of higher prey abundance (Fromentin and Powers 2005). Eddies and fronts therefore attract fish and squid which in turn are preyed upon by birds such as albatross (Xavier *et al.* 2003).

6.4 Energy inputs

6.4.1 Primary production

Photosynthesis in the oceans is limited almost exclusively to phytoplankton and Cyanobacteria. Floating macrophytes are important only in the Sargasso Sea, in the mid-Atlantic Ocean. Elsewhere, with the exception of the limited

6

primary production at hydrothermal vents (Section 6.7), the ocean ecosystem is dependent on the energy fixed in the photic zone by phytoplankton. Away from upwellings, highest ocean production occurs in regions of intermediate stability where nutrients are mixed across the thermocline but algae are not mixed down into deep waters. Nitrate is generally the limiting nutrient in oceanic waters, although diatoms may be limited by their requirement for silica.

Iron as a limiting nutrient

In the equatorial Pacific and the Southern Ocean, nitrate, phosphate and silicate are present in relatively high concentrations but primary production is low. In these regions, iron may be the limiting nutrient (Chavez *et al.* 1991), because although common in the Earth's crust and in inland and coastal waters it has low solubility and enters the ocean only in small amounts. The main source is wind-blown dust particles, but the distribution of global winds means that the equatorial Pacific Ocean and the Southern Ocean around Antarctica receive little of this dust.

In 1993 and 1995, the importance of iron in controlling primary productivity in the equatorial Pacific was investigated by releasing approximately 0.5 t of iron, in the form of ferrous sulphate labelled with an inert tracer molecule (so the water with the added iron could be distinguished from the surrounding ocean, even if all the iron was consumed). Within a few days, up to 30-fold increases in chlorophyll concentration and phytoplankton biomass were recorded in the water to which the iron had been added. This has led to speculation about the feasibility of adding iron to the ocean to stimulate primary production and hence removal of CO_2 from the atmosphere to offset anthropogenic increases and global warming.

6.4.2 Secondary production

There is a fairly constant ratio of carbon in the oceans.

Dissolved organic matter (DOM)	1000
Particulate organic matter (POM)	125
Phytoplankton	20
Zooplankton	2
Fish	0.02

Thus POM and DOM represent a large reservoir of organic carbon, whose dominant users are heterotrophic microorganisms. Transfer of energy to the deep ocean is by marine snow: detritus sinking from the photic zone. Once it has settled on the ocean bed, some POM is resuspended into the pelagic zone. Resuspension of even small amounts of bottom sediments by internal waves or near-bottom currents can provide a large input of nutrients and POM to the overlying waters. Resuspension is seasonal, as a result of the seasonal pattern of wave action, but is also limited by stratification, which both acts as a barrier to waves interacting with the bottom and effectively isolates the photic zone from any nutrients generated by resuspension.

Composition of POM

Much POM consists of recognizable fragments of phytoplankton and zooplankton, along with faecal pellets. A large proportion is however in the form of amorphous masses, up to 50 mm in diameter, containing bacteria, small algae and protists, along with much inorganic carbonate, or particles less than 5 mm in diameter, with no obvious origin, but possibly either breakdown products or building blocks of these amorphous masses. Surface waters contain semi-transparent flakes, up to 25 mm in diameter, which may be the result of films which form on air bubbles and remain after the bubble collapses.

Most POM is recycled in the top 300 m, so its concentration declines with increasing depth. A large proportion is associated with the surface film, and can be concentrated into slicks by the action of Langmuir circulation cells (**Box 6.4**). Typical concentrations are 50–1000 mg C dm^{-3} at the surface, 5–100 mg C dm^{-3} at 100 m depth

Box 6.4   Langmuir circulation

Langmuir circulation is the name given to the series of rotating cells set up by moderate winds blowing over the surface of the water. The cells run downwind and may extend over several hundred metres.

Under conditions of Langmuir circulation, plankton are passively accumulated into patches which may be several kilometres long but only 1 m or less wide. The figure shows how a steady moderate wind over the sea surface sets up Langmuir circulation cells. Neutrally buoyant particles remain randomly distributed (A), negatively buoyant particles are aggregated in the upwellings (B), while buoyant particles are concentrated at the downwellings—often forming visible slicks on the surface (C). Actively swimming organisms may position themselves in areas of high relative flows (D), lower velocity flows (E) or the low flow region out of the vortices (F). After Stavn (1971).

Accumulation of particles in regions of downwelling and upwelling results in the frequently observed phenomenon of greater variability in samples parallel to the wind than those taken in perpendicular to it. A net towed perpendicular to the wind will tend to integrate across the patches, while downwind samples will either be in or out of a patch and hence more variable.

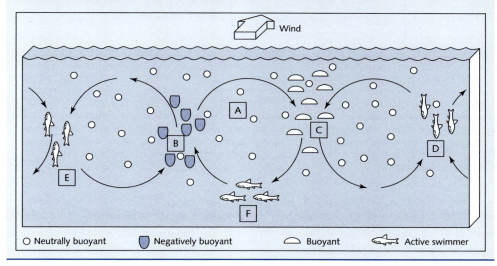

O Neutrally buoyant   Negatively buoyant   Buoyant   Active swimmer

6

and <10 mg C dm^{-3} at 250 m depth. From about 300 m to the seabed, POM concentration is fairly constant.

Biological processing of POM

Biological activity is extremely important in packaging POM. Small particles with a slow sinking rate are concentrated by water column detritivores, such as filter-feeding copepods, salps, and mysid shrimps, into large, rapidly sinking faecal pellets which, below about 1000 m, are a major component of sinking POM. This transfer is supplemented by vertical migration of organisms which feed at night near the surface, then migrate down at dawn (Section 6.8), often covering a vertical distance in excess of 1000 m (**Figure 6.5**), and release their faecal pellets at greater depths.

POM particles act as the basal resource in deep ocean systems. The biological repackaging of small particles into faecal pellets, along with the ease with which phytoplankton cells aggregate into 'flocs' several centimetres across,

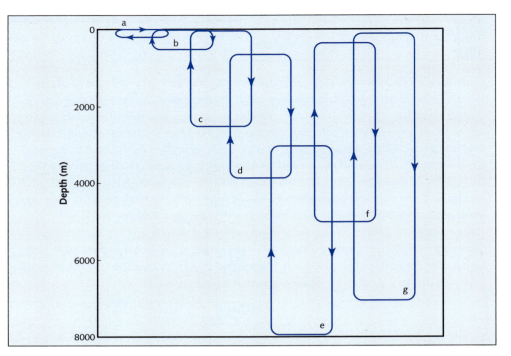

Figure 6.5 Diagrammatic representation of how vertical migrations of different amplitudes and mid-points overlap, allowing material consumed at the surface to be liberated at mid-depths, where it is reconsumed and later liberated at greater depths. (a) Diurnal vertical migrations in the photic zone; (b) diurnal and seasonal migrations in the photic and upper mesopelagic; (c) seasonal and ontogenetic migrations including the surface and bathypelagic realms; (d) ontogenetic migrations encompassing the bathypelagic and the abyssopelagic; (e) migrations wholly within the abyssopelagic; (f) migrations including the deep sea and the mesopelagic; (g) short-term irregular migrations covering most of the water column. From Raymont (1983) after Vinogradov *et al.* (1970).

6

mean that sinking rates are quite rapid, of the order of 300 m per day, and detritus reaches the ocean floor within weeks of leaving the surface. Many zooplankton are filter feeders, which do not distinguish between living and dead POM and will gain much of their nutrition from the microbial communities on POM particles, rather than from the detritus itself (see Section 4.4). The same processes occur on POM in the sediments, where deposit feeders strip the heterotrophs off the particles.

In surface waters, the ratio of C:N in POM is 6:1, whereas in deep waters this has shifted to 20:1, probably because nitrogen is used by bacteria for protein synthesis, while carbon is present in refractory forms. Soluble carbohydrates are rapidly lost from the particles so that, below 200 m, only insoluble carbohydrates are left.

Fats generally comprise less than 1% of POM, and they too are rapidly reused by bacteria in the top 200 m. Protein and amino acid concentrations decrease rapidly from the surface to 200 m, then more slowly to 300 m.

6.5 Community structure in the photic zone

6.5.1 Ocean plankton

Species richness in pelagic communities appears to be similar to that in many terrestrial communities, but the generally wider distributional limits of the pelagic biota mean that the total species pool of pelagic species is much lower than for terrestrial species. There is a latitudinal gradient in species richness from the tropics to the poles that mirrors that seen in terrestrial environments. This cline is not, however, constant and there appears to a sharp decline at 40° latitude, which has been attributed to the increased seasonality and greater productivity occurring at the higher latitudes.

The paradox of the plankton

Hutchinson (1961) first referred to the paradox of the plankton: how can so many species co-occur without any distinct subdivision of the resource base and no spatial segregation? The lack of spatial structure and heterogeneity in the pelagic environment sets it apart from all other environments, because there is no scope for niche partitioning based on microhabitat specializations. Below 1000 m, no light penetrates and only the gradual change in pressure with depth marks the position in the water column, but the majority of species lack specialist pressure-sensing organs, so cannot detect this gradient.

Given the lack of physical structure, the expectation would be for low species richness in pelagic communities. This is not the case. For example, McGowan and Walker (1985) showed that 175 species of copepod coexist in the North Pacific in a relatively uniform water mass and that the relative and absolute abundance of each species had remained constant for over 20 years. Is this the result of competition being reduced by specialization and niche segregation or are the communities in a state of non-equilibrium, continually disturbed by predation, patchy resources or some other factor? In benthic communities, such questions would be tested experimentally. However, the nature of the pelagic environment means that any attempt to enclose a portion, in order to manipulate it, introduces large artefacts which make interpreting the results of the experiment difficult. An alternative approach which does

not suffer from the problem of artefacts is the 'natural experiment', although the uncontrolled and non-replicated nature of such experiments brings its own limitations.

In the central North Pacific, two events have been observed which can be used as such 'natural experiments' (Dayton 1984). The first occurred in 1969, when vertical turbulence across the thermocline increased, resulting in increased mixing of nutrient-rich waters up to the photic zone. If competition was an important factor in structuring this community, then increased resources would have allowed those species with high reproductive rates to dominate. Phytoplankton productivity doubled (confirming that they were nutrient limited) as did zooplankton biomass, but the relative abundance of the various zooplankton species did not alter in spite of the change in total abundance. The second experiment occurred as the result of a relatively new tuna fishery, which reduced numbers of top predators in the system by a considerable degree, allowing their prey—planktivorous fish and squid—to increase, and therefore altering predation pressure on the copepods upon which they feed. It did not, however, result in any shifts in the structure of the copepod assemblage. All the consumers of plankton in this system—predatory copepods, chaetognaths, euphausiids, shrimps, squid and fish—are dietary generalists, so, although total predation pressure increased, it did so uniformly across the range of prey species and no shift in relative abundance occurred. The sum of the predation pressure exerted by this assemblage may be important for maintaining the diversity of the copepod component of the community.

Patchiness in plankton distribution

Changes in the physicochemical properties of the water mass, life history variations and short-term migrations account for the heterogeneous distribution of zooplankton in the vertical plane. Spatial distribution of zooplankton in the horizontal plane is also patchy. As early

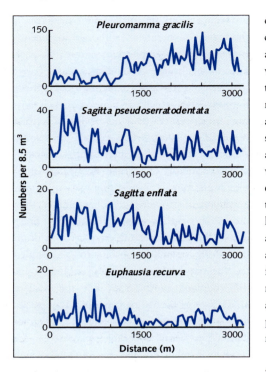

Figure 6.6 Horizontal patchiness of zooplankton occurs at a variety of scales, with large patches superimposed on small patches. From Wiebe (1970).

as the mid nineteenth century, it was recognized that plankton were not distributed randomly, but often occurred in dense swarms or blooms. By the 1950s (e.g. Barnes and Marshall 1951) it was established that zooplankton distribution is generally more clumped or aggregated than random. Cassie (1963) showed that the distribution of the diatom *Coscinodiscus gigas* was non-random, even within 1 m², while Wiebe (1970) described plankton patches on scales ranging from less than 20 m to several kilometres (**Figure 6.6**). The causes of this patchiness can be divided into those of physical and biological origin, although observed patches are likely to be a response to a number of physical and biological factors, and even to the sampling method used to make the observations!

Physical effects occur over a wide variety of scales ranging from kilometres of advection events, such as upwellings and downwellings, to centimetre scales of random turbulence. Other physical processes influencing spatial distribution of plankton include Langmuir circulation (**Box 6.4**), frontal systems (**Box 6.1**), and the upwelling and downwelling associated with cyclonic and anticyclonic systems, respectively (**Figure 1.5**). Phytoplankton tend to respond primarily to these physical processes and the supply of nutrients. Zooplankton are subjected to the same physical processes, but also undergo a range of biological processes which lead to spatial patchiness. There is some evidence that zooplankton are able to maintain their patch integrity against small-scale turbulence by active swimming. Zooplankton are attracted to regions of high phytoplankton, and are also able, given that patches may be stable for several months, to increase their numbers more rapidly in regions where phytoplankton are abundant. This is analogous to the ability of phytoplankton to develop dense patches in regions where nutrients are abundant.

Some phytoplankton produce inhibitory compounds, which reach their highest concentrations in the centre of a phytoplankton patch. Grazers, therefore, tend to forage from the edge of the patch inwards, resulting in the fragmentation of the large patch into a number of smaller patches as the zooplankton gain control (**Figure 6.7**).

Aggregations of copepods in tropical waters and euphausiids such as krill in the Antarctic appear to be active social groupings. Whereas many fish shoal to reduce predation pressure, it is difficult to apply this argument of safety in numbers to zooplankton, because many predators require dense patches to ensure that their prey is at a high enough density to sustain them. It has, therefore, been proposed that aggregating and attracting predators gives individuals away from the swarm a respite from predation. This is unlikely to apply to most zooplankton, but may be true for asexually reproducing species and for cladocerans when they are reproducing parthenogenetically.

Predatory plankters such as ctenophores and chaetognaths are capable of very high feeding rates and, in some cases, a clear correlation may be found between increase in abundance of

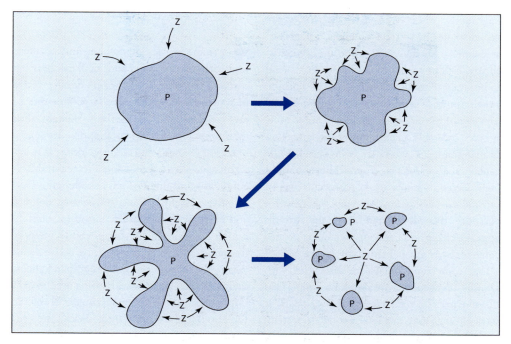

Figure 6.7 Dense patches of phytoplankton (P) attract grazing zooplankton (Z). However, the presence of any inhibitory compounds will be greatest in the patch centre, so grazing tends to occur from the edges. Any irregularities in the patch result in progressive incursions by the grazers, eventually leading to fragmentation of the original patch.

these predators and decline in copepod abundance. However, other studies show no such link and the role of zooplankton predators in controlling or structuring grazer assemblages remains to be resolved.

6.5.2 The role of oceanic predators

Low productivity and long food chains in the oceanic realm limit the biomass of the top predator guild and mean that predators have to forage over large areas to obtain their food. In the tropics and mid latitudes tuna, sharks and toothed whales are the dominant, natural, top predators.

Tuna

Tuna (*Thunnus* spp.) are fast-swimming fish—bursts of up to 20 body lengths per second have been recorded—which never stop swimming.

They are negatively buoyant and 'fly' underwater, generating lift from their extended pectoral fins. Their continual motion also allows them to ventilate the gills by forcing water against the gill membranes (ram-ventilation) as they swim with their mouths open, allowing them to meet the oxygen demands of their high metabolic activity. Uniquely amongst fish, they have a body temperature warmer than the environment, although whether this is simply due to their high metabolic activity generating heat more rapidly than they shed it, or is a case of active thermal regulation—a 'warm-blooded' fish—remains controversial. The oceanic distribution of tuna is probably driven by changes in water temperature (and thermocline position) and the need for well-oxygenated water masses. The average density of yellow fin tuna (*Thunnus albacares*) in the Pacific, without any human exploitation, would be one 10 kg fish every 2.8 km^2 of ocean, but tuna form large aggregations which undertake ocean-scale migrations. Three types of aggregation can be distinguished—tuna schools, porpoise-associated shoals and flotsam-associated shoals. Fish in porpoise-associated shoals tend to be larger than those associated with flotsam, which, in turn are larger than school fish. The size of fish in any given school tends to be similar, probably a reflection of the energetics of maintaining a particular swim speed, while genetic analysis of fish from schools shows a high degree of similarity in the occurrence of rare alleles—indicating that many school members are related. Despite this, the schools appear to form and break up, as smaller groups come together and subsequently separate.

That some tuna shoals are associated with schools of porpoises or dolphins emphasizes the similarity in their dietary requirements. It seems strange that, in a low production system, two potentially competing species would operate together. Our knowledge of the ecology of the open ocean is still so limited that it is not known what benefits the species may confer on each other to offset the potential negative impacts of competition. A possible explanation is that, while food density is low on average, both species require large aggregations of prey—social swarms, aggregations at fronts or cyclonic ring upwellings—within which food may not be limiting. If the species use each other as cues for good feeding conditions then this might represent an example of facilitation.

Sharks

The precise role of sharks in oceanic communities is poorly understood. Most sharks are dietary generalists with broader diets than other fish from the same region, although differing habits, mouth morphology and dentition leads to relatively low dietary overlap in co-occurring species. In larger species, the dietary overlap is potentially greater but competition may be avoided by spatial separation, either vertically in the water column or temporally, as many species show seasonal migrations between oceanic areas and shelf waters.

Toothed whales

While some species of dolphin and porpoise feed with tuna on squid and small fish near the surface, the majority of open ocean species of toothed whale also take large amounts of mesopelagic fish, particularly lantern fish (Myctophidae), for which they dive to depths of up to 250 m. There is some evidence that different species of oceanic dolphin feed at different depths and times of day, dividing the food resource. At close range, all oceanic dolphin and porpoise species utilize vision for hunting but, in addition, many species can locate prey by sound and, in some cases, by active echolocation. The animals emit broad band clicks, their well-developed hearing being used to gain information about the environment from the reflections and echoes of these clicks.

Sperm whales (*Physeter macrocephalus*) feed mainly upon squid, avoiding competition with other species through their large size (males up to 20 m, females up to 12 m in length) and by foraging at greater depths. They have been recorded at 1200 m depth and indirect evidence suggests that occasional dives may exceed 3000 m. Females and juveniles remain in low latitude waters, while males range throughout the oceans, providing an interesting example of a species which avoids intersexual competition for resources.

6.5.3 Baleen whales

In contrast to toothed whales and tuna, which feed predominantly in low and mid latitudes, baleen whales generally undergo migrations between high latitude feeding grounds and low latitude breeding and calving grounds, allowing them to exploit the high, but short-lived, productivity of the high latitudes while calving in warm waters—which may reduce neonatal mortality. Navigation during these migrations often involves movement parallel to a coastline by following bottom contours, supplemented by direct visual observation of the coastline, while oceanic species such as blue whale

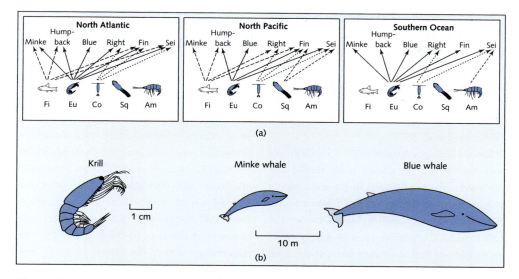

Figure 6.8 (a) Differences in food consumed by rorqual whales (Balaenopteridae: minke whale (*Balaenoptera acutirostrata*), fin whale (*B. physalus*), blue whale (*B. musculus*), sei whale (*B. borealis*), humpback whale (*Megaptera novaeangliae*) and right whales (Balaenidae: *Balaena* spp.) in different oceans. Fi, fish; Eu, euphausiids; Co, calanoid copepods; Sq, squid; Am, pelagic amphipods. (b) Krill (*Euphausia superba*) is the dominant euphausiid in the Southern Ocean and, along with the phytoplankton upon which it feeds, provides the basis for many Antarctic food webs. Although rarely exceeding 10 cm in length, it aggregates in vast swarms and provides the main food for all southern populations of baleen whales, from the 10 t minke whale to the 150 t blue whale.

(*Balaenoptera musculus*), which have good visual acuity, may navigate using the position of the sun and the stars.

Baleen whales are filter feeders. A gulp of water containing macrozooplankton prey is taken into the mouth and the tongue is pushed up and back, driving the water out through the filter of baleen plates in the roof of the mouth and moving the food mass to the back of the buccal cavity to be swallowed. The various species of baleen whale appear to specialize on different prey groups or in different locations (Figure 6.8). In the Southern Ocean, most species exploit krill (*Euphausia superba*), a large crustacean which aggregates into dense swarms. Humpback whales (*Megaptera novae-angliae*) actively enhance concentrations of prey by 'bubble netting', which may be carried out by a solitary individual or a group working in cooperation. The whale swims in a circle below the prey, while emitting bubbles of air from its blow hole. As these bubbles rise, they form a barrier, a bubble curtain. Rising up and swimming in smaller circles, the whale concentrates the prey within the bubbles before finally swimming down and then rising vertically—mouth agape—through the prey aggregation.

6.5.4 Ecological functioning

Carbon dioxide sequestration

While the level of primary production is low in most parts of the ocean, its vast overall size means that the total carbon fixed by the ocean ecosystem is large and this represents a removal of carbon dioxide from the atmosphere. The oceans draw down 1.5 to 1.8×10^{10} tons of carbon each year and it is estimated that about half the 800 billion tonnes of carbon dioxide put into the atmosphere by mankind since the start of the Industrial Revolution has been absorbed by the seas. The oceans hold around 38,400 Gt (10^{12} tonnes) of carbon. At any one time about 1 Gt is in the form of organic carbon (living and dead) and 670 Gt is dissolved in surface waters in various inorganic forms (CO_2, HCO_3^- and CO_3^{2-}). The remaining 36,730 Gt is held in the deep sea as dissolved inorganic forms (Falkowski *et al.* 2000).

While a large proportion of the CO_2 is rapidly recycled when an organism is consumed and the fixed carbon is respired by the consumer, a proportion is essentially removed from the system. This is a major contrast between terrestrial and marine ecosystems, as only a very small proportion of terrestrial vegetation and animal remains are not subject to rapid consumption/decay. Carbon removal in the ocean is to a large extent a consequence of the widespread use of calcium carbonate in structural materials, such as the shells of pteropods and cephalopods, and in the plates of coccolithophores. These structures are all relatively heavy and undigestable, and on the death of the organism they tend to sink rapidly; even when the organism is consumed, its carbonate-dominated components are left untouched and so end up in the deep sea sediment.

Modelling studies suggest that the level of productivity, and hence carbon sequestration, has declined as the oceans have warmed (Kamykowski and Zentara 2005). In addition, there is concern that increased atmospheric carbon dioxide levels are leading to increased acidity of sea water (see Section 1.3) and this will both increase the metabolic cost of producing calcium carbonate shells and cause them to dissolve after death and so reduce the drawdown of carbon into ocean sediments and out of the active biosphere (Engel *et al.* 2005).

Productivity and food chain length

Most of the ocean is oligotrophic, exceptions being high latitudes, which show a strongly seasonal pattern of productivity, and upwelling regions where high productivity is sustained by nutrients from beneath the thermocline. In rel-atively productive high latitudes, euphausiids such as krill directly exploit phytoplankton and, in turn, are exploited by large filter feeders such as baleen whales and seals. In oligotrophic waters, however, phytoplankton are smaller than in more productive regions, because the uptake kinetics for scarce nutrients favour small cells with high surface area to volume ratios. They are too small to be grazed directly by macrozooplankton, and extra microzooplankton and mesozooplankton links are inserted into the food chain. Ironically, therefore, low-energy waters have longer food chains, with the associated loss of energy, than high-energy waters. The lower inherent production and the greater number of steps in the food chain means that few large predators can be supported in oligotrophic areas of the ocean.

6.6 Community structure in the mid-depths

The largest, by volume, ecological region on the planet is the bathypelagic—the zone that extends from the bottom of the mesopelagic to just above the sea floor in the open ocean. Lying above the bathypelagic is the mesopelagic, which extends from the bottom of the photic zone (around 200 m in clear ocean water) to the limit of natural light (about 1000 m). The bathypelagic and mesopelagic share many characteristics.

The most dramatic defining feature of this region is the complete lack of defining features. There are no solid surfaces, other than the organisms themselves, there are no strong horizontal gradients in physical or chemical parameters and it is only in the mesopelagic that any natural light is detectable, and this forms a dim, diffuse field. However, biologically produced light, bioluminescence, is common.

While the absence of light or any strong chemical gradients preclude photosynthetic and chemoautotrophic primary production, the meso- and bathypelagic realms support diverse assemblages. These include free-living microbes and protists, along with others attached to falling organic particles, zooplankton including copepods, krill, and various forms of jellyfish, squid and fish. Deep-diving toothed whales and seals forage in the mesopelagic and upper bathypelagic zone.

6.6.1 Adaptations to life without boundaries

With no physical structures to use as reference points, the inhabitants of the mid-ocean depths must rely primarily on vibrational and chemical cues to sense their environment and drive their behaviour. Surprisingly for a region where natural light is absent or at extremely low levels, many have well-developed eyes in addition to well-developed chemosensory apparatus. In general vibration or motion detectors are only useful for identifying nearby events, the vibrations from activities further away being quickly lost in the general background of water movements. They do however play a key role for many organisms by eliciting an 'escape reaction', such as a sudden change in behaviour, a flash of light or release of ink.

Like most zooplankton, the plankton of this region have the ability to control, by swimming, their position in the vertical plane, but they have no control over horizontal dispersal. However, compared to shelf regions the strength of the mid-water currents is generally weak. Copepods and krill from mid-depths show, compared to shallow water relatives, enhanced levels of chemosensory apparatus including longer antennae and greater density of chemoreceptors.

Fish and squid, along with some of the larger crustaceans, are much more mobile and can actively move against current flows and so can swim up current following a 'scent' trail. Baited camera experiments, in which a time lapse camera is deployed with a bait placed in front of the lens, show that fish and crustaceans arrive very

quickly following deployment and typically from a downstream direction.

Given the lack of autochthonous primary production it has been a popular view that most large mesopelagic and bathypelagic organisms are extreme, opportunistic, generalists. While many of the fish species do have morphological adaptations to the mouth and gut that allow them to consume prey only slightly smaller than themselves, the limited data available on feeding ecology do show some degree of specialization. For example, Williams et al. (2001) sampled five species of mid-water fish from the lower mesopelagic and upper bathypelagic (700–1500 m) off southern Tasmania. Three of the fish were myctophids (lantern fish) and two were stomiiforms (hatchet fish). The myctophids fed primarily on plankton, including euphausiids and copepod crustaceans, one of the stomiiformes consumed fish, decapod crustaceans (shrimps) and euphausiids while the other species of stomiiformes fed exclusively on fish. These small fish are important prey items for larger fish including the commercially important orange roughy (*Hoplostethus atlanticus*). As the small fish are feeding at trophic levels 3–4, this implies orange roughy (and other large mid-water fish) are at trophic level 4–5 and so marine mammals and the fishery are at trophic level 6. This is a further example of a low productivity system having long food chains.

While chemoreception plays a key role in foraging for many species, there is also evidence for the use of chemical cues in aggregation and reproductive behaviours. The difficulty of observing natural behaviour for species living several kilometres down in the middle of the ocean has rather constrained such studies. For example, it is clear that male angler fish (*Lophius piscatorius*) use a chemical cue, which could be described as a pheromone, to locate females, while laboratory observations of deep water copepods (*Calanus marshallae*) show that chemical and mechanical cues are important during the mate search behaviour of the male (Tsuda and Miller 1998).

6

6.6.2 Bioluminescence and camouflage

When the submersible *Trieste* reached the bottom of the Marianas Trench, the deepest part of the ocean, its occupants observed from the observation port a flat fish and noted its bulbous eyes. Eyes on a bottom-dwelling fish over 10 km from the nearest sunlight may seem odd, but throughout their descent the explorers had noted that many of the organisms were equipped with lights. Bioluminescence works through the controlled oxidation of a substrate (luciferin) by an enzyme (luciferase). This causes a photon of light to be emitted. The oxidized luciferin can then be reduced, using metabolic energy, and used again. Thus metabolic energy is converted to light. Different organisms have slightly different versions of the substrate and enzymes, which means that the light can vary in characteristics but also highlights that bioluminescence has evolved independently a number of times. Most bioluminescent light when produced is at a wavelength of around 475 nm; blue light of 470–480 nm wavelength has the greatest penetration in clear sea water, so bioluminescence uses a wavelength that downwells furthest from the surface in the mesopelagic zone and exploitation of this wavelength also means that any lights used in signalling or to hunt prey will travel furthest.

Some free-living bacteria are bioluminescent, but most occur as symbiotic light-emitting bacteria in structures within other organisms that form light organs. There are species of flagellate protists, ctenophores, medusae (jellyfish), shrimps, sea cucumbers, squids, salps, and fish which have specialist light-emitting cells and in many species these are organized into light-emitting organs, which in the most complex can have lens and muscular controls to vary the colour, amount and direction of the light emitted. At least 12 phyla include bioluminescent species and in the North Atlantic more than 70% of the fish species and 90% of the individuals are bioluminescent. The

equivalent figures for shrimp-like crustaceans are 80% of species and 80% of individuals.

In the mesopelagic, invertebrates tend to be transparent, like the near-surface forms, while fish are silvered. Silvering takes advantage of the uniform distribution of downwelled light to effectively make an organism invisible from the side **(Figure 6.9)**. Whether mirrored or transparent, the shape is still visible from above, so that the back is pigmented (dark in fish and red in crustaceans). Most difficult to hide is the silhouette of the body against the milky luminescence of the downwelled light when seen from below, but both fish and crustaceans have evolved ventral counter-illumination, in which downward-directed bioluminescent lights emit light of the same intensity and colour as the downwelling light and so masking the silhouette **(Figure 6.9)**.

Around 600 m depth a change in colour schemes occurs. In the bathypelagic, most species are uniform matt colours. Most fish are black, crustaceans are red and medusae are brown or deep purple. *In situ* all these colours have the same attribute, in that they absorb blue light. Thus a stray flash of bioluminescence from a copepod will not reflect off and alert a passing predator to a potential meal.

While mesopelagic species can benefit from the counter-illumination attributes of downward pointing light organs, most light organs occur on organisms that live permanently below the depths to which natural light ever penetrates, and therefore are not downward directed. The evolutionary reasons for these are many and include bioluminescent 'headlights' located by the eyes. In some species these emit red light and so are specialized at identifying crustacean prey. In other species the lights are arranged over the body in patterns that differ between the sexes, implying a role in sexual signalling. In other species the light may be used in social interactions, such as schooling. Deep sea angler fish have luminous lures while some species when threatened emit a flash before swimming away, emit luminous

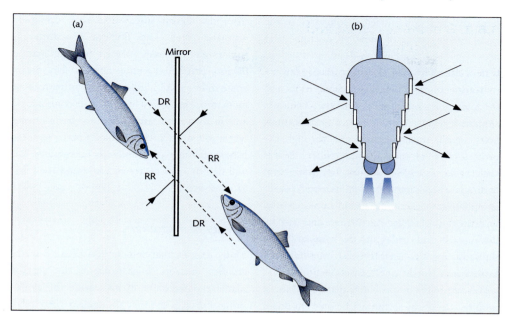

Figure 6.9 The symmetrical distribution of radiance in the ocean provides a means of camouflage by mirrors. (a) A predator cannot distinguish between the reflected rays (RR) from the mirror (M) and direct rays (DR) of light from above. (b) Mirrors following the curve of the body wall would not be effective and so the mirror (guanine crystals) in the scales of silver-sided fish need to be orientated vertically. In some species ventral photophores (light organs) match the downwelling light and so hide the silhouette.

slime/ink or detach luminous body parts when threatened.

6.6.3 Ecological functioning

Habitat provision

Given the lack of physical structures in the mid-depths of the open ocean, organisms themselves provide physical habitat. Amongst the most obvious of these relationships are the colonization of whales by whale barnacles (Nogata and Matsumura 2006) and the accompanying of large sharks and other large fish by remora (suckerfish). Whale barnacles appear to have little or no impact on their hosts, while some species of remora appear to hitch a ride on their host and consume their faeces while in other species the remora scavenge food scraps and may provide a service, cleaning teeth and removing parasites. In both cases the larger host provides habitat but not does not suffer as a result of this commensal association.

Another case of biogenic habitat in the open ocean and one whose influence can be traced throughout the water column is *Sargasso* weed. On his first transatlantic voyage, Columbus reported the existence of a large area of sea with dense areas of floating weed. These were initially thought to be drift algae detached from nearby shores but in fact the brown seaweed *Sargasso nutans* is a free-floating brown alga able to complete its life cycle without an attached phase. The region of low winds and low surface currents in the centre of the North Atlantic gyre is known as the Sargasso Sea because of the widespread occurrence of rafts of the algae.

The presence of the algae dramatically alters the ecology of this region, making it very different to the low productivity regions in the centres of the other ocean gyres. The algae are large and so retain nutrients, compared to the rapid turning over of 'new production' that occurs in the phytoplankton-dominated surface waters elsewhere. The floating rafts of seaweed to some extent reduce light penetration and so shade primary production below. The rafts support a diverse community of grazers (molluscs, crabs, shrimps), fish and, through the provision of physical habitat, filter feeders such as barnacles, sea squirts, bryozoans and hydroids. The assemblage contains many specialist endemic species, only found in association with the rafts of algae. The rafts are also used by juvenile fish and turtles and the Sargasso Sea is the spawning area used by eels (*Anguilla*) from Europe and North America (see Section 4.3).

This assemblage provides detritus (faeces, moults, carcasses) to the sea below. On death the *Sargasso* gas bladders lose their integrity and the plant sinks relatively rapidly, taking its cargo of associated fauna with it. Thus the abyssal region below receives regular large food falls rather than the rain of small particles typical over most of the ocean. Therefore this single species alters the functioning of the photic, mesopelagic and abysso–benthic zones of this region, making them distinct from those in the other ocean areas.

There has been a long history of observations of fish (particularly juveniles), amphipods and even crabs being associated with jellyfish. These are often assumed to provide transport, food resources via scavenging and protection. Salps are planktonic tunicates which feed by pumping water through their barrel-like bodies by rhythmic contractions of the body wall. Amphipods of the genus *Phronima* capture a salp and consume the internal organs before taking up residence in the now empty barrel, which is used as a shelter from which to launch attacks on other gelatinous plankton and eventually as a chamber for the deposition of eggs.

Large floating structures such as litter, fallen trees and mats of coastal vegetation washed out to sea act as ecological hot spots. They often develop a distinct fauna and flora of rapidly growing opportunistic species and clumps of vegetation may also harbour a relic community

from the coastal site where they originally occurred. Such large floating structures are sometimes referred to as 'fish-aggregating devices' (FADs) as a range of small fish species come to live in and under the structure, feeding on material produced by the sessile community, and in turn these are subject to predation by larger predators such as tuna. In many regions, fishers seeking to exploit large, far ranging, fish will deploy rafts to act as FADs and then fish around them.

Ladder of migration

Many species that inhabit the oceanic water column undergo migrations in the vertical plane (see Section 6.8) and while the evolutionary drivers for this are unclear **(Box 6.6)**, these movements play an important role in controlling the functioning of the deep ocean. The distances moved can be considerable, typically 10^4 to 5×10^4 body lengths, equivalent to 50–250 m for a copepod and 500–700 m for a large shrimp or small fish. These migrations can be seasonal, linked to the life cycle or completed daily.

In the Pacific Ocean, 43% of individuals and 47% of the biomass is below 400 m by day but migrates up to above 400 m at night. Calculations of the biological flux caused by these diel vertical migrations have estimated a vertical movement of 23 t km^{-2} day^{-1} from 250 m to the surface, equivalent to 10^9 t day^{-1} over the whole ocean (Longhurst 1976). Individual species (and individuals within a species) migrate differently so individuals overlap to differing extents with other species of predator and prey. Thus species which feed at their shallower depth and are then predated, or simply defecate, when at the deepest depth have moved organic matter down the water column. Each species can be thought of as a rung on a ladder linking the productive photic zone and the bathyal. It should be emphasized that the species are not migrating in order to form this 'ladder of migration'; each species is migrating in response to selective pressures acting upon it, but the emergent pattern of all these different migrations is the ladder that rapidly links surface to deep ocean (**Figure 6.5**).

6.7 Community structure on the ocean floor

Of the 270 million km^2 of the Earth's surface covered by abyssal plains, the total area quantitatively sampled by all the sampling programmes to date adds up to less than 50 m^2. They have, however, been relatively well sampled compared to trenches, for which most of our knowledge is derived from a handful of Russian studies in the Pacific (**Box 6.5**).

6.7.1 Benthic diversity

The macrofauna of abyssal plains is dominated by nematodes, polychaetes, bivalves and pericarid crustaceans. The Indian, Pacific and Atlantic are separated at depths of about 2000 m into three distinct basins, with a high degree of species endemism in each basin, although Indian and Pacific faunas are more similar to each other than either is to the Atlantic. There appears to be an observable faunal change at 6000 m, below which amphipods, polychaetes, bivalves, holothurians and echiurids increase in proportion, while echinoids, asteroids, sipunculids, ophioroids, coelenterates, bryozoans and fish decline in importance. Endemism is high in trenches, as much as 75% in some Pacific trenches, which further suggests a specialist fauna distinct from that on the adjacent abyssal plain.

Species diversity of macrofauna in deep sea sediments may be comparable to that of tropical rainforests and coral reefs. Diversity increases from the continental shelf down the slope to a maximum at about 3000 m, with a subsequent decrease down to 6000 m, beyond which it remains relatively constant. This pattern is not, however, universal. Two groups of theories, which are not mutually exclusive, have been proposed to explain the high diversity of the deep sea bed fauna.

Environmental stability and habitat complexity

Sanders (1968) originally developed the stability–time hypothesis, his argument being that, as the deep sea is a very stable environment and has been in existence in this form for many millions of years, the organisms have evolved to reduce competition, so individual niches are very narrow and many species can be accommodated. Recent evidence from video and camera studies shows that the deep sea floor is not the unstructured constant environment described by Sanders, but shows considerable heterogeneity. Thus, coexistence can be promoted without the need for extreme levels of niche specialization. Furthermore, whether the required dietary specialism occurs in the deep sea remains unclear, as gut analyses of the larger fauna indicate that generalists and non-selective sediment feeding dominate.

6

Non-equilibrium dynamics and biological disturbance

Dayton and Hessler (1972) advanced the view that deep sea diversity could be maintained by the activities of a large number of unselective predators holding populations below their carrying capacity. Thus resources were not limiting, allowing additional species to invade the community. The subsequent realization of the importance of disturbance in generating small-scale patchiness in marine communities (Section 2.4) has furthered the case for considering deep sea communities to be non-equilibrium ones. The maintenance of a spatiotemporal mosaic, through the action of foraging predators, grazers, bioturbators and detritus falls, is responsible for continually creating patches of underexploited resource available for colonization. Each patch then follows a succession towards the climax community but,

Box 6.5   The ecology of the hadal

Depths below 6000 m only occur in the ocean trenches, a zone often referred to as the hadal to distinguish it from the abyssal zone. Trenches only form in areas where oceanic crust is being subducted, so they are all in seismically active regions and the fauna have to cope with earthquakes and slumping of seabed sediments. Worldwide, there are only 37 trenches, 28 of which occur in the Pacific.

No mid-water samples have ever been taken from hadal depths with a closing net, so nothing is known about the fauna of the water column below 6000 m. Samples have been obtained by bottom trawl nets, corers, by baited traps and remote cameras and in 1960 the bathysphere *Trieste* made a unique manned descent to the hadal seabed. Thus, we have limited knowledge of the infauna and epifauna of the seafloor.

More than 600 species have been recorded from hadal depths, of which more than 50% are known only from these depths, the remainder having also been recorded in abyssal samples. While there appears to be an overlap zone with the abyssal fauna between 6000 and 7000 m, below this the hadal benthic fauna changes little with depth. Initial sampling showed a high degree of endemism, with species often being restricted to a single trench, but increasing sampling effort now suggests this might have been a sampling artefact and while endemism is high it is comparable with the hydrothermal vent fields, which also present similar challenges for dispersal and outbreeding.

Polychaete worms dominate the infauna and holothurians (sea cucumbers) the epibenthos—diversity and abundance of holothurians are remarkable, and below 8000 m depth they make up 98% of the individuals and more than 99% of the wet weight in samples from the Kuril-Kamchatka Trench. Decapod crustaceans, bryozoans, cumaceans, fish and echinoderms (apart from holothurians) are all poorly represented in the hadal fauna compared to their role in the abyssal region.

The *Trieste* on the surface of the Pacific Ocean.

meanwhile, new patches have been created. So the system at any one time contains a mosaic of patches at different stages of the trajectory from newly disturbed to climax state and has a high overall diversity.

6.7.2 Ecological processes at hydrothermal vents

Hydrothermal vents (Box 6.2) have been compared to oases in a desert. They differ from the surrounding abyssal plane in the availability of hard substrata but, more importantly, in the autochthonous production occurring in response to the emissions being produced (Section 1.4). Chemoautotrophy proceeds based on both reduced sulphur compounds, particularly sulphides, and reduced carbon sources such as methane. Sulphide-based systems are more common at hot plumes and methane-based systems at cold seeps, but both occur to some extent at most locations.

Vent plumes contain around 10^6 bacterial cells l^{-1}, and this represents productivity in excess of the overlying mid-ocean surface waters. The bacteria generally occur as clumps approximately 100 μm in diameter, at the interface between the two water types, where temperatures are not too high and oxygen is plentiful. They are preyed upon by attached filter-feeders, tube worms, bivalve molluscs and, in Atlantic systems, swarms of shrimps. Bacterial mats also build up on the hard surfaces of the rocks and chimneys and these are fed on by 'grazers' such as limpets, other gastropod snails, squat lobsters and polychaete worms. Predators and scavengers are also present, usually in the form of crabs and fish.

Many of the organisms present at the vents contain symbiotic bacteria. In some cases the symbionts are maintained extracellularly (exosymbionts) but in many cases they are hosted within the cells and are therefore endosymbionts. At least 10 phyla contain species that host chemoautotrophic symbionts. The vent mussel *Bathymodiolis thermophilus* possesses enlarged gill tissues which contain chemosyn-

thetic bacteria, but the mussels retain the ability to filter feed and so can exist away from the vent plume. The entire nutritional requirement of the clam *Calyptogena magnifica* can be met by endosymbiotic chemoautotrophic bacteria. One of the most dramatic discoveries at vents in the east Pacific is large colonies of the tube-living worm *Riftia pachyptila*, which lacks a mouth and gut but has a highly vascularized set of gills and whose body is packed, up to 50% of its mass, with endosymbiotic bacteria. The worms live in the mixed flow conditions close to the vent, where they can extract carbon dioxide, sulphides and oxygen from the water, and contain haemoglobin which enables them to withstand periods of anoxia. Time lapse photography shows that the worms regularly retract into their tubes, to reduce exposure to anoxic conditions as the flow regime fluctuates, and to reduce predation, often in the form of gill nipping, from fish and crabs.

Vent fields are transient phenomena, as changes in flow regime, precipitation of minerals in the channels and collapse of chimneys can all stop the hydrothermal flow and hence the supply of energy to the chemoautotrophic system. The specialist fauna must therefore be able to colonize new vents and so avoid extinction when the vent expires. Initially it was thought that vents were highly isolated phenomena and that colonization of new vents would be a major challenge. Early studies on the genetics of vent organisms, however, revealed high levels of out-breeding and genetic exchange. Subsequent survey effort has revealed that hydrothermal vents occur at most locations on the ocean spreading centres where conditions are appropriate; on the mid-Atlantic Ridge, for example, the average distance between fields is of the order of tens of km. Given the distance between vents, dispersal must be via the water column, and two possible routes exist. First, the hot rising plume of vent fluid could be utilized to lift larvae high into the water column, where the lateral shear of the currents distributes them over a wide area. Secondly, there are currents flowing along the axis of most mid-ocean ridge

6

systems, into which the release of larvae will disperse them to new vent sites on the ridge. While the latter method is more targeted to suitable habitats, vent sites are not distributed evenly along ridges and the former method will distribute larvae over a wide area—although much of it will be unsuitable environment. Whale carcasses may also act as a temporary stepping stone between vents (see below).

The transient nature of hydrothermal vents and the ability of the fauna to colonize them is illustrated by observations on the East Pacific Rise. A vent field that was thriving in 1989 was dead when visited again in 1991, as a result of an anoxic event. Clumps of tubeworms (*Tevnia*) were, however, present at the site within 12 months, reproducing vestimentiferan worms (*Riftia*) were recorded 21 months later and mussels had colonized within 4 years (Shank *et al.* 1998).

While individual vent fields are transient phenomena, as a habitat they have been present for many millions of years, as evidenced by fossil structures, and this has allowed the evolution of the specialist fauna and the unique ecology that they support. While productivity at vent fields can be extremely high (for the ocean), with biomasses of 10–50 kg m^{-2}, they account for only around 0.03% of global oceanic primary production and 3% of the total carbon flux to the deep sea floor. However, the ecological impact of the vents, along with similar systems, is important not just at the site where they occur but probably also to production and ecological processes in the surrounding area, through spill-out of production and the export of biomass by mobile species and in the form of larval production.

6.7.3 Other reducing environments

Methane gas seeps

Hydrothermal vents are not the only reducing environments in the ocean, and the discovery of chemoautotrophic food webs at vents has

stimulated research in environments such as methane gas seeps. At these sites, both free-living and endosymbiotic chemoautotrophic bacteria have been observed and the presence of methane gas seeps supporting chemoautotrophic communities may contribute significantly to the productivity of some areas of the ocean. The fauna at some of these sites contains species similar to those at vents, but their geographical separation from the vents means that they are unlikely to act as 'stepping stones' between vent fields for dispersing fauna.

Whale carcasses

The discovery in 1987 of a partially decomposed whale carcass in the deep ocean off the California coast allowed observations of the impact of large carcasses on the local environment. The input of organic matter from this 21 m blue or fin whale (*Balaenoptera* sp.) carcass had significantly altered the local sedimentary environment—raising the organic content and reducing the sediment oxygen content. In places, the bones and sediment were covered in a mat of the bacterium *Beggiatoa*, a chemoautotrophic species utilizing sulphides and oxygen. Anaerobic decomposition of lipids—which are stored in bones for buoyancy and can account for as much as 60% of the skeletal weight of a living whale—released reduced organic compounds and hydrogen sulphide which were utilized by free-living and endosymbiotic chemoautotrophs to fix organic carbon (Smith 1992).

The bones of the whale had been colonized by clams, mussels, gastropods, polychaete worms and amphipod crustaceans. While some of these species were members of the background abyssal fauna, others were species recorded previously only from hydrothermal vents, including the limpets *Pyropelta corymba* and *P. musaica*, which graze sulphur bacteria. It is, therefore, possible that whale carcasses may act as a stepping stone for dispersal between seeps. Approximately 1000 grey whales (*Eschrichtius robustus*) die each year. If 50% of their carcasses arrive at the deep-sea floor and they are distributed randomly, then the distance between carcasses, within their limited coastal distribution in the north Pacific Ocean, would be about 9 km. A similar density of carcasses throughout the oceans would probably ensure effective dispersal for vent-dependent species.

6.7.4 The role of patchiness in community dynamics

The fauna of the deep sea floor exhibit patchy distributions at a wide range of spatial scales although, given the problems of sampling—both the logistical problems of remoteness and the sparse nature of the fauna requiring large numbers of large volume/area samples—there remains some debate over the degree of patchiness and the causal mechanisms. Analysis of photographs and sediment samples shows that the deep sea floor has variations in the physical environment on scales from millimetres to metres, including mounds, tracks, pits/burrows and faecal deposits, all of which arise from biological activity. Such biogenically produced features interact with the weak bottom currents in the deep ocean to produce further heterogeneity as a result of deposition in the lee of mounds. Superimposed on this physical heterogeneity is the variability associated with the arrival of food material, ranging from aggregations of a few phytoplankton cells to whole whale carcasses, and that derived from the variability in the settlement of propagules. Figure 6.10 illustrates the observed patterns of spatial and temporal heterogeneity. They range from short-term aggregations for reproduction to assemblages persisting for many years on hydrothermal vents and, in size, from a few square centimetres to several hundred square metres.

Organic matter on the sea floor

A key component of the patches identified in Figure 6.10 is the importance of detrital inputs; only vents and seeps are independent of allochthonous inputs from above. It is difficult

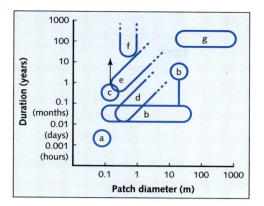

Figure 6.10 Spatial and temporal scales of deep sea faunal patches. (a) Small-scale social/reproductive aggregations, lasting no more than a few hours. (b) Slow-moving detritivores, ranging from those feeding on very patchy resources to larger 'herds' feeding on more generally distributed resources. In any one locality, these aggregations will only last for a few days, but the large mobile herds may stay together as a group for very much longer periods. (c) Meiofaunal patches based on small organic resources. These are very small in size, reflecting the size of the organisms and the resource, but will persist for as long as the resource remains. (d) Scavenging necrophages attracted to animal carcasses. The size and duration of these patches is directly proportional to the size of the carcass and mobility of the organisms; in the case of whale carcasses (Section 6.7), they may persist for months or years. (e) Aggregations on plant material. These are similar to those in (d) but longer lasting, as plant remains are generally more refractory and so attract less active scavengers. (f) Patches of sessile filter feeders, such as sponges, which are limited to areas where currents bring food to them but, once established, persist for many years. (g) Patches associated with seeps and hydrothermal vents. Modified from Rice and Lambshead (1994).

accurately to estimate the frequency of arrival of organic matter at the sea floor, but there is probably a steady drizzle of material derived from faecal material and moults from animals living in the water column, in addition to pulses of higher levels of deposition due to the arrival of material following blooms in surface waters. This material may arrive only days or weeks after the surface bloom and may be predictable from year to year, giving a seasonality to the deep sea environment.

In some areas, close to the ocean margin, as much as 20% of the carbon in the deep sea sediments is of terrestrial origin. Large pieces of plant material, including tree trunks, saltmarsh plants and macroalgae, are relatively frequently observed, but this is probably largely a function of their refractory nature. This material tends to be utilized by a variety of small infaunal species which colonize the sediment under and adjacent to the plant remains. Offshore (with the exception of the centre of the North Atlantic gyre, below the Sargasso Sea) large falls of plant material are rare and animal carcasses are the main input of carbon in large parcels. Baited camera observations show that animal material rapidly attracts mobile scavengers, particularly amphipods, decapods and fish, which consume and disperse the material within a few days. Inputs of large detrital masses, such as whale carcasses, are irregular and infrequent but, once present, may support a localized community for several months, or even years (**Figure 6.10**).

6.7.5 Sea mounts

Sea mounts are undersea mountains that rise steeply from the floor of the ocean but do not reach the surface. Those that reach the surface form islands. Most sea mounts are the result of undersea volcanoes and while many rise steeply to a point others—known as guyots—reached the surface during periods of lower sea level and have had their tops eroded to form a flat surface. Sea mounts occur in all the oceans but are most common in the Pacific.

Ecologically sea mounts are important for a number of reasons. They provide habitat diversity in the deep sea with areas of slope, hard substrate and physical features such as crevices. By rising above the abyssal plain they interfere with the flow regime, setting up eddies and potentially areas of enhanced deposition of organic matter. Deflections of deep current up the slope can lead to upwelling and boost surface productivity (**Box 6.1**). Sea mounts rising to near the surface can influence surface flows and create meso-scale eddies which can stimulate upwelling.

The availability of hard substratum for attachment and the high levels of organic material in the entrained flow, in part derived from increased surface primary productivity, support a diverse assemblage of filter feeders including large sponges, tube worms, sea fans, and cold water corals. This habitat diversity and productivity also supports, for the deep sea, relatively high densities of bottom-associated fish. There is also evidence that the pelagic and mesopelagic regions near sea mounts support enhanced levels of large predators such as fish and whales, while feeding by oceanic sea birds also seems to be correlated with sea mounts.

As sea mounts are isolated phenomena, their biota face major challenges with respect to initial colonization and subsequent maintenance of populations. Unlike hydrothermal vents the longevity of a sea mount as a habitat is measured in geological time, so the balance between long range dispersal, to find a new sea mount, and ensuring propagules find the right environment is different. Studies of sea mounts in the Pacific that rise to near the surface suggest that majority of the fauna may have arrived by rafting from coastal sites and that the sea mount assemblage is dominated by species with direct development or short larval durations. The tendency for a sea mount to set up eddies in the surface flow may allow a large proportion of short duration larvae to be retained in its vicinity. Recent studies in the south-west Pacific suggest that up to 20% of the species found at sea mounts may be endemic either to individual sea mounts or to a local group of sea mounts suggesting relatively recent speciation and localized species distributions, a rare event in the deep sea (de Forges *et al.* 2000).

In recent years the fish populations associated with sea mounts have been targeted by fisheries (see Section 6.8). While the presence of productive and diverse assemblages on sea mounts has led to their being considered a high priority for nature conservation initiatives, the location of the majority in international waters has prevented effective protection.

6.7.6 Ecological functioning

Nutrient regeneration

In coastal waters and shelf seas organic matter breakdown in marine sediments releases a significant proportion of the nutrients required to meet the annual needs of the phytoplankton assemblage in the overlying waters (see Section 5.4). Similar processes occur in the sediments of the deep ocean but with a number of ecological differences. Compared to shelf sediments:

- the rates of breakdown are slower due to the lower temperatures;
- the efflux of nutrients is lower as the organic matter is, on average, more refractory;
- the nutrients are not mixed back up to the overlying surface waters but advected horizontally by bottom currents, re-entering the photic zone in regions of ocean upwelling.

Benthic production

The deep sea floor acts as a physical barrier to the sinking particles of organic matter and so accumulates everything from particles of marine snow, through flocs of phytoplankton debris, faecal pellets and exuvia to sinking whale carcasses. The majority of benthic organisms would be characterized as being deposit feeders, grazing on this organic material either selectively or via ingestion of the substratum and subsequent digestion of the organic portion. Many species will however also feed opportunistically on each other, the key determinant being whether or not they can overpower them. The biomass of the deep sea benthos declines with increasing depth but also correlates with the productivity of the overlying waters (Thurston *et al.* 1998).

In a manner analogous to the demersal fish component of the continental shelf ecosystem, the deep sea floor has associated with it a benthopelagic fauna. This comprises mainly fish but also includes cirrate octopus, decapod crustaceans, isopods and even benthic siphonophores that use their tentacles to attach to the bottom so that they float like hot air balloons just clear of the bottom. The benthopelagic fauna have access to the rich food of the benthic environment but also have much more mobility to search for food aggregations than the true benthos. They are however also more susceptible to larger mobile predators.

6.8 Temporal change in oceanic systems

Oceanic communities are in a continual state of flux, the position of organisms changing in response to processes acting at the scale of minutes to those which function over several decades. Much of the short-term variation in plankton distribution can be attributed to physical processes such as wind-driven mixing and 'eddies', while that in the nekton is usually caused by behavioural responses, including predator avoidance, food gathering and shoaling.

6

6.8.1 Vertical migration

As the only reference frame is the vertical one provided by the gradient of light intensity, it is perhaps not surprising that the dominant behaviour shown by many zooplankton is movement in the vertical plane (Figure 6.5). Dramatic changes in distribution can occur, many of which can be explained by active behavioural mechanisms—vertical migration in response to seasons, life cycle or even on a daily basis.

Ontogenetic vertical migration

Ontogenetic vertical migration refers to a change in depth at different stages of development. It occurs in two basic forms: breeding at the surface, the female rising to lay the eggs, and breeding at depth, following which young animals swim up, not necessarily all the way to the surface. The latter is more common, the copepods *Calanus plumchrus* and *C. cristatus* in the temperate North Pacific providing good examples (Section 5.4). Females lay their eggs in winter and early spring at 500–3000 m depth. Eggs and nauplii float or swim up, reaching the surface in spring as copepodite larvae, which exploit high spring and summer phytoplankton populations to achieve rapid growth. In July, the copepodites, now in their final larval stage (cV), begin to move down, ceasing feeding once at depth. The mature females are non-feeding and die after laying. All samples from depth contain both adults and cVs, so it appears either that some mature after one season while others delay for another year, or that a proportion never mature.

Diel vertical migration

The detailed analysis of ocean plankton distributions from many regions has shown a general pattern of plankton moving downwards at dawn and upwards at dusk, or, occasionally, the reverse. Diel vertical migration has been observed in polar, temperate and tropical seas and many lakes (Section 7.5). By the 1920s it was fairly well established that the main cue for many species was a particular isolume, or level of light intensity **(Figure 6.11)**. This explains the phenomenon of 'midnight sinking': the population moves up as a distinct group at dusk as light levels fall and the isolume rises. After dark, without the light cue, the plankton are dispersed over a wide depth range by random movements. At dawn, the isolume reappears, so the plankton re-group near the surface and then move deeper as light intensity increases. While the optimal light level may explain the cue used by many species, others obviously do not use this cue as they start to migrate upwards hours before sunset; nor does it explain the evolutionary advantage of reverse migration or of moving the equivalent of up to 100,000 body lengths per day **(Box 6.6)**.

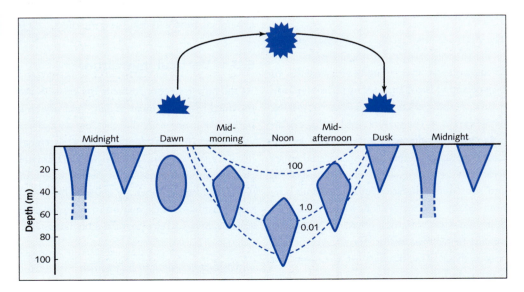

Figure 6.11 Diurnal vertical migration of zooplankters following a particular level of light intensity (an isolume). At midday, the sun is high in the sky and a large amount of light passes through the sea surface, so the isolume is at its deepest. As the sun sinks towards dusk, more light is reflected off the sea surface and the depth of the isolume becomes increasingly shallow. After dark the isolume disappears and random movements by zooplankters now lacking a behavioural cue lead to them being dispersed over a wide depth range, a process referred to as 'midnight sinking'. At dawn, the isolume reappears near the surface and zooplankters reorientate themselves.

Box 6.6   Evolutionary mechanisms driving diel vertical migration

The two main contenders for the evolutionary cause of diel vertical migration (DVM) are the 'more efficient energy use' or growth rate hypothesis and the 'predator avoidance' or death rate hypothesis.

The growth rate hypothesis

Migration confers a bioenergetic advantage by reducing maintenance costs. The basal metabolic rate is lower in the cold deep water than at the surface. So, if mechanics of feeding and digestion are such that an organism cannot feed continually, then there is an advantage to spending time in a region of lower costs. It may also lead to increased lifetime fecundity, increasing the lifespan of individuals by slowing growth.

Criticisms of the growth rate hypothesis

- Non-feeding life stages migrate, even though they would gain most from reduced metabolic costs by staying at depth.

- The intensity of migration often increases with age. Young, rapidly growing individuals often do not migrate, even though they would gain most from the energetic benefits of migration.

- It cannot account for the generally greater migratory activity of larger individuals, which are most obvious to a visual predator.

- In polar waters, migration ceases during midsummer, when there is no night. Under the growth rate hypothesis, there would be no advantage to staying in cold deep waters, yet the plankton do (as predicted by the predator avoidance model).

The death rate hypothesis

If the main predators hunt by sight, then predation rates will be lower at depth than at the surface during the day. At night, susceptible organisms can migrate up to exploit the higher productivity of surface waters, avoiding predation under cover of darkness.

Criticisms of the death rate hypothesis

- Many migratory species do not go deep enough to avoid visual predators.

- In the Antarctic Ocean, the main predators are not visual, but are filter feeders, yet migration still occurs.

- Animals in very deep waters migrate, even though they would never be in the light, even at the top of the migration.

- Many fish feed near surface at night to exploit the plankton which have moved up.

- Salps, medusae and ctenophores are not palatable yet still migrate.

- Transparent forms should migrate less, as they are less visible; this is not the case.

6

6.8.2 Seasonality

At latitudes greater than 40°, ocean productivity is highly seasonal (Section 5.4). Seasonal changes in the tropics and subtropics may be driven by ocean-scale phenomena—the monsoons over the Indian Ocean and Indo-Pacific region, for example. Seasonal migrations of species such as whales and tuna may bring seasonality to otherwise non-seasonal areas. At high latitudes, seasonality is driven by the production cycle. The flocculation of phytoplankton cells and production of faecal material by zooplankton exploiting the phytoplankton bloom all aid in the rapid transfer of this material to the deep sea, whose faunas also experience seasonal production at high latitudes. Our knowledge of the response of meso- and abyssopelagic faunas to this seasonality is very limited. In the deep sea, time lapse photography, used to follow the fate of settled material derived from the phytoplankton bloom, shows that while pelagic detritus arrives as a relatively uniform covering, it is moved by bottom currents to accumulate in depressions and the lee of mounds. These patches then persist for many months and possibly years, and may be an important source of spatial heterogeneity in the deep sea benthic environment.

6.8.3 Interannual changes: ENSO

In addition to cyclic seasonal changes, there is a series of aperiodic changes which cause interannual variation in the ocean ecosystem. One of the most well known of such phenomena is the El Niño southern oscillation (or ENSO) event. ENSOs are climatic fluctuations centred in the Pacific region. They occur every 2–10 years and are indicated by the occurrence of unusually warm water off the coasts of Peru and Ecuador. The term El Niño is the local name for this phenomenon and literally means 'the Christ child', making reference to the timing of the warming of the coastal waters just after Christmas.

ENSOs are a disruption of the coupled atmosphere–ocean system in the Pacific and it remains unclear as to whether it is the ocean which drives a change in the atmosphere or vice versa. During a normal year, hot air over Indonesia rises, and the resultant low pressure pulls air west from the high pressure region in the subtropical South Pacific, creating southeast trade winds **(Figure 6.12a)**. During an ENSO, however, atmospheric pressure over Indonesia remains high and these trade winds over the central Pacific weaken, or may even reverse (becoming westerly) in the western Pacific. The reduction in the trade winds causes the sea surface slope across the Pacific to disappear and the thermocline also becomes horizontal **(Figure 6.12b)**, allowing a large body of warm mixed layer water to flow eastwards, bringing warm water to the coastal areas of Peru and Ecuador and suppressing the normal upwelling. The warm mixed layer water carries with it the nutrient-limited plankton community typical of the central Pacific, replacing the normal coastal pelagic community which is dependent upon the nutrient-rich upwelling. This, in turn, causes a failure of the fishery, which is based on planktivorous fish such as the anchoveta (*Cetengraulis mysticetus*).

The ramifications of ENSO are much more widespread than a change in the upwelling dynamics and fishery off Peru. Weather patterns over the whole Pacific are altered. Droughts affect normally wet regions, while arid areas suffer torrential and destructive rain storms. Tropical cyclones are also more frequent and track further east in ENSO years, while there is evidence of ENSO events being linked to fluctuations further afield, with, for example, extremely cold winters in Eurasia and North America tending to coincide with ENSO years.

6.8.4 Long-term changes in ecosystem function

The North Atlantic zooplankton

The zooplankton of the North Sea and the North Atlantic are surveyed every month by the

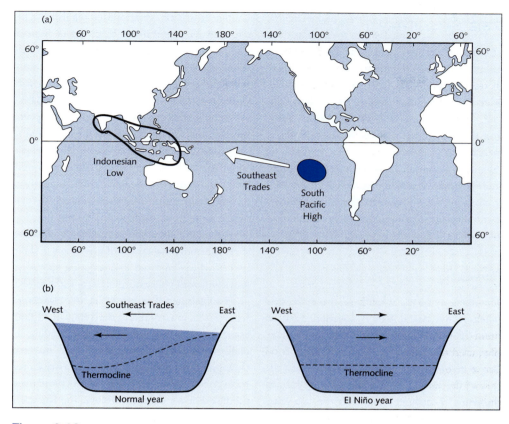

Figure 6.12 (a) The 'normal' position of the Indonesian Low and the South Pacific High and the path of the southeast trade winds produced by the pressure gradient. (b) A section along the equator in the Pacific in a 'normal' (left) state and during an El Niño (right) event, with prevailing wind and current directions shown by arrows. Note the loss of sea surface slope and the horizontal thermocline during the El Niño event.

continuous plankton recorder (CPR) survey. Analysis of the species composition shows that since 1960 there have been marked changes in the zooplankton community in the North Atlantic (Beaugrand *et al.* 2002). Comparisons of the distribution of groups of species indicative of certain conditions, for example 'temperate oceanic species typical of warm water masses' or 'sub-arctic species' show large-scale shifts, occurring mainly since 1980 in the north-east Atlantic and a little later in the adjacent shelf seas. Overall the pattern consisted of a move north of over 10° of latitude for the warm water species groups and a commensurate retreat of Arctic and sub-Arctic species. This can be interpreted as being the result of increasing northern hemisphere temperatures

and associated changes in the dynamics of the North Atlantic Ocean.

The Antarctic

The Antarctic pelagic food web is centred on krill, which is consumed by baleen whales, seals, penguins, fish and squid. At the beginning of the twentieth century, there were large populations of baleen whales in Antarctic waters each summer, consuming around 190 million tonnes of krill each year. Commercial whaling has since decimated these whale populations. Estimates of the consumption of krill for the mid 1980s (Laws 1985) indicated that baleen whales only accounted for 40 million tonnes and the amount consumed by crabeater seals (*Lobodon carcinophagus*) and penguins had more than doubled. Since the moratorium on hunting, whale populations have increased more slowly than population models had predicted they should, suggesting that the balance of the ecosystem has shifted, the present seal- and penguin-dominated system being stable and resistant to invasion by whales. If this is the case, and it is by no means certain that it is, then it is an example of an ecosystem exhibiting multiple stable states (see Section 7.5). It is also a clear message about the scale and unpredictability of the changes that overexploitation can bring about in the natural world.

Harvesting

The majority of the world's marine harvest is taken from coastal waters (Section 5.9), as these are more productive and tend to have shorter, and therefore more efficient, food chains than oceanic regions. In addition, the further the catch has to travel to market from the point of capture, the greater the sale value has to be to cover costs. Therefore, oceanic harvesting concentrates upon exploitation of high-value oceanic whales, tuna and squid. Recently, however, 'prospecting' fisheries have developed for species which may potentially yield enough to offset costs, such as krill and deep sea fish. Test harvesting of Antarctic krill by the Japanese and Russians has led to estimates of a sustainable annual harvest of 100 million tonnes (some estimates actually being double this). One of the barriers to exploitation of krill is how to utilize it: there are no direct routes for human consumption and most proposals for its harvest advocate processing into a protein meal, for use in other products or as a food for livestock. The costs of catching krill in the inhospitable Antarctic, transporting it to consumers in the northern hemisphere and processing it make these proposals far from economic at present. If large-scale harvesting goes ahead, however, further major changes in Southern Ocean ecosystems can be expected.

While most harvesting of deep sea fish can be regarded as exploratory, there are a limited number of established fisheries based on deep water species. Off Madeira there is a traditional long-lining fishery for scabbard fish (*Aphanopus carbo*) from depths of 750–1000 m, using baited hooks. The Pacific supports fisheries, on the continental slope to approximately 1500 m depth, for the sable fish or black cod (*Anoplopoma fimbria*), while around Australia and New Zealand there is a trawl fishery on the continental slope and on the tops of sea mounts, at depths of 750–1000 m, for orange roughy. In addition to their food value, these fish contain a variety of wax esters in their tissues which are used in cosmetics, specialist lubricants and pharmaceuticals, and they are a good substitute for sperm whale oil. However, from peak capture rates in the 1980s of 42,000 t yr^{-1}, catches have been declining, leading to concerns that the fishery was based on old, slow-growing fish and will not, therefore, recover. Similar concerns have been voiced over the best-known deep sea fishery, the Russian-dominated harvest of roundnose grenadier (*Coryphaenoides rupestris*) and roughhead grenadier (*Macrourus berglax*) in the northwestern Atlantic and sub-Arctic. Between 1968 and 1978, catches averaged 83,800 t yr^{-1}, but have since declined, although whether this is a result of exploitation or of long-term cooling of the water masses is unclear.

Ocean predators

Human use of living marine resources is highly selective—we choose the species and often the size of the individuals we take. Thus while harvesting has had an effect across many parts of the ecosystem, the group which has suffered most are the top predators. Man is both an additional trophic level above the natural top predator and a competitor for lower trophic levels. In a global analysis of the distribution of two top predators, tuna and billfish (Istiophoridae such as marlin, and Xiphiidae, the swordfish), Worm *et al.* (2005) have shown that their natural diversity is greatest near thermal fronts, in oxygen-rich waters and in warm seas (around 25°C). Unfortunately this study also showed that diversity of these oceanic predators had decreased by between 10 and 50% in every ocean in the last 50 years, these declines being associated with overfishing and climate fluctuations (ENSO).

6.8.5 Regime shifts

While it is clear that marine ecosystems vary widely on a variety of space and timescales, it is also clear that human activity has caused some of the changes, for example by fishing and whaling. In some cases it also appears that the changes have caused the ecosystem to settle into a different configuration. Dramatic changes in ecosystems have now been observed many times and in a wide variety of locations and these are often referred to as 'regime shifts'. The term was first used in terrestrial communities and then applied to marine ecosystems to describe concurrent fluctuations in anchovy (*Engraulis encrasicolus*) and sardine (*Sardina pilchardus*) populations in several regions of the world. The term has since been applied to apparent shifts in oceanic and climatic conditions and marine community structure around the world (Collie *et al.* 2004).

Changes in oceanic systems often become apparent only when they impact coastal seas, because these are zones in which most monitoring is occurring. Thus a shift in the climatic regime in the north-east Atlantic was linked to a change in the patterns and quantity of phytoplankton and zooplankton in the North Sea. This in turn has been linked to altered recruitment of a number of commercial fish (Reid *et al.* 2001).

Lees *et al.* (2006) developed a rigid definition for marine regime shifts and then considered the extent to which reported changes in various parts of the ecosystems actually met the criteria for a regime shift. Their criteria emphasize the sudden onset of the change, the magnitude of the change (relative to previous levels of variability) and the inclusion of both environmental and ecological components of the system. Of the regime shifts identified in both the Atlantic and Pacific Oceans, climatic drivers were implicated in all and overfishing in most of those that involved changes in the biological components of the system. However, care must be used in interpreting correlations between time series such as weather data and zooplankton abundance or fisheries data, as statistical artefacts can mislead (Rudnick and Davis 2003).

6.8.6 Ocean scale changes

The oceanic system has persisted as a continuous entity throughout the period during which life has been present on Earth. Its individual components are not, however, fixed, and all oceans go through a 'life cycle' as a consequence of plate tectonics.

The Earth's surface consists of a series of plates, which move relative to each other. Movement is generated by sea floor spreading, the creation of new sea floor at spreading centres, which form an almost continuous ridge system, resembling a mountain chain, throughout the major ocean basins. Creation of new sea floor forces the rest to move away and, as the Earth's surface does not change in area, creation must be matched by destruction. Continental crust is less dense and therefore at a higher elevation than oceanic crust so, where ocean and continent are forced together, the continental plate rides over the oceanic plate, which is subducted beneath it, creating an oceanic trench (Figure 1.6). Continental plates, therefore, remain relatively unchanged in size and shape and can be considered to 'float' over continually changing oceanic plates.

Evolution of ocean basins

The average lifespan of an ocean is about 200 million years. Thus, as the Earth's oldest rocks

6

Table 6.1 Six stages in the evolution of an ocean			
Stage	**Examples**	**Dominant motions**	**Characteristic features**
1 embryonic	East African rift valleys	Crustal extension and uplift	Rift valleys
2 young	Red Sea, Gulf of California	Subsidence and spreading	Narrow seas with parallel coasts and a central depression
3 mature	Atlantic Ocean	Spreading	Ocean basin with active mid-ocean ridge
4 declining	Pacific Ocean	Spreading and shrinking	Ocean basin with active spreading axes; also numerous island arcs and adjacent trenches around margins
5 terminal	Mediterranean Sea	Shrinking and uplift	Young mountains
6 relict scar	Indus suture in the Himalayas	Shrinking and uplift	Young mountains

were formed about 3800 million years ago (Ma), the oceans have turned over 15–20 times. Individual ocean basins grow from an initial rift, reach a maximum and then shrink, ultimately closing completely. The Atlantic and Indian Oceans, which contain active spreading centres and no major subduction zones, are expanding, while the Pacific is contracting because subduction zone activity outpaces spreading. The rates of movement, averaged over time, are of the order of several centimetres per year.

Several stages in the life cycle of an ocean can be identified (Table 6.1). Rift valleys develop along the line of continental separation. Basaltic magma wells up between the continental blocks, creating crust which is thinner and more dense than continental crust and so lies below sea level. The young ocean basin is shallow and, if influxes of sea water are partially or totally evaporated, then salt deposits build up on the sea floor, along with the more normal deposition of sand and mud (Stage 1). Sediment from the adjacent coasts builds out into the rift, eventually creating a continental shelf-slope system. As the spreading axis builds away from the coast, the sediment supply dwindles and the continents become increasingly distant from the spreading ridge (Stage 2). The crust contracts as it cools, and abyssal plains are produced, allowing the continental shelf and rise system to become fully developed. At this stage (Stage 3), the continental margins are approximately parallel to the spreading centre. In due course, one or more destructive plate margins will be formed, as a result of continental collision or new continental rifting (Stage 4) and the ocean begins to disappear (Stage 5).

Figure 6.13 shows the relative position of land masses at three times during the past 170 million years. During the Jurassic period (Figure 6.13a), about 170 Ma, the continental plates were joined together as a single land mass, called Pangaea, and there was a single ocean, Panthalassa, with an arm called the Tethys Ocean. By the Cretaceous period (Figure 6.13b), about 100 Ma, the continental plates had separated into two aggregations, Laurasia and Gondwanaland. Gondwanaland was beginning to fragment, causing contraction of the Tethys Ocean. By the Eocene period (Figure 6.13c), around 50 Ma, the Atlantic and Indian Oceans had started to form, Panthalassa, now identifiable as the Pacific Ocean, had shrunk considerably in size and the Tethys Ocean was confined to a relatively small area covering the present day Mediterranean basin and southwest Asia.

Birth of an ocean: the Red Sea

The clearest example of a young ocean today is the Red Sea. Extensive evaporite deposits were produced when the embryo Red Sea was

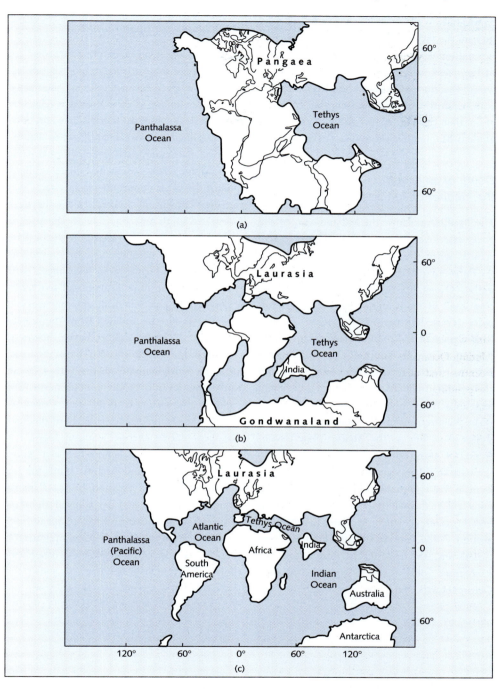

6

Figure 6.13 Reconstruction of the distribution of the continents (a) 170 Ma, (b) 100 Ma and (c) 50 Ma. Note that coastal shelf seas are not differentiated from continental land masses.

connected intermittently to the Mediterranean. Since this closed and the connection to the south opened, there has been considerable infilling with terrigenous and pelagic material.

In the southern part of the Red Sea, a ridge system very similar to the Mid-Atlantic Ridge has developed, while to the north, new ocean floor is forming in the areas greater than 1000 m in depth, but no rift system is apparent. The sea floor in the vicinity of the southern rift system is 5 million years old, while that in the vicinity of the deeps is 2 million years old or less. It would therefore appear that the Red Sea axial zone is a northward-propagating zone of separation that opened up properly in the south about 5 Ma, but has yet to do so in the north.

Death of an ocean: the Mediterranean Sea

The Mediterranean is the last remnant of the once extensive Tethys Ocean **(Figure 6.13)**, created when the African plate collided with the Euro-Asian plate, cutting off a section of the Tethys Ocean. Its floor is a complex realm, composed of many minor plates. In general, older crust occurs in the east of the Mediterranean and there is evidence for a subduction zone to the south of Cyprus. Thus the patterns of crust removal are complex and no clear correlation between the age of the sea and tectonic activity appears to exist.

The Mediterranean is an ocean in the final stages of its life, as the African plate is consumed under the Euro-Asian plate. Thus, unless the pattern of movement changes, Africa and Eurasia will eventually collide and new mountains will be thrown up. The sea has, however, had one reprieve: when the plates first collided around 7 Ma, the Straits of Gibraltar were closed and the isolated Mediterranean Sea evaporated, only to refill from the Atlantic Ocean about 5.5 Ma.

Implications for ocean biota

The main consequence of plate tectonics for marine organisms is from a biogeographical perspective. Deep ocean species are unaffected by surface climate, as their habitat is relatively constant throughout the world. They are, however, fragmented by oceanic ridges and shallow seas and particularly by severance of oceanic connections when continental plates collide. One of the most important recent effects was the development of a land bridge between North and South America, isolating the Atlantic and Pacific Oceans. Prior to the merger of these two continents, oceanic currents moved between them, allowing biotic exchange. Now the only oceanic links are in cool temperate and polar regions, which means that tropical Atlantic species are isolated from the tropics elsewhere. Atlantic biota have, however, colonized the Mediterranean basin because, although it originated from the Tethys Ocean to the east, its only modern connection to the main oceanic system is to the west. Construction during the nineteenth-century of the Suez Canal, a marine channel without locks, has allowed a direct interaction between warm water Indo-Pacific and Atlantic biota for the first time since the creation of the Panama Isthmus over 2 Ma, and has allowed an influx of species into the eastern Mediterranean from the Red Sea (so-called 'Lessepsian migration'), though few species have gone the other way (Por 1978).

Bulleted summary

- The majority of the oceanic realm is a low productivity environment. Surface productivity is nutrient limited at low latitudes and seasonally limited by low light at higher latitudes. Important exceptions to this occur at regions of upwelling. The sea floor and the water column below the photic zone are dependent upon allochthonous production and the cycling of organic matter. Transfer of detritus to the ocean depths is enhanced by detritivores, both by packaging it into faecal pellets, which sink relatively rapidly, and by vertical migration, often over great distances.

- The oceanic water column can be regarded as a series of layers varying in their salinity and temperature. The boundaries between these water masses may act as a barrier to the movement of plankton and so provide some structure to this otherwise homogeneous environment. The abyssal plains are covered by soft sediments, mainly of biological origin.

- In the water column, most heterotrophs appear to be generalist and opportunistic in their behaviour—food webs are structured by size. No single species or group of species, including the larger mobile predators, appear to be functional 'keystone' species. Low productivity and patchiness of resources means that large nekton are generally very mobile, foraging over great distances.

- Species diversity at the sea floor appears to be promoted by small-scale—often biologically produced—disturbances, such that these systems are non-equilibrium ones. The patchy, unpredictable arrival of detritus inputs accentuates this patchiness. These communities are species-rich compared to most terrestrial ones, but the total species pool in the oceans is much smaller than on land, each species having a much larger geographical range. Deep ocean trenches, in contrast, appear to support a fauna which is both distinctive from that of the abyssal plains and often shows a high degree of endemism within a given trench.

6

- Hydrothermal vents represent areas of primary production in the deep sea and support a specialist fauna, including grazers, filter feeders, scavengers and predators, many of which are restricted to vent environments. The problem of dispersal between vent fields may be reduced by the use of the reducing environment associated with large carcass falls as stepping stones.

- Outside the tropics, the oceans are seasonal, with the seasonality in surface production giving a seasonal pattern of organic flux into the deep sea. In addition, the oceans undergo other temporal changes on time scales of months to decades. Some of these patterns, such as ENSO, have major climatic consequences over large geographical areas.

- Despite their size, oceans are not permanent, but follow a clear life cycle of formation, spreading and then contraction, powered by plate tectonics.

Further reading

Gage J.D. and Tyler P.A. (2002) *Deep-sea biology: A natural history of organisms at the deep-sea floor.* Cambridge, Cambridge University Press.

Harris G.P. (1986) *Phytoplankton ecology: Structure, function and fluctuation.* London, Chapman and Hall.

Herring P. (2002) *The biology of the deep ocean.* Oxford, Oxford University Press.

Lalli C. and Parsons T. (1995) *Biological oceanography: An introduction.* Oxford, Butterworth-Heinemann.

Mann K.H. and Lazier J.R. (1996) *Dynamics of marine ecosystems: Biological–physical interactions in the oceans.* Oxford, Blackwell Publications.

Van Dover C.L. (2000) *The ecology of deep sea hydrothermal vents.* Princeton, NJ, Princeton University Press.

Assessment questions

1 What mechanisms exist in the open ocean to transfer productivity from the photic zone to the deep sea floor?

2 What are the main constraints on primary production in the open ocean?

3 What events other than creation of hydrothermal vents provide environments suitable for chemoautotrophy in the deep sea?

4 What mechanisms generate temporal variation in the abundance of zooplankton in the open ocean?

5 What major morphological adaptations do abyssal and mesopelagic species show?

6 How does marine snow contribute to the ecology of the deep ocean?

7

Lakes and Ponds

■ ■

7.1 Introduction

A lake may be defined as a body of water, encircled by land, whose outflow, if present, is small relative to its volume. Lakes fed by rivers can be distinguished from water bodies fed only by rainfall, groundwater or occasional inundation, such as ponds, oxbow lakes and, in coastal areas, lagoons, but ecologically these water bodies share important similarities and so all are considered in this chapter (Figure 7.1). Pools, backwaters and other permanent dead zones in rivers also share functional similarities with lakes, emphasizing that the boundary between a river and a lake can often be very blurred.

Lakes, ponds and other enclosed water bodies are often referred to as standing waters, or lentic systems, as opposed to rivers, which are flowing, or lotic, systems. This is a convenient division but, functionally, is a gross oversimplification because all but the smallest, most sheltered ponds will contain moving water masses. The differences between lentic and lotic systems relate to the form this movement takes: flow in rivers is unidirectional, proceeding downhill under the influence of gravity, whereas direction of flow in lakes and ponds is less consistent, being affected by, amongst other things, wind direction and speed, and by convection currents generated by differential heating and cooling in different parts of the water body. Furthermore, the turnover time for water in lentic systems is slower; a water particle in a river is continually displaced downstream, whereas in a lake it may recirculate many times before being lost to the outflow or to evaporation. Absence of a strong unidirectional current allows pelagic communities to flourish so that, ecologically, large lakes in particular are more analogous to the open sea than to rivers.

7

(a)

Figure 7.1 Two contrasting lakes. (a) A cirque lake, formed by scouring at the head of a former glacier and fed mainly by run-off from a small, high-altitude catchment. The catchment is dominated by slowly weathering rock, with little vegetation, and the lake is deep and oligotrophic: Lochnagar, Scotland.

(b)

Figure 7.1 Continued (b) An oxbow lake, formed when a river cut off a former meander and now fed by groundwater and occasional floods from the parent river. The catchment is well-vegetated alluvium and the lake is shallow and eutrophic: River Bollin, northern England. Photos M. Dobson.

7.2 The abiotic environment

7.2.1 Water level

Formation of a lake depends upon the presence of a depression and a supply of water to collect in it (Box 7.1). Generally, a lake will fill the entire basin and feed an outflow river, draining from the lowest point of the basin's edge, but in arid areas, where water input is matched or exceeded by rate of evaporation or loss to groundwater, the depression may be only partially filled and a saline lake without an outflow will form. The presence of an outflow demonstrates that a basin is overflowing, and therefore the water level will remain relatively constant, changes in inputs being matched by changes in outflow. Fluctuations will, however, occur in response to prolonged or marked changes in rainfall. Equatorial lakes in East

7

Africa, after experiencing low levels during the first half of the twentieth century, experienced a sharp rise during the 1960s; the surface of Lake Victoria rose by 2.5 m between 1960 and 1964, doubling its output into the River Nile. More impressive, however, are fluctuations in saline lakes, which lack an outlet to limit the degree to which they can rise. Mar Chiquita in Argentina has a surface area which fluctuates between 2000 and 5000 km^2, depending upon the rainfall in its catchment. Great Salt Lake in Utah, whose mean depth is variable but generally in the range 4–7 m, has fluctuated in depth by over 5 m between its lowest recorded level in 1963 and its highest in 1873, with a corresponding variation in surface area of 2600–6500 km^2 (Figure 7.2); this, in turn, is a tiny remnant of Lake Bonneville, a pluvial lake formed by increased precipitation during the Late Pleistocene, whose maximum area was 51,640 km^2, when its shoreline was 335 m above that of the present day lake. More spectacular still are arid zone saline lakes which are

Box 7.1   Types of lake basin

Geomorphologists recognize around one hundred different lake types, based upon the origin of their basins, but they can generally be placed into three structural categories.

Depressions in bedrock

Suitable depressions include those caused by glacial ice scouring out a depression, forming long, thin lakes in glacially eroded valleys, or more rounded cirque lakes near the tops of mountains, marking the head of a former glacier. Other sites are formed by tectonic activity, raising the ground to create shallow basins, such as that now containing Lake Victoria, or lowering the ground to produce rift (or graben) lakes, collapse of cave systems in limestone areas, volcanic activity and even the impact of meteorites. These include the largest lake types, and can reach great depths; glacial scour lakes include most of the larger lakes in Canada and northern Europe, notable amongst these being Great Slave Lake in north-west Canada (maximum depth 614 m, extending to 464 m below sea level; the deepest lake in North America), Mjösen in Norway (720 m; the deepest lake in Europe) and Loch Morar (310 m) and Loch Ness (230 m) in Scotland, as well as the Great Lakes of eastern North America. Volcanic crater lakes can be very deep: Crater Lake in Oregon fills a large volcanic crater and, despite having a surface area of only 55 km^2, extends to a maximum depth of 608 m. These depths are easily exceeded, however, by the deepest rift lakes: Lake Baikal (1637 m) in Russia and Lake Tanganyika (1480 m) in the East African Rift Valley.

Depressions in sediment

Large areas of sediment left behind following glacial or periglacial activity may contain depressions in which water will collect, creating kettle lakes, which are often short-lived; the mosses and meres of Cheshire and Shropshire in England were formed in this way: meres still contain open water, whereas mosses have infilled to become wetlands (Section 8.6 and Figure 8.8). Other lakes occur on

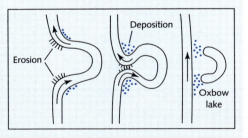

fluvial sediments, and may be formed as a river changes course or shifts sediment around. These include oxbow lakes, formed when a river cuts off one of its own meanders (Figure 7.1b).

Barrier lakes

A barrier thrown across a valley and acting as a dam will create a lake. Barriers include landslides, lava flows (creating so-called coulee lakes), wind-blown sand and, in areas of glacial activity, ice barriers. Among the most common, however, are those formed by glacial moraine. Glaciers push out scoured rock fragments, or moraine, and aggregations of this sediment mark the furthest extent of a glacier. After the glacier has melted, these can form effective dams, behind which a lake will develop.

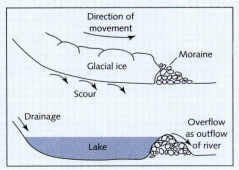

7

Box 7.1   Continued

The Finger Lakes of New York are unusual in that they are impounded at both ends by glacial moraine, the southern ends by the debris from an older ice advance, the northern ends by more recent glacial activity.

Barrier lakes form along coasts when longshore drift creates a sand or gravel barrier; complete severance from the sea allows a freshwater lake to develop; otherwise, a brackish lagoon will be created.

Biologically created water bodies

Biological processes create some lentic water bodies, generally small in size. Beaver ponds, formed by active damming of a river, are barrier lakes; *Sphagnum* bog extending across a valley mouth can, similarly, create a barrier behind which a lake will form. *Sphagnum* bog, too, is typically full of small ponds created by depressions in the hummock–hollow sequence of such structures. Ephemeral water bodies which can, nevertheless, be important local habitats for aquatic organisms during dry periods are those created by large animals: alligators in the Florida Everglades, for example, create scrapes which remain water filled during the dry season. Water which collects in hollows in living plants, such as tree holes, supports a specialized community dependent upon the leaf litter which collects within them. The simple but variable communities which occupy these phytotelmata, as they are called, have been used to study the effects of resource availability on food chain length (Pimm and Kitching 1987).

An exception to the general pattern of biotically created water bodies being small is those produced by human activity. By damming river valleys to create reservoirs, people have created many thousands of lakes, varying in size from those covering a few square metres to reservoirs which extend over several thousand square kilometres. Human activity has also created many flooded depressions in sediment, generally as an accidental by-product of mining or quarrying activities, although in many parts of the world, tiny ponds are dug in large numbers for ornamental or fish-rearing purposes. Gravel pit lakes, for example, simply fill up the depression left after gravel extraction. The Norfolk Broads in eastern England were created when mediaeval peat cuttings were flooded. The soft glacial sediments in Cheshire in northern England contain several ponds formed in depressions created by bombs dropped during the Second World War.

Canals are interesting examples of anthropogenically produced water bodies which, although clearly linear in shape and flowing like a river, generally flow so slowly that they support biota more typical of lakes.

temporary, drying up completely for extended periods; there are thousands of such lakes in drier parts of the world, most of them small but including Lake Eyre North, fed by intermittent rivers in the desert of South Australia, which can reach a maximum area of 8430 km².

7.2.2 Thermal stratification

Lakes share many physical characteristics with oceans, modified by their size and depth.

Recognition of these similarities is, unfortunately, confused by the different terminology used by marine and freshwater ecologists. All but the shallowest lakes will undergo stratification similar to that in marine systems, with the development of a thermocline (Box 1.3) and consequent nutrient depletion (Section 5.4). In such lakes, as in the oceans, distinct layers can be distinguished, separated by the thermocline. The differences between layers are, however, relative: Lake Victoria in tropical Africa develops

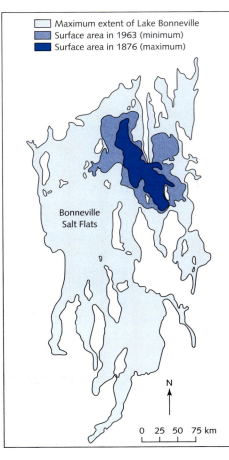

Figure 7.2 Extent of the Great Salt Lake (Utah) and its Pleistocene predecessor, Lake Bonneville. The Great Salt Lake fluctuates markedly in volume and surface area: the maximum (1876) and minimum (1963) water surface areas recorded in historical times are marked. From Burgis and Morris (2007).

different patterns of stratification in different places. This has important implications for nutrient availability, because stratification of the entire water body reduces the opportunity for nutrients to be replenished by processes such as deep circulation and upwelling (Section 1.2).

Patterns of stratification

The pattern of stratification in lakes is determined by latitude, as in marine systems, and altitude. A generalized stratification sequence is illustrated in **Figure 7.3**, other patterns being variations on this theme. Warm temperate lakes, like temperate oceans, can be described as monomictic, as they have a single season of mixing—often referred to as the overturn—during the winter and a season of stratification during the summer. Cold temperate lakes are

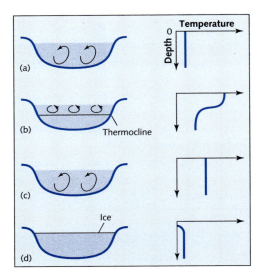

a thermocline from January to May, but its hypolimnion, although completely dark, has a temperature of 23°C, only 2°C cooler than the epilimnion.

An important difference between the physical characteristics of lakes and marine environments is that whereas the ocean, as a single unit spanning all latitudes, undergoes different stratification patterns in different parts of its water mass, a lake, as a discrete unit, will undergo the same pattern of stratification throughout its area. Only very few lakes large enough to span several climatic zones show

Figure 7.3 The stratification cycle of a temperate dimictic lake, showing zones of mixing and the temperature profile with depth during different times of the year. Compare with ocean stratification patterns (Figure 6.1). (a) Spring: complete mixing occurs; (b) summer: development of a thermocline; (c) autumn: the thermocline breaks down and mixing occurs throughout; the lake's situation is similar to that during the spring, but the temperature is generally higher; (d) winter: stratification develops as the surface layer freezes, resulting in a reverse thermocline, with the warmer water at depth.

dimictic, stratifying twice per year, once during the summer and again, with warmer water at depth, when they freeze at the surface during the winter (Figure 7.3). Warm tropical lakes are polymictic, stratifying during the day but with a temperature difference between epilimnion and hypolimnion so slight that mixing occurs overnight. Deep tropical lakes may be continually stratified, like tropical oceans, although severe storms may induce occasional, irregular mixing; these are therefore classified as oligomictic. Certain polar or high altitude lakes are amictic—permanently frozen, their waters never fully mix; the presence of a layer of ice on their surface can result in the development of a reversed thermocline in which temperature rises with increasing depth beneath the ice, often then to drop again, sometimes appreciably below 0°C, in very deep lakes.

Depth also has an influence on mixing. Relatively deep lakes may undergo mixing and stratification in their upper layers, while the deep water remains permanently stratified, equivalent to the deep ocean (Section 6.2); such lakes are referred to as meromictic. Lakes which undergo mixing throughout the entire water body are referred to as holomictic. Fluctuations in water level can cause saline lakes to change from meromictic to holomictic and vice versa (Box 7.2). Some tropical lakes are monomictic not because of seasonal temperature changes, but because they undergo seasonal variations in depth. Lake Calado is a seasonally flooded lake on the Amazon flood plain, whose maximum depth varies from 1 m in October to 11 m in July, related to the rise and fall of the Solimoes River. When Lake Calado is less than 3 m deep, it mixes fully to the bottom, with no

---

Box 7.2 Depth, mixing and productivity—the case of Mono Lake

Mono Lake is a hypersaline alkaline lake in California, on the western fringe of the North American Great Basin. It covers 186 km^2 and has a mean depth of 18 m. Its current salinity ranges between 72 and 87 g l^{-1} and high concentrations of bicarbonate and carbonate give it a pH of around 10. It is highly productive, supporting two common metazoan species—the planktonic brine shrimp *Artemia monica* and the littoral benthic alkali fly *Ephydra hians*. These, in turn, provide food for large numbers of birds, particularly California gulls (*Larus californicus*) which breed at the lake, but also migratory phalaropes (*Phalaropus* spp.) and eared grebes (*Podiceps nigricollis*).

Mono Lake is fed by a series of freshwater streams, much of whose water was diverted from 1941 to supply the city of Los Angeles. Since diversion started, the lake level has dropped by about 14 m (from 1956 to 1942 m above sea level [asl]) and between 1941 and 1982 its salinity doubled. Increasing salinity reduces standing crops and diversity of benthic algae and increases physiological stress on brine shrimps and alder flies, whose reproduction is then impaired. Declining reproduction of these invertebrates, in turn, influences food supply for the birds. Furthermore, several species formerly present in the lake, including two rotifer species, became extinct once salinity rose above their physiological tolerance.

A management strategy for the lake is to reduce or cease diversion of the lake, and this is currently being implemented. However, a series of natural fluctuations in rainfall has demonstrated that such an approach may be counterproductive to some of the biota, at least in the short term. A large input of fresh water induces meromixis, with the less saline inputs sitting on top of the denser, more saline water and a clear halocline developing between the two. Nutrients continue to be lost from the upper layer, particularly through defecation by brine shrimps, and the water above the halocline becomes

Box 7.2 Continued

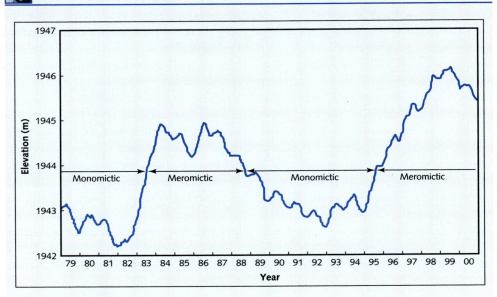

From Melack *et al.* 2002.

nitrate-limited. This occurred during the mid-1980s as a consequence of high precipitation, and again during the late 1990s; only as drought conditions and water transfer resumed did the upper layer become sufficiently saline to resume mixing and restore algal productivity.

The management of Mono Lake involves raising its surface level from 1942 m asl to 1948.3 m asl. The experience of the late twentieth century suggests that this may be counterproductive, but a higher lake level would eventually induce meromixis because freshwater inputs would be a smaller proportion of the total volume of the lake. Furthermore, higher water levels are beneficial to alkali flies, which require solid substratum on which to develop as larvae; as the lake level rises, the proportion of the littoral zone that is solid bedrock increases, allowing their numbers to grow. Finally, higher water levels mean that overall salinity is relatively low (although still considerably higher than that of sea water), allowing the rotifers to recolonize; these did reappear in the late 1990s, during a period of meromixis, but were most abundant in nearshore pools whose salinity was less than 53 g l^{-1}.

thermocline, but once it has exceeded this depth stratification occurs. The thermocline depth is variable but never exceeds 7 m.

Mixing across the thermocline

Mixing in lakes can be induced by wind displacing surface waters. A strong, steady wind will tilt the thermocline by pushing surface waters towards the downwind end of the lake, allowing the hypolimnion to rise at the upwind end (**Figure 7.4**). Mixing across the thermocline can then occur in three ways: (a) the wind displaces the epilimnion so far that deep waters upwell to the surface (**Figure 7.4a**); (b) continual lateral movement of surface waters causes friction along the thermocline, leading to localized mixing across its boundary (**Figure 7.4b**); (c) when the wind stops, the two layers of water revert to equilibrium but, in the process, the

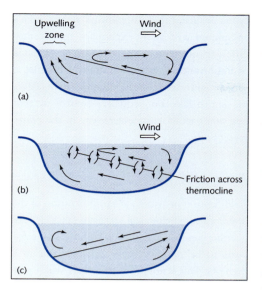

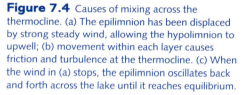

Figure 7.4 Causes of mixing across the thermocline. (a) The epilimnion has been displaced by strong steady wind, allowing the hypolimnion to upwell; (b) movement within each layer causes friction and turbulence at the thermocline. (c) When the wind in (a) stops, the epilimnion oscillates back and forth across the lake until it reaches equilibrium.

thermocline rocks backwards and forwards in a series of decreasing oscillations (known as seiches), leading to friction and mixing across its boundary (**Figure 7.4c**).

In temperate zones, decreasing solar radiation during the autumn leads to gradual cooling of the epilimnion and breakdown of stratification, but the process will be hastened by stormy weather, inducing active mixing between the two layers.

Horizontal stratification

It is tempting to see stratification across the thermocline/nutricline as the only structural break in the otherwise homogeneous structure of the water body, but this is a misleading impression of systems whose patchiness, even within such an apparently homogeneous environment, can be pronounced. Lakes can generate horizontal flows through differential heating and cooling of various parts of the water body.

Shallow shoreline regions, particularly semi-enclosed bays, change temperature more rapidly than deeper offshore areas and, if heated by solar radiation, their water will flow over the cooler offshore water, displacing it laterally.

7.2.3 Oxygen stratification

A second difference between oceans and lakes relates to oxygenation of deep waters. The deep oceans are kept oxygenated by a global circulation system in which cold, oxygen-rich currents descend near the poles and flow along the ocean bed (Section 1.2). Lakes, lacking the combination of both cold and warm surface zones, have no such currents and replenishment of deep water can be very slow, which in combination with bacterial respiration at depth leads to oxygen depletion beneath the upper mixed layers. Deeper lakes can be completely anoxic in the hypolimnion, a condition typical of meromictic tropical lakes. Lake Tanganyika in East Africa is interesting in this respect in that it extends across 5 degrees of latitude: its northern end, at 3°S, is subject to an equatorial climate and oxygen penetrates to a depth of 70 m; its southern end, at 8°S, is subject to a slightly greater degree of seasonality, increasing mixing and allowing oxygen to penetrate to 200 m depth. In contrast, Lake Baikal in Russia, whose maximum depth exceeds 1600 m, is unique in that it has a complex and, as yet, incompletely understood, water circulation system which ensures oxygenation throughout the water column.

Oxygen stratification can be generated or accentuated by photosynthetic organisms. Eutrophication of Lake Victoria since the 1960s (Section 7.7) has increased the biomass of phytoplankton; as a consequence, the upper layers of the lake have a higher concentration of oxygen than previously, due to enhanced photosynthetic rates. Concentrations of oxygen at depths greater than 40 m are, however, appreciably lower and anoxia, formerly very rare, is common and widespread, probably due to increased inputs of dead and decaying algal

cells into the hypolimnion. At a smaller scale, submerged macrophytes can increase oxygen concentrations in their vicinity and, through diffusion from their roots, can oxygenate bed sediments (Section 7.3), but dense mats of floating leafed macrophytes can decrease oxygen concentrations by impeding exchange with the atmosphere (Carpenter and Lodge 1986).

layers ensuring that different parts of a lake's epilimnion will rarely experience significantly different nutrient status (Box 7.3). This is of importance to pollution control because a single point source of pollutant may affect the entire lake but, equally, control at that one point will facilitate recovery in the entire water body.

7.2.4 Water chemistry

Vertical stratification ensures that marked differences in water chemistry can occur between the hypolimnion and the epilimnion, but horizontal variations are much less profound, the enclosed nature and complete mixing of upper

Acidification

Acidification provides a clear illustration of the interconnected chemical status of lakes. A lake is either acid or it is not, with little scope for parts to be acidified while others maintain a high pH. This is not to say that there will not be

Box 7.3   Defining nutrient status

The terms oligotrophic and eutrophic are widely used to describe lakes, although precise definitions are difficult, not least because they refer to relative rather than absolute states.

An oligotrophic lake is one with low nutrient levels. Phytoplankton production is low and therefore water clarity is high. Low bacterial production in the hypolimnion means that decomposition rates are slow and use up oxygen at a rate more slowly than it can be replaced by mixing from the surface, so the water remains well oxygenated.

A eutrophic lake has high nutrient levels and therefore high rates of phytoplankton production. High densities of algal cells reduce water clarity and decomposition of detritus in the hypolimnion uses up oxygen more rapidly than it can be replaced, leading to reduced concentrations, or even anoxia.

Assessment of trophic status can be made using one or more of the following methods.

- *Nutrient concentration (TP)*. The mean annual total phosphorus concentration (i.e. including that within plankton cells within a water sample). Normally expressed as micrograms per litre (μg l^{-1}).

- *Phytoplankton biomass (Chl a)*. Normally the mean or peak annual biomass is quoted. This is difficult to measure directly, so concentration of chlorophyll *a* (Chl *a*), a component of all phytoplankton whose concentration is directly related to photosynthetic activity, is generally used instead. Normally expressed as micrograms per litre (μg l^{-1}).

- *Rate of primary production (PP)*. Measured as mass of carbon assimilated (i.e. converted from inorganic carbon dioxide to organic compounds) per m^2 surface area per day (mg C m^{-2} d^{-1}).

- *Water transparency*. This is a measure of light attenuation (Box 1.2). Normally expressed as Secchi depth, the depth measured with a Secchi disc. This is a disc 20 cm in diameter and divided into four equally sized segments, alternately black and white in colour. It is lowered horizontally into the water and the depth at which it just ceases to be visible is the Secchi depth.

Box 7.3 Continued

Although there is no universally accepted definition of oligotrophic and eutrophic states, various attempts have been made to quantify them. The table below gives useful approximate values for the parameters listed. They allow inclusion of several further categories of trophic status—ultra-oligotrophic, mesotrophic and hypertrophic.

	TP (Chl *a*) ($\mu g\ l^{-1}$)	mean (Chl *a*) ($\mu g\ l^{-1}$)	max ($\mu g\ l^{-1}$)	PP ($mg\ C\ m^{-2}\ d^{-1}$)	Secchi depth mean (m)	maximum (m)
Ultra-oligotrophic	<4	<1	<2.5	<30	>12	>6
Oligotrophic	4–10	1–2.5	2.5–8	30–100	6–12	3–6
Mesotrophic	10–35	2.5–8	8–25	100–300	3–6	1.5–3
Eutrophic	35–100	8–25	25–75	300–3000	1.5–3	0.7–1.5
Hypertrophic	>100	>25	>75	>3000	<1.5	<0.7

This classification works on the principle that phytoplankton biomass and production change in tandem with phosphate concentration and, in turn, have a direct effect upon water clarity. This is undoubtedly true in a great many cases, but should be applied with caution, as there are many exceptions. Section 7.3 describes the example of Loch Ness, oligotrophic in terms of TP and PP, but with a low Secchi depth, not a result of algal cells but of humic substances in the water. Conversely, lakes with high nutrient inputs can be very clear if these nutrients are absorbed by macrophytes (Section 7.5).

fluctuations: acid streams, for example, will cause localized elevation of acidity at their mouths, but mixing processes will dilute this input throughout the water body and such patches will only be maintained by constant inputs. Acidification can be countered by using lime as a buffer; this can be added to the lake at a single point and will raise the pH throughout the lake. At Loch Fleet in southern Scotland, lime added to the catchments of several of the small streams flowing into the lake had the same effect: it was washed into the lake at several discrete points but raised the pH to a constant level throughout the entire water body.

Salinity

Saline lakes are defined as those with salinity (see Section 1.3) of at least 3 g l^{-1}, although typically lakes in arid zones have salinities in excess of that of seawater and are referred to as hypersaline. Inputs of water, however fresh, contain some dissolved salts and a lake lacking

an outflow will lose water mainly through evaporation, leaving the salt behind. Salinity is therefore determined by the volume of water within which these salts are dissolved, which, in turn, is dependent upon inflow and evaporation rates, and can be very variable. Mar Chiquita was relatively small for most of the twentieth century, with a salinity of 200–300 g l^{-1}, but increased rainfall since the 1970s raised the volume of the lake such that salinity had dropped to 30 g l^{-1} during the 1990s. The Caspian Sea, the world's largest lake, has a mean salinity of just under 13 g l^{-1}, with only minor seasonal variations, but it is significantly diluted by inflow from the Volga River to the north, which accounts for around 80% of inputs to the lake and creates a marked salinity gradient (Figure 7.5). A consequence of this gradient is the presence in the northern fringe of the lake of a fully freshwater fauna, including the zebra mussel (*Dreissena polymorpha*), whereas the central and southern parts of the lake have a brackish water fauna.

7

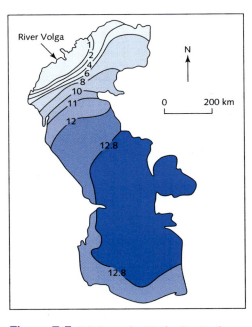

Figure 7.5 Salinity gradient in the Caspian Sea during the summer. The figures refer to salinity. Inflow from the Volga River, in combination with increased temperature and aridity towards the south of the lake, creates a marked salinity gradient. From Kosarev and Yablonskaya (1994).

less well understood, although equally dramatic (Section 7.3).

7.3 Ecological effects of the abiotic environment

Within a given lake, the component of the abiotic environment which fluctuates with the greatest frequency is mixing within the water column. This, in turn, affects light and nutrient availability for primary producers, with consequent effects upon energy budgets for the entire lake. Therefore, this section concentrates upon these determinants of primary production.

7.3.1 Light intensity

The absorption of light, described in Box 1.2, is as relevant to fresh waters as to marine systems. High latitude lakes have strongly seasonal inputs of light, with long summer days and deep penetration into the water column creating an optimum light environment, whereas low penetration and reduced day length in winter ensure that light is severely limiting. Absorption of light is further determined by water clarity: a very clear, clean lake may allow photosynthetically active radiation to penetrate several tens of metres into the water column, but coloured humic solutes (such as those originating from peat), suspended sediment and even a high density of algal cells themselves will significantly reduce penetration. A dense algal bloom, for example, can limit light penetration to the top few centimetres.

In clear shallow lakes or the edge of deep lakes, light penetrates to the lake bed, allowing the growth of macrophytes and benthic algae. Confusingly, this zone is often referred to as the littoral zone, a term used in marine biology to describe the intertidal (Section 5.2). Here, the term 'lake littoral' is used, to ensure that the two are easily distinguishable.

The salinity of many of the World's saline lakes is being raised by diversion of inflowing water for human use (Box 7.2). The most dramatic example of this is the Aral Sea in central Asia, whose two inflowing rivers have since 1960 been almost completely diverted for irrigation, with the result that by 1990 its volume had dropped to less than one third of that in 1960, and its salinity had risen correspondingly from 10 to 29 g l^{-1}. In 1989 continuing shrinkage caused the Aral Sea to split into two lakes: the Small Aral in the north and the Large Aral in the west; the Large Aral is further subdivided into an eastern and a western basin, with relatively little exchange of water between the two, and by 2002 the salinity of the western basin had reached 75 g l^{-1}, whilst in the eastern basin it reached 150 g l^{-1} (Mirabdullayev *et al.* 2004). The human tragedy associated with the loss of the Aral Sea has been well documented, but effects upon its ecology are much

7.3.2 Nutrients

The nutrient status of lakes is of fundamental importance in determining their community structure and productivity. There are certain environments, such as carbonate-precipitating marl lakes, in which nutrients, though abundant, are in insoluble forms unavailable to photosynthetic organisms but, in general, the higher the nutrient concentrations, the higher the rate of photosynthetic activity (**Box 7.3**).

Stratification limits the availability of nutrients by minimizing their transport across the thermocline (see Section 5.4 for a description of the same process in marine environments). Photosynthetically active organisms are confined to the epilimnion by their requirement for light and will normally deplete nutrients to the extent that they become limiting, whilst at the same time the hypolimnion may contain high concentrations, enhanced by sinking of detritus and even of living algal cells from the epilimnion. Therefore, any process that adds nutrients to the epilimnion in a stratified water body can enhance primary production, and point sources of nutrients from catchment inputs can induce patches of high phytoplankton biomass (**Figure 7.6**). Wind-induced upwelling from the hypolimnion or mixing across the thermocline will have similar effects but, in contrast to major zones of oceanic upwelling (Section 1.2), is dependent upon strong or steady winds and will be temporary or intermittent. Wind-induced mixing in shallow lakes can also cause substantial nutrient exchange between sediment and water.

Nutrient concentrations will also undergo temporal variation, although this is generally a result of seasonal variation in uptake by phytoplankton (see Section 7.4).

Vertical mixing within the epilimnion

The epilimnion is a layer of mixed water. Some of the larger phytoplankton, such as dinoflagellates, are able to propel themselves through the water and will endeavour to remain in patches

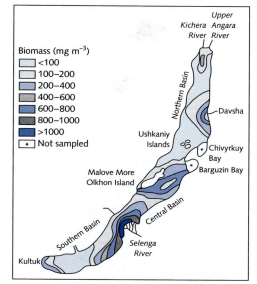

Figure 7.6 Horizontal variations in biomass of phytoplankton in the surface layers of Lake Baikal. The highest concentrations occur around the mouth of the Selenga River, the main inflow, in response to the nutrient load it discharges into the lake. From Bondarenko *et al.* (1996).

of optimal conditions, but their rate of movement upwards is less than the velocity of vertical currents created by surface winds. Rotation of water through the mixed layer is, therefore, too rapid for even motile phytoplankton to counter, so if light penetrates through only a part of the epilimnion but water is being mixed throughout, photosynthetic activity will be limited (**Figure 7.7**).

Loch Ness, in Scotland, is an interesting example of a lake in which light penetration is reduced relative to the depth of mixing, limiting primary production. It has a summer thermocline at a depth of a least 30 m, but the water is clouded by high concentrations of humic substances, limiting light penetration to a depth of 6–7 m. Therefore, even during daylight hours phytoplankton may only spend around 20% of the time in the photic zone, reducing their ability to photosynthesize effectively. During the winter, when solar radiation is reduced both in terms of strength and day length, vertical stratification is absent and mixing occurs to the

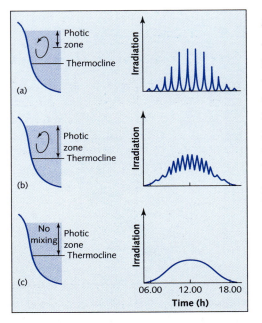

Figure 7.7 The effect of internal mixing within the epilimnion upon photosynthetic activity of a phytoplankter. (a) Mixing occurs throughout the epilimnion, but the photic zone extends only half way to the thermocline: algae are being constantly carried into and out of the photic zone, resulting in photosynthetic activity marked by major peaks and troughs. (b) Mixing occurs throughout the epilimnion, but the photic zone extends to the thermocline: algae are mixed between higher and lower light intensities, but can photosynthesize throughout daylight hours. (c) No mixing occurs. Algae are subjected only to the daily pattern of varying light intensity. Adapted from Reynolds (1987).

bed (whose mean depth is 137 m), reducing photosynthetic potential even further. Thus, even though it is oligotrophic, primary production in Loch Ness is constrained not by nutrient availability but by light. It is, in this respect, unusual: oligotrophic lakes are normally clear, allowing deep light penetration, and phosphate is the limiting factor. Nutrient-rich lakes, in contrast, can support such a high biomass of phytoplankton that the algal cells themselves reduce light penetration, confining photosynthesis to a narrow vertical zone close to the surface (Box 7.3).

Stratification within the epilimnion

The epilimnion is defined as the upper mixed layer, but calm weather conditions will reduce vertical mixing, leading to the development of internal stratification, with measurable differences in nutrient concentrations, phytoplankton biomass and phytoplankton species assemblages at different depths even above the thermocline. Such conditions favour motile phytoplankton, which are able to move vertically within the water column to the zone of optimum light intensity, but are detrimental to heavier or non-motile species, which sink to the bottom of the epilimnion. In sheltered lakes, this stratification can be persistent, but more exposed lakes will be subjected to weather conditions which periodically break it up.

7.3.3 Salinity

Moderately saline lakes (<10 g l^{-1}) are normally occupied by halotolerant freshwater species, but above this salinity they will be replaced by specialist saltwater species. These require adaptations to tolerate the osmotic stress of saline environments and the fluctuations in salinity typical of most salt lakes, and in temporary salt lakes they need adaptations to withstand complete desiccation (Williams 2002). The biota of saline lakes is therefore taxonomically restricted, as few organisms are adapted to the generally high and fluctuating salinities. However, several taxa specialize in these environments. Brine shrimp (*Artemia* spp.) are rare in salinities below 45 g l^{-1} and several species, such as *A. franciscana* in Great Salt Lake, reach their physiological optimum at a salinity of around 140 g l^{-1}, four times that of seawater. These animals have two modes of reproduction: in stable environments, they produce ovoviviparous eggs which hatch immediately into nauplii; in unstable conditions, however, they produce resistant cysts. This second mode of reproduction is a strategy to survive events such as drying or extreme hypersalinity, to both of which saline lakes in arid areas are prone. In Great Salt Lake, however, cyst production has a second function in that it allows the species to persist during the winter, when shrimps would normally be killed by the cold. *Artemia franciscana* has two major periods of reproduction in the lake: spring, when eggs dominate, and late summer/autumn, when cysts dominate (Wurtsbaugh and Gliwicz 2001).

As diversion of inflow waters causes the salinity of many saline lakes to increase, so changes in their biota are expected. These are greatest when the original salinity is low. An increase of salinity in Australia's Lake Corrangamite of 35 to 50 g l^{-1} since the 1960s has resulted in almost complete disappearance of most of its native fauna, whereas the salinity of the Dead Sea in the Middle East has increased from 200 to 450 g l^{-1} since 1980, with little effect on its biota.

The fate of the Aral Sea (Section 7.2) demonstrates the impacts of persistently increasing salinity. Species richness of phytoplankton in the Large Aral Sea declined from 375 in 1925 to 159 in 2002, while that of zooplankton was even more dramatic, from 42 species in 1971 to four species in 2002. Brine shrimps were unknown in the Aral Sea until 1998, when *Artemia parthenogenetica* was recorded for the first time; by 2000 it comprised over 99% of zooplankton biomass. Zoobenthos and fish species richness initially increased, following introduction of marine species such as ragworm (*Hediste diversicolor*), flounder (*Platichthys flesus*) and herring (*Clupea harengus*), particularly in the 1960s; however, most of these species, along with the native fauna, have disappeared. The Small Aral Sea has fared somewhat better, because the restoration of inputs from the Syr Darya (River), combined with the intermittent presence of a dam restricting drainage to the southern basin, reduced its salinity from a high of 29 g l^{-1} in 1990 to below 20 g l^{-1} since 1998. Pelagic and benthic productivity is high and the fauna still relatively diverse, retaining many of those species now extinct in the Large Aral Sea (Aladin *et al.* 2005).

7.4 Energy inputs

Primary production in deep lakes is dominated by phytoplankton, whose productivity is subject to the same constraints of light penetration, nutrient status, mixing and grazing as is that in the sea (Section 5.4). Attached macrophytes are confined to the lake littoral, generally a region close to the shore although, in shallow lakes, it may extend across much of its bed, whereas in deeper lakes the dominant benthos will be detritivores, relying upon allochthonous inputs from above. In very deep or nutrient-rich lakes, lack of oxygen ensures that the only living organisms are anaerobic bacteria. Lake Baikal contains hydrothermal vents (Section 6.7), with an associated concentration of bacterial and benthic faunal biomass; this raises the possibility that, with so many of the world's lakes in areas of volcanic activity, and even in volcanic craters (Box 7.1), lake chemoautotrophy may be more widespread than has hitherto been considered.

This section emphasizes the pelagic system, although most of the constraints upon phytoplankton production affect macrophytes and attached algae as well. Macrophytes are considered in some detail in Chapter 8, and potential competition between phytoplankton and macrophytes is returned to in Section 7.5.

7.4.1 Limiting nutrients

Phytoplankton typically require nitrogen and phosphorus in the relative proportions of 7:1 (the Redfield Ratio), so if inputs deviate from this ratio, limitation will ensue. Normally, the major limiting nutrient in fresh waters is phosphorus, and its addition to water bodies, both experimentally and as a pollutant, has been demonstrated on many occasions to increase primary production. Indeed, Schindler and Fee (1974) showed that addition of less than 0.5 g phosphorus per m^2 of surface area in small Canadian lakes resulted in algal populations

50–100 times greater than those in control lakes. Nitrogen limitation occurs commonly, although not exclusively, in the tropics, and productivity in some tropical lakes may even be limited by sulphate availability. Silicon can become depleted by high diatom biomass.

Total nutrient input determines biomass of primary producers, but the timing of their addition determines community diversity. Continual input of small doses favours competitive species, resulting in a constant biomass and domination by very few species, whereas addition of the same total concentration of nutrients in a series of discrete pulses favours opportunistic species: biomass may fluctuate widely, peaking after each addition and then crashing, and different species may become abundant after each input, with no one species being able to maintain domination. Neill (1994) demonstrated that frequent inputs of nutrients to experimental patches of a small oligotrophic lake led to domination by small flagellates, whereas occasional large doses resulted in continually altering species composition in which small flagellates remained rare, though eventually large gelatinous algae dominated.

Nutrient recycling

Nutrient limitation in the epilimnion ensures that any free inorganic nutrients that become available are quickly recycled. Biotic processes such as excretion by zooplankton or fish produce micropatches of nutrient enrichment whose persistence can be measured in seconds before they are destroyed by diffusion, but algal cells that happen to be caught in such a patch may gain their entire daily nutrient requirement even within this short space of time.

Autotrophic bacteria are efficient at assimilating dissolved nitrogen and phosphorus at very low concentrations, and in doing so are in direct competition with algae. Their extremely small size, however, ensures that they do not sink into the hypolimnion so nutrient stock that has been incorporated into bacterial biomass remains in the epilimnion. Bacteria are heavily grazed by protists—rotifers, flagellates and ciliates—and even by very much larger water fleas (*Daphnia* spp.), whose consumption and subsequent excretion remineralize and recycle nutrients back into the water column.

Particulate organic matter will eventually end up at the lake bed, where it provides the basis for benthic communities. Benthic invertebrates and benthos-feeding fish may release large quantities through excretion (Table 7.1), thus contributing significantly to nutrient recycling within the water body.

Eutrophication

Eutrophication is the name given to enrichment of waters by plant nutrients (Box 7.3). This process occurs naturally, but inputs of very high levels of nutrients from human activities, such as fertilizer run-off and discharge of partially treated sewage, can lead to hypertrophy, the presence of nutrients at excessive concentrations. With nutrient availability removed

Table 7.1	Fluxes of ammonium and phosphate in Acton Lake, Ohio	
Flux process	**Ammonium flux (μmol N m^{-2} d^{-1})**	**SRP flux (μmol P m^{-2} d^{-1})**
Benthic invertebrate excretion	770	46.7
Direct release from sediments	1357	198.7
Fish excretion	2370	185.0
Catchment input: normal year	806	194.7
wet year	1706	810.2
dry year	61	25.1

SRP: soluble reactive phosphorus.
From Devine and Vanni (2002).

as a limiting factor, photosynthesizers will increase in biomass until they are constrained by space or by shading. At such densities, they can disrupt the entire lake ecosystem, most notably through oxygen depletion. During the day, shallow water may be supersaturated by photosynthetic oxygen production, but at night the combined effects of respiration and decomposition of detritus may lead to complete deoxygenation, and the consequent death of other organisms living within the lake.

Hypertrophic water bodies are dominated either by macrophytes or by phytoplankton, each flourishing in the presence of high levels of nutrients, but rarely, if ever, do they support high densities of both. It is a common feature of many water bodies that clear water and a high density of macrophytes is replaced by turgid, algal-dominated water as nutrient levels rise. However, even at high concentrations of nutrients, the macrophyte-dominated clear water conditions may persist if conditions are appropriate; this phenomenon is considered further in the discussion of alternative stable states in Section 7.5.

The impact of nutrients on a water body is not a simple consequence of nutrient loads, but is also determined by its morphology and by the residence time of water. Therefore, application of figures such as those given in **Box 7.3** needs to be interpreted in the context of the individual water body. For deep lakes, the Vollenweider model (Vollenweider 1975) describes this relationship as:

$$P = L/z(s + r)$$

Where P is the concentration of phosphorus; L refers to inputs from the catchment, atmosphere and sediment, z is the mean depth of the lake, s is the net rate of phosphorus exchange with the sediment (positive if phosphorus is lost to the sediment; negative if it is released from the sediment) and r is the residence time of water in the lake. As there is usually a strong positive correlation between total phosphorus concentration and algal biomass, managing eutrophication aims to reduce inputs, thus lowering total phosphorus loads.

7.4.2 Primary production as a seasonal phenomenon

Primary production generally follows seasonal patterns, although the causes of this seasonality differ between temperate and tropical zones.

Seasonality in temperate zones

In temperate zones, low light intensities and temperatures restrict winter production. A peak is reached during the spring in response to improving physical conditions, but is curtailed in late spring by nutrient depletion following stratification and by zooplankton grazing, leading to a midsummer dip in productivity. During late summer there is generally a second peak, reflecting a change in algal species composition and occasional weather-controlled breakdown of the thermocline. Algal biomass declines rapidly during the autumn, as light intensities and temperatures again fall, despite the influx of nutrients as seasonal stratification ceases. This pattern may be compared with the temperate North Atlantic pattern of seasonality observed among marine plankton (Section 5.4). It is illustrated by Lake Esrom in Denmark, a lake in which primary productivity is limited by nitrogen and silicon, the latter being required by diatoms; concentrations of nitrates and nitrites in the water drop from over 300 µg l^{-1} in March to 0 µg l^{-1} by June, whereas phosphorus accumulations remain very high throughout the year. A small spring bloom is therefore followed by a drop in production until late summer, when it is warm enough for nitrogen-fixing Cyanobacteria, which are temperature limited, to proliferate. Mixing of the water column stimulates a second algal peak in the autumn, before lack of light and low temperatures become the limiting factors once more **(Figure 7.8)**.

Seasonal changes in the relative abundance of different phytoplankton taxa may occur. Lake Teganuma in central Japan illustrates this well: it is a shallow, hypertrophic lake, in a warm temperate climatic zone, whose planktonic

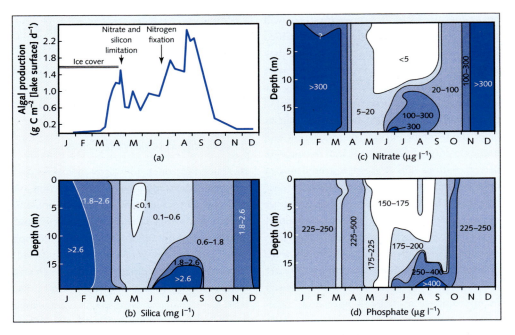

Figure 7.8 Seasonal patterns in primary production and nutrient availability in Lake Esrom, Denmark. (a) Primary production and relative temperature and light intensity. (b)–(d) Change in nutrient levels. Nutrient concentrations are displayed as contours on depth–time diagrams, in which depth is along the vertical axis and time along the horizontal axis. For a given date, the concentration at different depths is plotted and then points of equal value are joined together. After Jónasson (1996).

primary production, although showing a peak in August and September, is high throughout the year. There is however a marked change in the species composition of its phytoplankton: the green alga *Micractinium* dominates between February and June, but disappears over the summer, to be replaced by Cyanobacteria—*Sprirulina* in July and August and *Microcystis* in September—while a diatom, *Cyclotella*, dominates over the winter.

Seasonality in tropical zones

In the tropics, light intensity and temperature are relatively constant, but phytoplankton production follows seasonal patterns related to hydrology. Low flow rates in rivers and lakes, or periods of water storage in reservoirs, create a period of stability, allowing phytoplankton to flourish, whereas during high flow or water release periods their numbers decline. In larger African lakes, where such rainfall-related pat-

terns are diminished by long water retention times, even small variations in temperature translate into patterns of stratification and mixing which can, in turn, affect phytoplankton production.

7.4.3 Secondary production

Grazing

Phytoplankton are consumed by a range of grazing zooplankters, predominant among them being cladocerans such as water fleas (*Daphnia* spp.). Zooplankton can consume 10–30% of algal biomass per day in summer, but different species have different vulnerabilities, grazers preferring small, unicellular or unarmed species to large, colonial forms. *Daphnia* has a very high grazing rate and is able to generate 'clearwater phases', even when algal productivity is high, following which its numbers will collapse through food limitation.

Exact mechanisms by which algae persist in the face of such grazing pressures are not fully understood, but include daytime recovery to counter nocturnal grazing and modification of community structure in response to predation pressure, along with a range of morphologically adaptive responses such as colony formation, development of spines and generation of gelatinous mucus. Open water, in the presence of fish, supports larger or less motile species such as diatoms, whereas macrophyte beds are dominated by small motile forms, which are more difficult for grazers to catch. A possible explanation for this pattern lies in water movement: motile species are able to counter sinking and are therefore favoured in the relatively still waters around macrophytes, whereas diatoms require constant mixing to counter sinking, so will be confined to open waters. In macrophyte-free lakes with high zooplankton grazing pressure, however, small flagellates dominate even the open water, suggesting a role for grazing in determining this pattern.

In nutrient-limited lakes, grazing may stimulate algal production by recycling nutrients, the digestive activity of grazers releasing remineralized nutrients back into the water column. The destructive activity of grazers also increases the loss of DOC from algal cells into the water column, and this DOC is incorporated into a microbial loop (Section 2.2), whereby it is consumed by heterotrophic bacteria which, in turn, are consumed by protists. Thus organic carbon which would otherwise have been lost is recycled within the pelagic zone, the bacterial grazers being consumed by larger zooplankton.

Detritus

Much phytoplankton biomass eventually settles to the lake bed where, along with inputs of detritus from rivers and the lake littoral zone, it provides food for benthic detritivores. Benthic macroinvertebrate communities are dominated by chironomids, oligochaetes and filter-feeding bivalves, mainly feeding upon fine particulate organic matter.

Benthic chironomids can play an important role in organic matter dynamics in eutrophic lakes, because larval growth in sediments is followed by a period of emergence, when pupae rising to the surface are available as food for free-swimming pelagic predators. In temperate lakes, maximum chironomid growth and emergence often coincides with spring and autumn overturn periods, when primary production is high and oxygen is freely mixed to the lake bed. In Japanese lakes, a single species, *Tokunagayusurika akamusi*, can account for over 70% of total benthic macroinvertebrate biomass. This species emerges as an adult in very large numbers over a period of approximately two weeks in late autumn, providing a valuable food resource for pelagic predators at this time, and around three times as many individuals are predated as emerge. Adults which successfully emerge are however effectively transferring energy out of the lake and into surrounding terrestrial systems.

7.5 Community structure

7.5.1 Competition

The influence of addition of extra nutrients on algal biomass demonstrates the importance of resource limitation in most lentic water bodies which, in turn, leads to competitive interactions among species. Under the relatively predictable physicochemical conditions encountered in freshwater lakes, competitive interactions of this type are often strong, and some of the classic examples of competitive interactions derive from studies of lentic organisms (Brönmark and Hanssen 2005). Two examples of competitive interactions are given below.

Light versus nutrients

Photosynthesizers require light and nutrients. Light is limiting away from the water surface,

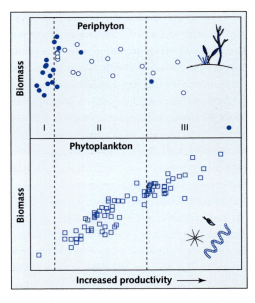

Figure 7.9 Changing biomass of periphyton and phytoplankton along a gradient of increasing productivity. From Brönmark and Hansson (2005).

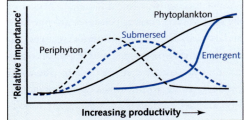

Figure 7.10 Relative importance of different primary producers along a productivity gradient. From Brönmark and Hansson (2005).

7

while nutrient release is greatest near the lake bed due to activity of decomposers at the sediment surface. Therefore, in low productivity lakes, phytoplankton have a light advantage but periphyton have a nutrient advantage. In oligotrophic lakes, all algae are normally nutrient limited, so biomass is low (**Figure 7.9**, area I). As nutrient levels increases, all algae benefit and biomass increases, but periphyton increase more rapidly because they have greatest access to limiting nutrients (**Figure 7.9**, area II). At high nutrient levels, however, periphyton numbers decline, despite high nutrient levels, because they are being shaded by large numbers of phytoplankton (**Figure 7.9**, area III). The competitive advantage of being close to the source of nutrients is therefore overcome by the disadvantage of being relatively far from the source of light.

Addition of macrophytes into this scenario creates a complex pattern of competitive interactions. Submerged and floating leaved macrophytes can absorb nutrients from water and sediment, giving them a competitive advantage over both periphyton and phytoplankton.

Emergent macrophytes absorb nutrients from sediment but photosynthesize above water; therefore they are able to outcompete all other primary producers.

Figure 7.10 adds macrophytes to the productivity–biomass relationship. Periphyton do best at low productivities, but are shaded out at higher productivities. Submerged macrophytes can outcompete them by shading, and do better than phytoplankton at moderate productivity because they derive their nutrients direct from the sediment; at high nutrient levels, however, phytoplankton can reach densities high enough to shade and therefore suppress submerged macrophytes. Emergent macrophytes only become important at high productivity because they need high sediment nutrient levels to support their large growth forms. They are only limited by depth.

Competition among fish

Changing nutrient levels can indirectly influence competitive interactions among predatory fish species. Increasing phosphate levels across the gradient of lakes studied by Jeppesen *et al.* (2000) demonstrated the interaction between competition and predation among fish. As nutrient levels increased, numbers of piscivorous perch (*Perca fluviatilis*) declined, releasing zooplanktivorous cyprinid species from predation pressure. Therefore, numbers of cyprinids rose with increasing total phosphorus concentrations, although their mean body size declined, probably a consequence of greater competition for food. Two species—young perch and ruffe (*Gymnocephalus cernuus*)—feed on benthic invertebrates. Greater eutrophication and therefore greater turbidity favoured ruffe as, unlike perch, it is not dependent on light to forage. However, ruffe, in turn, competed with bream (*Abramis brama*) which dominated at high nutrient levels because it is able to forage in the pelagic zone as well as the benthic zone. Therefore, ruffe numbers peaked at moderate phosphate concentrations and then declined, whereas bream reached maximum abundance at high concentrations.

7.5.2 Predation

Predation is one of the primary determinants of community structure in lakes and ponds, fish in particular having a demonstrably strong effect upon numbers of their prey. A complex example of a predation effect was reported by Cryer *et al.* (1986) from Alderfen Broad, one of the Norfolk Broads in eastern England, a shallow, eutrophic lake in which the main zooplankton predator was roach (*Rutilus rutilus*). In years when fish recruitment was poor, a high biomass of large cladocerans predominated in the zooplankton but, when roach underyearlings were abundant, copepods and small rotifers dominated, cladoceran density being significantly reduced at the time they became important components of the diet of young roach; the smallest (*Bosmina longirostris*) declined first, followed by the larger species (*Ceriodaphnia quadrangula* and *Daphnia hyalina*) a few weeks later, reflecting the gradual increase in size of the young fish. Experimental exclusion of fish from parts of the lake resulted in an increase in both density and individual body size of *Daphnia*. During the 1970s and early 1980s, roach in Alderfen Broad followed a two-year cycle of abundance, which Cryer *et al.* speculated was related to their predatory influences. A year of high recruitment produced large numbers of young fish depleting the food supply, so that growth and fecundity were low;

this resulted in a low number of eggs and therefore few fish present the following year. In the absence of strong predation pressure from fish, zooplankton reached high densities, providing plenty of food for roach, which grew rapidly, leading to higher fecundity the following year and repeating the cycle. Interestingly, this pattern was maintained by another predator, pike (*Esox lucius*), which severely depleted older, and therefore larger, roach; as a result, reproduction was predominantly by roach under three years old, at which age they were feeding upon the same zooplankton resource as were underyearling fish.

Other examples of effects of fish predation abound. In Lake Kinneret (Sea of Galilee) in Israel, distribution patterns of fish and cladocerans coincide, both being more abundant on the eastern side of the lake where gradients of temperature and oxygen concentration are highest. At a smaller spatial scale, however, fish aggregation leads to localized reduction in zooplankton biomass, maximal fish densities being matched by minimal cladoceran densities. The fish in question, Kinneret sardine (*Mirogrex terraesanctae terraesanctae*) congregate in dense aggregations which form and disperse every 4–5 days, possibly as a consequence of local over-exploitation of their prey. In several of the Norfolk Broads, predation by roach and bream (*Abramus brama*) reduces densities of benthic chironomids and oligochaetes. Reduction of fish densities in Tuesday Lake, Michigan, resulted in an increase in density of larvae of phantom midges (*Chaoborus* spp.) which, in turn, reduced densities of smaller zooplankton, midge larvae having reached densities towards the end of summer at which they could potentially consume the entire rotifer population within a single day.

7.5.3 Strategies to avoid predation

Faced with such intense predation pressures as those described above, zooplankton have developed a range of strategies to reduce their

7

vulnerability to predation. These may be classified as morphological, life cycle and behavioural, although many species adopt a combination of these strategies.

Morphological strategies

Cladocerans are important components of the grazing community in many lakes. If subjected to heavy predation pressure by invertebrate predators, they develop structural features which reduce their susceptibility to predation. *Daphnia pulex* defence strategies in the face of predation by larvae of the midge *Chaoborus* (Figure 7.11a) include increased body size, growth of neck spines (Figure 7.11b, c) and development of larger neonates so that adult size can be achieved more rapidly. Defences

such as these are developed in response to chemical cues originating from predators. They make cladocerans more difficult to handle when caught, increasing their chance of escape, but development of a crest by the Australian *D. carnata*, heavily predated by notonectids, apparently increases manoeuvrability and therefore reduces the number of times they are caught in the first place. Each of these strategies, however, has an energetic cost (Dodson 1989). Neck spines reduce overall growth rates, larger neonates require larger, and therefore fewer, eggs, while increased body size, although reducing capture efficiency of *Chaoborus*, increases encounter rates and therefore the probability of being caught. Therefore, in the absence of predators, they do not develop. Adult *Daphnia* are generally too large to be eaten by *Chaoborus* and lose the defensive structures that they developed as larvae, although those that coexist with the large-bodied *Chaoborus trivittatus* in North America will retain their spines into adulthood (Reissen and Young 2005).

Defensive structures are not confined to plankton. Many dragonfly larvae develop backward-pointing abdominal spines, which make them less susceptible to fish predation. The northern European *Leucorhinia albifrons* has long abdominal spines (Figure 7.11d) and is only successfully consumed by fish if attacked from the front.

Life-cycle strategies

Cladocerans are eaten not only by various size classes of invertebrate predators but also by much larger fish, which selectively take larger individuals. The complexity of interactions between these groups is illustrated in Figure 7.12. In the absence of fish, medium-sized individuals are most at risk from predation (Figure 7.12a), so juveniles grow rapidly until they are too large to be eaten by midge larvae, after which they will reproduce. In the presence of fish, however, they will never grow too large to be eaten, but small individuals, less attractive to

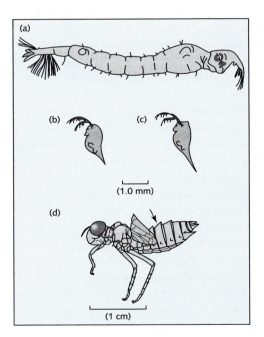

(1.0 mm)

(1 cm)

Figure 7.11 Development of structural features in response to predation. (a) Larva of the midge *Chaoborus americanus,* a major predator on grazing zooplankton. (b) Young *Daphnia pulex* grown in the absence of *Chaoborus.* (c) Young *Daphnia pulex* grown in the presence of *Chaoborus*; note the extended tail spine and the tooth on the back of the neck. From Dodson (1989). (d) Larval damselfly *Leucorhinia albifrons,* showing abdominal spines. From Mikolajewski and Johansson (2004).

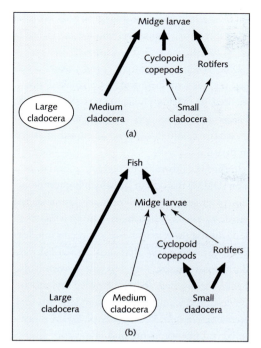

Figure 7.12 Effect of predation on size of dominant cladocerans. The width of each arrow is proportional to the intensity of predation and the size range least impacted by predation is encircled. See text for more details.

fish, are more likely to survive **(Figure 7.12b)**. As **Figure 7.12** demonstrates, this is not simply an interaction between fish and cladocerans, but also incorporates smaller predators, which feed on very small cladocerans but, in turn, are eaten by midge larvae. Abundant midge larvae keep numbers of these small predators down, reducing predation pressure on very small cladocerans, so the optimal reproductive strategy adopted by cladocerans in the absence of fish is to produce many small offspring. Where fish are present, however, cladocerans concentrate on producing relatively large offspring to reduce the risk of predation by small predators (Lynch 1980).

Behavioural adaptations

Fish are predominantly visual predators, whose foraging is effectively confined to the photic zone during the hours of daylight. Therefore, invertebrates which are susceptible to fish predation adopt strategies that reduce the probability that they will be caught by visual cues.

In the open water, diel vertical migration (DVM) is common, with zooplankters descending at dawn and rising to feed at dusk. This, in turn, may induce their prey species, which are too small to be at risk from fish, to adopt the opposite strategy. Fish, too, will migrate if they are subject to predation by piscivorous fish. Juvenile sockeye salmon (*Oncorhynchus nerka*) in western North America spend the night 20–30 m below the surface but descend during the day to depths of 80–120 m. In Babine Lake, British Columbia, this induces a complex pattern of vertical migration: sockeye eat the cladoceran *Daphnia longispina* but also the larger calanoid copepod *Heterocope septentrionalis*. *Daphnia* does not migrate but *Heterocope* avoids salmon predation by descending at night. This, in turn, induces the small cladoceran *Bosmina* to rise at night, to avoid *Heterocope*, its main predator.

In European lakes the copepod *Eudiaptomus gracilis* descends deeper than *E. graciloides*, with which it coexists, as it is larger bodied and therefore more likely to be targeted by fish predators. Furthermore, females tend to descend more deeply than males, as those with egg sacs are larger and more conspicuous than males. Both of these species are, however, smaller than *Daphnia* and under intense predation pressure it will migrate beneath *Daphnia*, which fish preferentially select, thereby using them as a 'living shield'; in the absence of fish predation *Eudiaptomus* will remain above *Daphnia* (Jamieson 2005).

Avoiding predators is not the only adaptive value of vertical migration. For example, descending to the cold hypolimnion for part of the day reduces metabolic activity and therefore expenditure of energy, so less food is required (Lampert 1989). However, Neill (1990) showed that, when *Chaoborus* was eliminated by trout from Gwendoline Lake in British Columbia, the calanoid copepod

7

Diaptomus kenai ceased to migrate vertically, but remained in the upper layers, with the phytoplankton upon which it fed, throughout the day. Three years after the removal of *Chaoborus*, its reintroduction induced reinstatement of vertical migration by *Diaptomus* within four hours. Interestingly, *Chaoborus* itself will not cease to migrate if fish are removed, although the intensity of migration will be significantly reduced.

Occasionally food supply may be greatest at depth, inducing a reversed migration. Clear oligotrophic lakes occasionally have maximum algal concentrations below the thermocline, where light penetration is adequate for some photosynthesis and nutrient availability is relatively good. In Oberer Arosasee, a Swiss alpine lake with a deep chlorophyll maximum, juvenile growth rates of *Daphnia galeata* are higher in the deeper water where food supply is greatest, and yet the animals migrate at night towards the surface, where food supply is poor **(Figure 7.13)**. The reason for this migration is unclear; oxygen concentrations and predation pressure are apparently similar,

so *D. galeata* may migrate upwards to take advantage of higher temperatures (Winder *et al.* 2003).

Diel vertical migration, often over much greater vertical distances, is also well documented in marine environments (Section 6.8, **Figure 6.5**). The two main hypotheses proposed to explain its occurrence are summarized in **Box 6.6**.

In the lake littoral zone diel horizontal migration (DHM) may occur, in which highest densities of grazing zooplankters are found in the vicinity of macrophyte beds, among which they hide during the day to avoid fish predation, foraging in open water after dark, when fish predation is reduced. Fish fry will however also benefit from the shelter afforded by macrophytes, and can be very abundant in such environments. As a consequence, DHM is not a universal phenomenon, and laboratory studies on subtropical *Daphnia obtusa* in Uruguay demonstrated that it actively avoided macrophytes, and that avoidance behaviour became more pronounced in the presence of alarm signals from predated conspecifics; avoidance of macrophytes was not simply stimulated by plants, but also by chemical cues derived from plants, whereas cues from a zooplanktivorous fish—*Cnesterodon decemmaculatus*—had no effect on their behaviour (Meerhoff *et al.* 2006).

7.5.4 The trophic cascade

The example of cladoceran response to changing predatory pressures illustrates the way in which a top predator, in this case the fish, can have a direct effect on two trophic levels by consuming both cladocerans and their main invertebrate predators. The fish also have an indirect effect, however: although they do not eat rotifers, they can modify their abundance indirectly by consuming midge larvae. Indirect effects such as these can be translated indirectly throughout the entire food web, in a process known as the trophic cascade (Carpenter *et al.* 1985). It is best illustrated by using an example in which two trophic levels are increased to

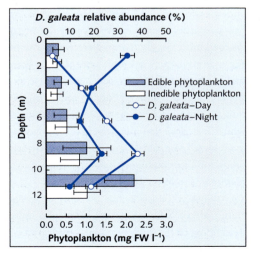

Figure 7.13 Vertical distribution of adult *Daphnia galeata* and edible and inedible phytoplankton biomass in the Oberer Arosasee, Switzerland, during summer 1998. Error bars indicate standard errors. FW, fresh weight. From Winder *et al.* (2003).

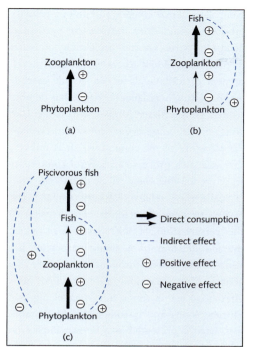

Figure 7.14 The trophic cascade, showing changes as two trophic levels are increased to four. The width of each arrow is proportional to the intensity of predation and positive and negative effects of each interaction are marked. See text for more details.

7.5.5 Species diversity

Dispersal abilities

Dispersal abilities of lake species can be contrasted with those inhabiting rivers (Section 3.7). The relatively short existence in geological time of most lakes (Section 7.6) ensures that their component species are well-adapted for dispersal. Small planktonic species, including algae, bacteria, fungi, rotifers and cladocerans, have resting stages such as spores and cysts which are both highly tolerant to desiccation and small and light enough to be wind dispersed. These species are widely distributed, many of them being cosmopolitan and limited in their distribution by water quality at a given site rather than by geographical barriers.

In contrast to river species, many of the dominant insects in lentic waters derive mainly from groups with strong flying abilities, such as dragonflies and water beetles, which have little difficulty in dispersing to new sites. Dragonflies and aquatic birds such as ducks carry with them algae and small zooplankton, contributing to the dispersal of these taxa. Other lentic specialists, such as amphibians, have a terrestrial phase in their life cycle during which they can walk to a new site.

Lentic water bodies which require the most advanced adaptations are those which are temporary in nature. Species occupying temporary or intermittent lakes and ponds require three main adaptations: stages in their life cycle that are resistant to desiccation, such as terrestrial adults or the ability to aestivate during the dry season; good powers of dispersal; and the ability to be adaptable in their use of the resources available. Organisms which possess these adaptations benefit from their ability to exploit an environment in which resources may be plentiful but specialist predators absent. Anuran amphibians (frogs and toads) are typical of temporary ponds; they are generally excluded from permanent water bodies by fish predation upon the tadpole stage of their life

four **(Figure 7.14)**. In **Figure 7.14a**, phytoplankton biomass is kept in check by zooplankton grazing; zooplankton, in turn, are limited by food availability. Addition of a planktivorous fish **(Figure 7.14b)** reduces the biomass of zooplankton, which are now limited by predation; the fish have had a direct negative effect on biomass of zooplankton which therefore leads to an indirect positive effect on phytoplankton biomass, as grazing pressure is reduced. Addition of piscivorous fish **(Figure 7.14c)** reduces biomass of planktivorous fish but, indirectly, releases zooplankton from predation pressure and allows them to increase in numbers, thus having a negative effect on phytoplankton biomass. Addition of a new trophic level in each case has a direct negative effect on the one immediately below it but, at the same time, the effect cascades down the food chain, with alternating positive and negative effects.

cycle but, to counter this, possess a highly mobile and essentially terrestrial adult stage.

Endemism and ancient lakes

The well developed dispersal abilities described above ensure that isolated water bodies are rapidly colonized. Geographically close lakes will share a common pool of species, and endemism is rare, because few lakes are stable enough for the geological timescales that speciation requires, and continual dispersal of individuals reduces the potential for genetic isolation.

It is for this reason that the ancient lakes of the world are special. Their age has allowed speciation to occur among groups with poor dispersal abilities, the most famous examples being the gammarid shrimps in Lake Baikal (around 1000 species) and the cichlid fishes of the East African Great Lakes (600 species in Lake Malawi alone, plus the haplochromines of Lake Victoria discussed in Section 7.7). For endemic species to develop in any numbers, however, stability is also required. Lake Kivu in Rwanda and the Democratic Republic of Congo is one of the world's oldest lakes, but has a low species richness and an immature plankton community because of volcanic activity within the last 10,000 years which literally cooked the lake and destroyed all life within it; presumably before this event it supported a very diverse assemblage of species, including many endemic forms, but recolonization has been by widespread, opportunistic species and those tolerant of conditions in rivers which were able to find refuge in its catchment during the volcanic event.

7.5.6 Ecological functioning

The nature of a lake or pond, as an enclosed body of water, ensures that it functions essentially as a single community. Traditionally, community structure in lakes was assumed to be a function of nutrient availability, the primary force acting upon it being 'bottom-up' control, in which competition among primary producers, limited by nutrients in particular, determines structure and productivity of higher trophic levels. More recently, however, development of ideas such as the trophic cascade and predation avoidance strategies has demonstrated an alternative, 'top-down' control, in which predation by tertiary consumers controls the structure of the community. It is unlikely that one such process can control community structure without some effects of the other, but certainly their relative importance can vary, as can be demonstrated by consideration of the perturbation of lake systems in which the modification of a single parameter has wide-ranging effects throughout the entire lake system. Two examples are presented here, alternative stable states following changes in nutrient loads and introduction of predatory fish. Each is presented as a human impact on lake ecosystems, but they illustrate the role of organisms in regulating community structure and production.

Alternative stable states

Typically, eutrophication will lead initially to increased densities of macrophytes and algae, but then phytoplankton crops become very large and submerged macrophytes are lost. This is not a straightforward case of algae outcompeting submerged plants, however, because there is a range of possible causes, none of them complete explanations. The most obvious is that floating algae shade submerged plants, but plants such as water lilies (*Nymphaea* spp. and *Nuphar* spp.), which start their growth when algal densities are low and grow to the surface, and floating plants such as *Lemna minor*, which overwinter at or near the surface or among debris at lake edges, also disappear. Other potential reasons relate to increased sediment instability, or even relative shortage of nitrate or carbon dioxide, all of which would potentially favour phytoplankton over macrophytes. Balls *et al.* (1989), using a series of experimental ponds in the Norfolk Broads

region of east England, demonstrated clearly that it was not simply a case of increased nutrient loading: fertilization to a level above that experienced by nearby lakes from which macrophytes had been eliminated resulted in no displacement of aquatic plants, and phosphate levels in the water did not increase, despite the inputs. Irvine *et al.* (1989) suggested that a hypertrophic lake will be dominated by one of two alternative states—aquatic plants or phytoplankton. Each is buffered against the other: macrophytes by absorbing extra nutrients and by suppressing the wave action that would otherwise re-suspend bottom sediments and lead to shading, algae by adding to the turbidity that shades potential vegetation growth. Although eutrophic lakes are generally turbid relative to their oligotrophic counterparts, macrophyte coverage exceeding 30% of the total surface area will usually maintain high water transparency, irrespective of nutrient concentration. This is because the turbidity is caused by the presence of algal cells, whose densities become so great in extreme cases that the water takes on a thick, soupy appearance.

The maintenance of either a macrophyte or a phytoplankton system is affected by secondary consumers, whose activities will serve to buffer whichever system is in place. In macrophyte-dominated systems, grazing pressure on algae is intense as cladoceran grazers, using macrophytes as predation refugia, are abundant, whereas in algal systems fish will keep numbers of grazers low, allowing algal numbers to remain high.

From a management point of view, the macrophyte system is usually more desirable than the algal system. Macrophytes disrupt access for boat movements, fishing, etc., but stabilize nutrients by retaining them for long periods within the plants' tissues; in temperate regions, an annual nutrient cycle is set up, incorporating absorption into plant tissues during the spring and then slow release during winter decomposition. Algae, with their short life cycles, cycle nutrients rapidly, leading to a succession of blooms. They can also respond very

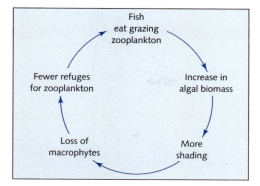

Figure 7.15 Eutrophication—the vicious circle which maintains high phytoplankton biomass to the detriment of benthic macrophytes.

rapidly to increases in nutrient supply and, in temperate regions, constant death and decomposition of algal cells can exacerbate problems of oxygen availability during the warm summer months, whereas macrophyte decomposition is mainly confined to the winter, when rates of decomposition are slow. Algal systems may become dominated by toxic Cyanobacteria, which have been known to kill livestock that drink contaminated water.

Unfortunately, the macrophyte system is easier to destroy and replace with the algal system than vice versa. Any major perturbation to macrophytes, such as chemical weedkillers, mechanical damage by boats or excessive grazing, will open up the still nutrient-rich water and allow algae, which can respond very quickly, to move in. Once their numbers increase, turbidity is raised and the opportunities for recolonization by macrophytes are reduced (**Figure 7.15**). This is well illustrated by Lake Apopka in Florida. Until the 1940s, this shallow lake, with a surface area of 125 km^2 but a mean depth of less than 2 m, was exceptionally clear, with a lush growth of macrophytes. In 1947, a hurricane uprooted macrophytes, causing massive mortality and, through their decomposition, releasing nutrients into the water. Within one week of the hurricane, the first algal bloom was reported, and phytoplankton quickly became the dominant primary producers (Carrick *et al.* 1993). When

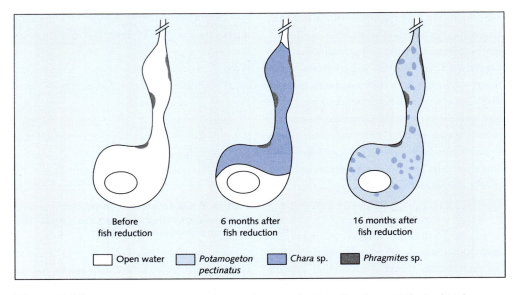

Before
fish reduction

6 months after
fish reduction

16 months after
fish reduction

☐ Open water ☐ *Potamogeton pectinatus* ☐ *Chara* sp. ■ *Phragmites* sp.

Figure 7.16 The changing distribution of macrophytes in Lake Bleiswijkse Zoom (Netherlands) in response to fish removal. From Meijer *et al.* (1990).

7

such a transformation occurs, restoration of the macrophyte system will almost certainly require the removal of planktivorous fish, or severe reduction in their densities, in order to release grazers from predation and allow them to attain high densities. This may be extremely difficult to achieve but, if successful, its effects upon macrophytes can be remarkable (Figure 7.16).

Introduction of fish

The trophic cascade demonstrates the top-down control by top predators of lake ecosystems. That top predators can have more fundamental effects, completely restructuring food webs, rather than simply altering the relative abundance of other species, has been documented many times through accidental or deliberate introduction of predatory fish to lakes, of which the example of Gatun Lake in Panama is one of the most celebrated. Prior to 1967, Gatun Lake supported a complex food web with eleven common species of fish. Introduction of the piscivorous cichlid *Cichla ocellaris*, a native of the Amazon basin, simplified the food web immensely, predation by this species eliminating seven species and redu-

cing numbers of three others. Only one species, the planktivorous cichlid *Cichlasoma maculicauda*, increased in abundance, probably because *Cichla* eliminated species which had formerly fed upon its fry. Of particular significance was removal of the atherinid *Melaniris chagresi*, formerly an important component of the diet of seasonally present consumers—Atlantic tarpon (*Tarpon atlanticus*) and black tern (*Chlidonias niger*)—both of which disappeared from areas where *Cichla* was present. Black tern was not the only bird affected, kingfishers and herons also being reduced in abundance by the arrival of *Cichla* (Zaret and Paine 1973).

In contrast to the experience of Gatun Lake, an apparent increase in species diversity was recorded from Lake Nakuru in Kenya following the introduction of the tilapine *Oreochromis alcalicus grahami* to the previously fishless saline lake. This species quickly became abundant, feeding on the cyanobacterium *Spirulina platensis*, but had no observed detrimental effect upon other components of the fauna because *Spirulina* was so abundant and prolific that total consumption by the fish was negligible. What did result, however, was an extension of the food chain in this lake to fish-eating birds, particularly great white pelican (*Pelecanus onocrotalus*), but including more than fifty species in total, whereas, prior to fish introduction in the 1950s, lesser flamingo (*Phoeniconaias minor*), itself a consumer of *Spirulina*, had been the only abundant bird species (Lévêque 1995).

7.6 Succession

Lake formation can be very rapid. As soon as a basin is created in the presence of a supply of water, it starts to fill with water (Box 7.1). The creation of reservoirs by artificially damming valleys demonstrates the speed at which this can occur, most filling within a few months of dam closure. One of the most rapid was Lake Kainji on the River Niger in Nigeria, which covers a maximum area of 1280 km^2 and has a volume of up to 15.7 km^3, yet reached its full size just 2.5 months after dam closure in 1968, but even the enormous Lake Volta, the largest man-made lake in Africa, reached its full extent of 8845 km^2 and volume of around 160 km^3 within five years. The Salton Sea in an arid region of California was created by accidental diversion of water from the Colorado River, via irrigation canals, into a desert depression. The leak was repaired within two years, but not before a lake covering 891 km^2 and with a volume of 7.8 km^3 had been created. Natural lakes can be assumed to fill with this, or perhaps even greater, rapidity. Reelfoot Lake, on the Kentucky–Tennessee border, is reputed to have come into being almost overnight after a series of earthquakes in 1811 created depressions into which the Mississippi River spilled.

The fate of all lakes is the same: as soon as formation is complete, they start to fill in. This is because, being concave depressions and inputted by rivers carrying sediment, they are natural zones of deposition: a river entering a lake loses its momentum, stops flowing and deposits its load. Even ponds, without river inputs, will accumulate sediment, whether wind blown or transported by biological agents. An inevitable consequence of this is eutrophication (Box 7.3), which will stimulate productivity and, in turn, accelerate the further build up of organic matter.

Lakes are relatively barren at first, with few animals and plants and limited sediment and nutrients. Initially, therefore, productivity is low, most nutrients coming from the catchment. Lack of sediment and nutrients keep macrophyte growth low and plankton losses may match or even exceed production if there is a large outlet. This is a relative condition: a pool formed by a beaver dam over rich forest soil will be sediment-rich compared to one formed in a glacial hollow, but typically the nutrient load of a lake will increase over time. Shallow lakes with a large littoral zone will, however, become highly productive once nutrient balances are established, and will rapidly acquire organic matter. Even large lakes, once production becomes high relative to losses, rapidly infill and will eventually become wetlands, a process explored further in Section 8.6.

The process of succession to a wetland can be surprisingly rapid: the East Anglian Fens, for example, had lost most of their open water a few thousand years after formation. Deeper, cleaner lakes persist much longer, but even the largest may persist only for tens or hundreds of thousands of years; the majority of lakes in northern Europe and North America, however large and deep, were formed within the last 20,000 years, with the retreat of glacial ice at the end of the Pleistocene. The size and solidarity of their basins makes bedrock lakes the longest lived and most stable lakes, although volcanic lakes, if formed within the crater of an active volcano, will be destroyed very rapidly in the event of an eruption. The oldest and deepest lakes in the world are those in tectonically active areas, filling the depression made as plate boundaries slip relative to each other. These include Lake Baikal and Lake Tanganyika, both around 20 million years old, which are maintained by the constant subsidence of the

basins they fill. Lake Baikal, for example, overlies sediment 8 km deep, enough to have filled it in several times over.

7.7 Complex interactions in lakes

Processes such as competition, predation and nutrient inputs have so far been presented in relative isolation, but they interact within a single lake, such that it is difficult to disentangle their individual impacts. A parallel problem relates to management of impacted lakes, as multiple impacts may be occurring simultaneously. Large size does not make a lake immune from such impacts as, however large, a lake functions as a single ecosystem. This is illustrated below by the example of Lake Victoria.

7.7.1 Lake Victoria—an example of complex perturbation

Lake Victoria provides an example of a complex interaction between alien species and eutrophication in modifying community structure, and demonstrates that even a water body of this size—at 68,000 km² the world's second largest freshwater lake in terms of surface area—can be completely modified by human activity. Lake Victoria has suffered from eutrophication, overfishing and the introduction of exotic aliens, all of which have contributed, often synergistically, to its current state. Between the 1920s and the 2000s Lake Victoria changed from a mesotrophic to a eutrophic state, with algal biomass increasing up to tenfold, primary production doubling and algal composition shifting from diatom to cyanobacteria domination (Balirwa *et al*. 2003). Initially these changes were gradual, but during the period 1961–1964, heavy rain brought exceptionally high lake levels (see Section 7.2), which flooded adjacent papyrus swamps, inputting much organic matter and

accelerating the eutrophication of the water which had been occurring gradually for the previous 40 years. At the same time as these high water levels, several exotic fish species were introduced to replace overfished native species, including Nile perch (*Lates niloticus*), a large predatory species, typically 35–50 kg in weight when full grown. *Lates* was first officially recorded in Lake Victoria in 1960, although it may have been present since 1954, so further introductions were made in Uganda in 1962. Its increase was initially very gradual, but during the late 1970s and early 1980s its population exploded and it spread throughout the lake. Lake Victoria formerly supported over 300 species of haplochromine fish, all of them endemic, of which around 200 disappeared or became severely endangered following the irruption of *Lates*. The pelagic zone of the lake formerly supported a large fish biomass (more than 200 g ha⁻¹), of which 40% was accounted for by grazing or detritivorous haplochromines and a further 40% by zooplanktivorous or piscivorous haplochromines; by the late 1980s fish biomass had dropped to around 100 g ha⁻¹, of which 90% was *Lates*. Haplochromine detritivores and grazers were replaced by a crustacean —the atyid prawn *Caridina nilotica*—and zooplanktivorous haplochromines by the native cyprinid *Rastineobola argentea*. *Lates* was feeding upon *Caridina*, *Rastineobola*, dragonfly nymphs and juveniles of its own species **(Figure 7.17)**.

It is tempting to see the situation in Lake Victoria as a simple case of a dominant predator restructuring the community, but *Lates* persisted at a low density for almost 20 years before suddenly irrupting to dominate the ecosystem. A possible explanation relates to eutrophication and consequent reduction in oxygen levels at depth (Section 7.2). The oxycline is typically around 20–30 m deep in the open lake, and can rise to as little as 5 m depth close to the shore, leading to fish kills and enforced migration of fish inshore where they are more susceptible to netting. Juvenile *Lates* may initially have been kept in check by large

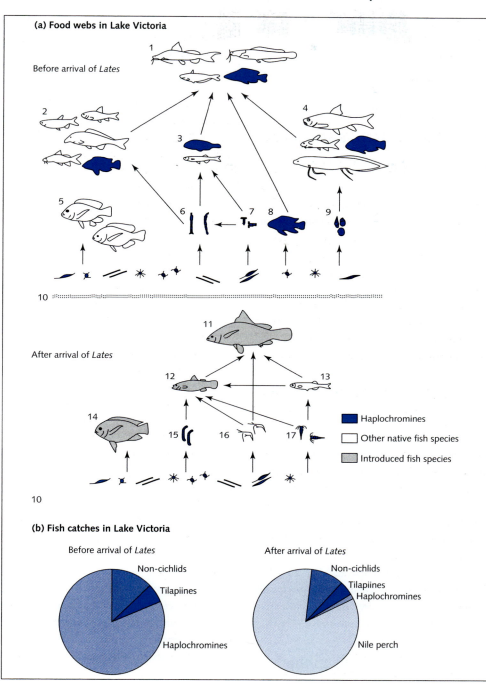

Figure 7.17 The effect of introduction of Nile perch (*Lates niloticus*) on the ecosystem of Lake Victoria. (a) Food web before and after establishment of *Lates*. (1) piscivorous fish; (2) insectivorous fish; (3) zooplanktivorous fish (including *Rastrineobola argentea*); (4) molluscivorous fish; (5) algivorous fish (native species of *Oreochromis*); (6) planktonic insect larvae; (7) planktonic crustaceans; (8) detritivorous and planktivorous fish; (9) molluscs; (10) phytoplankton and detritus; (11) adult *Lates*; (12) juvenile *Lates*; (13) *Rastrineobola argentea*; (14) *Oreochromis niloticus*; (15) chironomids; (16) *Caridina nilotica*. From Witte *et al.* (1992). (b) Percentage composition of fish catches in the Nyanza Gulf, Kenya, before and after establishment of *Lates*. From Lowe-McConnell (1994).

piscivorous haplochromines, themselves the prey of adult *Lates*. The reduction of haplochromines through overfishing may have released *Lates* from predation until the adult population was high enough for it, in turn, to become the dominant predator. Once common, *Lates* consumed large numbers of grazers and detritivores which had formerly kept algal biomass in check; their loss led to more algal detritus sinking into the hypolimnion, reducing oxygen levels. Eutrophication had started— and accelerated—before the irruption of *Lates*, as a consequence of land use changes, but its effects, in terms of algal blooms and deoxygenation, may have been kept in check by the large numbers of native haplochromines until *Lates* became well enough established to reduce their numbers. Alternatively, increasing deoxygenation of bottom waters may have forced benthic species inshore, providing more food for *Lates* and enhancing its population growth. Confounding the situation was the replacement of zooplanktivorous haplochromines by *Rastineobola*, a species whose population has been enhanced by the arrival of *Lates*. *Rastineobola* predation upon zooplankton is more intense than that of the species it replaced, so densities of larger zooplankters declined, with a corresponding reduction of grazing pressure on algae. Primary production was, therefore, much greater than grazing capacity and excess production descended into the hypolimnion. *Caridina*, a species tolerant of very low oxygen concentrations (less than 1 mg l^{-1}), proliferated as a consequence, as the practically anoxic hypolimnion provided it with a predation refuge from *Lates* (Gophen *et al.* 1995).

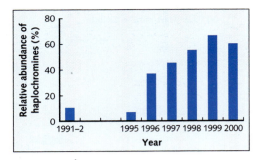

Figure 7.18 Evidence for the haplochromine resurgence. Numbers of haplochromines as a percentage of the total catch along experimental transects in Lake Nabugabo, a satellite of Lake Victoria in Uganda. From Chapman *et al.* (2003).

7.7.2 Recent changes to the Lake Victoria ecosystem

Since the early 1990s, several changes have occurred to the Lake Victoria ecosystem. Some of the haplochromines have started to reappear in large numbers (the 'haplochromine resurgence'), including several previously thought to be extinct. Those species persisted in refugia,

including satellite lakes and wetlands, rocky shores, rocky offshore islands and deeper waters close to the oxycline. Evidence for the haplochromine resurgence comes not only from increased catches (Figure 7.18), but also from analysis of the stomach contents of *Lates* and of other piscivorous fish species, whose diets are returning to haplochromines after a period when they were dominated by other food sources, particularly invertebrates (Table 7.2). Their recovery is probably a consequence of overfishing for *Lates*, most of whose catch is now immature individuals. Reappearance of haplochromines is beneficial to *Lates*, which grow better on a fish diet than one dominated by invertebrates. Effective haplochromine conservation may, therefore, help to maximize *Lates* production and create a sustainable fishery (Balirwa *et al.* 2003).

The lake ecosystem will never return to its original structure. Some species are probably completely extinct, the introduced species will not be eliminated and the haplochromines that persist are those that are tolerant of eutrophication, deoxygenation and predation by *Lates*. They are mainly representatives of zooplanktivore and detritivore feeding groups, the other feeding groups having shown little recovery in numbers. Furthermore, the species that persist are probably hybrids of the original forms: at such an early stage of speciation, fertile hybrids

Table 7.2 Percentage frequency of occurrence of major prey types in the stomachs of the catfish *Bagrus docmac* captured in Lake Victoria in 1958, 1991–92 and 1997

Prey type	Percentage of fish stomachs in which it occurred		
	1958	1991–92	1997
Fish			
Haplochromines	57.2	7.7	30.0
Lates niloticus	0	18.0	0
Rastineobola argentea	1.4	41.0	2.9
Tilapiines	1.9	0	4.3
Clarias spp.	0	0	7.1
Fish remains	–	30.8	20.0
Insects			
Mayfly: *Povilla adusta* (larvae)	7.5	15.4	42.9
Chironomidae (larvae)	3.8	12.8	11.4
Chironomidae (pupae)	0	12.8	15.7
Trichoptera (larvae)	0.7	10.3	11.4
Odonata (larvae)	9.3	61.5	1.4
Others			
Shrimp: *Caradina nilotica*	4.6	20.5	32.9
Molluscs	4.6	25.6	61.4
Crabs	1.3	0	17.1
Other invertebrate remains	0	0	20.0

Adapted from Olowo and Chapman (1999).

are possible among many species flocks and the characteristics used by females to distinguish males of different species are primarily visual; however, in the increasingly turbid waters of Lake Victoria, these visual cues break down and interbreeding occurs (Seehausen *et al.* 1997).

Another recent change to the Lake Victoria ecosystem is the appearance of water hyacinth (*Eichhornia crassipes*), an aggressively invasive floating plant originally from South America. First recorded in the late 1980s, it quickly dominated large areas of the coast, and then disappeared almost completely in the late 1990s, as a result of natural processes combined with vigorous control programmes. Despite the many problems associated with water hyacinth, it was beneficial to several fish species that used its physical structure as a predation refugium; these included lungfish (*Protopterus aethiopicus*) and the catfish *Charias gariepinus*, two native non-haplochromine species which had formerly been important in the Lake Victoria fishery and whole catches rose again significantly during the period of water hyacinth explosion (Njiru *et al.* 2005).

There is a fourth human impact on the Lake Victorian ecosystem: the almost complete elimination during the mid twentieth century of crocodile (*Crocodylus niloticus*). The effect upon fish assemblages of the loss of this species is unknown, but it is interesting to speculate upon the outcome of *Lates* introduction if this major piscivore had still been present.

Bulleted summary

- Lakes are created in depressions in the Earth's surface within which water collects. An outflow river will dampen large fluctuations in volume, but saline lakes, which generally lack an outflow, can fluctuate widely in surface and volume, in some cases disappearing altogether.

- Physically, lakes share much in common with the open sea, undergoing thermal stratification, often seasonally. Lack of mixing between the surface layer (epilimnion) and deep layer (hypolimnion) can result in oxygen depletion at depth; deep lakes are, with the exception of Lake Baikal, permanently anoxic at depth. Horizontal variations in water chemistry are less marked than vertical variations.

- Primary production in deep lakes is dominated by phytoplankton, which are subject to the same constraints of light penetration, nutrient status, mixing and grazing as occur in the sea. Primary production in temperate zones is seasonal, limited by temperature and light in the winter and by nutrient depletion and grazing in the summer. In the tropics, seasonal variations in hydrology affect primary production. In shallow lakes, primary production by benthic macrophytes is important, although eutrophication may lead to dominance by phytoplankton.

- Predation is the dominant biotic interaction in lakes. Strategies to avoid predation include development of structural defences, life-cycle strategies and avoidance behaviour. Predatory fish are often key components of lake ecosystems.

- Endemism is rare in lakes, as most species have effective dispersal abilities, in response to the geologically short lifespans of lakes, and the temporary nature of some lentic water bodies. Ancient lakes, in contrast, have high levels of endemism.

- Lakes act as sinks for sediment and will inevitably infill over time. The successional process involves gradual eutrophication.

- Anthropogenic eutrophication and introduction of alien fish species demonstrate that lakes, however large, act as single ecological units.

Further reading

...ark C. and Hanssen L.-A. (2005) *The biology ...kes and ponds*, 2nd Edn. Oxford, Oxford ...ity Press.

...d Morris P. (2007) *The world of lakes.* ...reshwater Biological Association.

...(1988) *Complex interactions in* ...New York, Springer-Verlag.

...and Coulter G. (eds) (1994) ...kes*. Stuttgart, Archiv für ...e der Limnologie.

O'Sullivan P.E. and Reynolds C.S. (eds) (2004) *The lakes handbook volume 1. Limnology and limnetic ecology.* Oxford, Blackwell Science.

Reynolds C.S. (2006) *Ecology of phytoplankton.* Cambridge, Cambridge University Press.

Talling J.F. and Lemoalle J. (1998) *Ecological dynamics of tropical inland waters.* Cambridge, Cambridge University Press.

Assessment questions

1 How does stratification influence biotic processes in lakes?

2 How do changes in nutrient status impact animal species in the pelagic zone of lakes?

3 What are the main morphological and behavioural adaptations employed by zooplankton to reduce predation pressure?

4 What are the main seasonal changes that occur in temperate lakes?

5 What general lessons about lake management can be learned from the example of Lake Victoria?

7

8 Wetlands

8.1 Introduction

A wetland is here defined as an area of land whose characteristics are determined by the presence of water, either permanent waterlogging or through regular, usually seasonal, flooding. If permanently inundated, this is to a depth shallow enough to allow the growth of emergent vegetation or rooted macrophytes. Open water bodies within the vegetated area, such as pools and oxbow lakes, are intimately connected to the wetland proper, and therefore considered to be part of the wetland ecosystem. Within this broad definition are many options for subdivision, reflecting the diverse nature and modes of origin of the world's wetlands, but all share important ecological similarities as a result of the high water table, hence their consideration together here.

There have been many attempts to classify wetlands, based on such features as morphology, source of water, hydrology, nutrient concentrations and vegetation characteristics. This complexity of classifications reflects the complexity of wetland types throughout the world. They do, however, fall into two fundamental types, for which the terms aquatic marginal and mire are here used.

Aquatic marginal wetlands

Aquatic marginal wetlands are created by open water bodies—rivers, lakes and the sea—which act as the source of water (Figure 8.1a). They can be conveniently subdivided into fringe wetlands and flood wetlands. Fringe wetlands have a continuous or very frequent hydrological connection with the parent water body (Figure 8.2a) and are typical of the shallow edges of lakes and slow-flowing rivers. Flood wetlands are, for most of the time, hydrologically separated from the parent water body, becoming connected only during high water (Figure 8.2b). Typical examples are river floodplain marshes, inundated frequently or for long periods during the wet season. Coastal saltmarshes incorporate both wetland types: areas subjected to daily inundation ('low marsh')

(a)

8

(b)

Figure 8.1 Examples of different wetland types. (a) Aquatic marginal wetlands: a fringe wetland along the edge of a river (foreground); behind it is a flood wetland receiving its seasonal input of water from the river: Danube Delta, Romania. (b) Mires: a fen, sitting in natural depression in the ground and fed by groundwater; Knockbane, Northern Ireland. Photos M. Dobson.

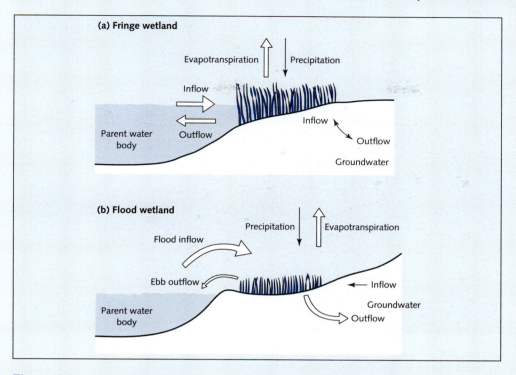

Figure 8.2 Inputs and outputs of water in (a) a fringe wetland and (b) a flood wetland. The width of each arrow is proportional to the volume of water following that route.

are flooded with a frequency and regularity adequate to describe them as fringe wetlands, whereas infrequently flooded areas ('high marsh') are flood wetlands (Figure 8.3). During the interval between flood events, flood wetlands will lose their water through drainage and evapotranspiration, and may, in some cases, dry out completely. The terms 'riverine wetland' and 'riparian wetland' have been applied to flood wetlands, as here defined, but these terms imply application only to freshwater wetlands associated with rivers, overlooking the important functional and ecological similarities these systems share with infrequently flooded coastal systems.

Fringe wetlands are not generally differentiated according to water quality, but occasionally such a separation can be useful. The flooded forests of the Amazon, covering up to 60,000 km², have been divided into the várzea and the igapó. The várzea receives 'white water'—turbid, nutrient- and sediment-rich waters from eroding head-waters; the igapó receives nutrient-poor water from the lower Amazon basin, either 'black water' if containing high concentrations of humic matter, or 'clear water' if humic concentrations are low. The várzea and igapó can be further subdivided, depending upon whether inundation is seasonal, daily (tidal) or irregular.

Mires

Mires persist independently of a parent water body, being fed by groundwater, overland run-off or precipitation. They, too, can be divided into two types, minerotrophic (or rheotrophic) mires and ombrotrophic mires, often referred to as fens and bogs, respectively. Fens are valley mires, receiving water and nutrients from groundwater and run-off from the surrounding catchment. They are generally nutrient-rich, unless fed by very pure groundwater (Figures 8.1b, 8.4a). Bogs, in contrast, occur above the water table and

Figure 8.3 A typical high marsh, densely vegetated with salt-tolerant plants. The channel is flooded with salt water during high tide: Mawdach Estuary, Wales. Photo M. Dobson.

8

are fed solely by rainwater and aerial deposition, so are nutrient-poor **(Figure 8.4b)**. Both bogs and fens are normally permanently waterlogged, ensuring reduced rates of decomposition and allowing partially decayed plant matter to accumulate in the form of peat; they are thus sometimes collectively known as peatlands. Bogs, which have acidic surface water, generally support vegetation characterized by mosses of the genus *Sphagnum*; if the pH rises above 4.5, fen vegetation, characterized by sedges (*Carex*) and woody plants, will take over.

Transitional wetlands

As with all aquatic systems, rigid definitions lead to difficulties with habitats which are clearly transitional. An area of floodplain may be fed by groundwater during the dry season, thereby remaining permanently waterlogged and taking on the characteristics of a fen. Whether it is then classified as a fen or a flood wetland, or simply a hybrid between the two, is a matter of subjectivity,

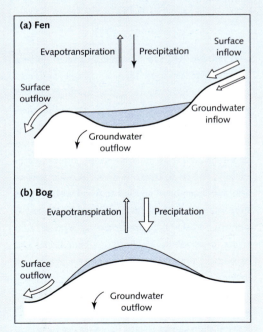

Figure 8.4 Inputs and outputs of water in (a) a fen and (b) a bog. The width of each arrow is proportional to the volume of water following that route. Note that, if the bog is growing, it will retain water and inputs will exceed outputs.

although if its soil is dominated by peat it is probably a fen which occasionally is flooded by a river, whereas if dominated by mineral soils, such as riverine silts, it is a flood wetland with fen-like characteristics. Similarly, as a fen forms over the site of a lake or pond, it passes through a transitional stage, and a distinction must be drawn between a small lake which supports a large fringing wetland, and an open pool formed in a hollow within a mire (Figure 8.8).

Swamp and marsh

This classification of wetland types is one of many which have been devised in an attempt to categorize this very diverse group of habitats. A problem encountered with many classifications is the use of terms that have different meanings in different parts of the world, along with others that are very geographically restricted in their use. We deliberately avoid using the terms 'swamp' and 'marsh' in a general sense, as both have been employed in several, rarely compatible, ways. Fojt (1994), for example, defines swamps as wetlands where the water table is normally above ground level and marshes as wetlands where it is normally at or just below ground level, whereas the normal use of 'swamp' in the United States is to refer to a wetland dominated by trees, while 'marsh' is reserved for wetlands dominated by non-woody vegetation. These terms are, however, valuable as accepted names for well known wetland types, such as saltmarsh in coastal areas, or reedswamp to describe single species stands of wetland grasses such as common reed (*Phragmites australis*).

8.2 The abiotic environment

8.2.1 The water table and oxygen availability

The abiotic environment of wetlands is dominated by the position of the water table relative to the substratum, and whether waterlogging or flooding is a permanent or a seasonal feature. The water table, in turn, determines the concentration of oxygen in the soil. Whereas well-drained soil contains many air spaces, these are filled with water in flooded soil. Furthermore, decomposition of organic matter in the soil by microbial action rapidly depletes oxygen. The rate of diffusion of oxygen through water (0.226×10^{-4} cm s^{-1}) is ten thousand times slower than its rate through air (0.209 cm s^{-1}), so the rate at which oxygen is used in respiration easily outstrips the rate at which it can be replaced. Absence of oxygen is, in itself, a problem which wetland organisms need to overcome, but it also has implications for soil chemistry.

8.2.2 Soil chemistry

It is difficult to separate abiotic from biotic processes in wetland soils because, although the anaerobic conditions are, in themselves, important in determining the ecology of wetland communities, they are compounded by indirect effects upon soil chemistry which result from microbial activity.

8

Nutrients

Permanently waterlogged soil is normally very deficient in nitrate, because rates of decomposition of organic matter, and therefore of remineralization of nutrients, are severely reduced in the absence of oxygen, and because of denitrification by anaerobic bacteria (Box 8.1). Seasonally flooded wetlands, on the other hand, go through a drying period when decomposition and nitrification can occur, releasing nutrients back into the system. Fringe wetlands

Box 8.1   Denitrification in wetlands

Creation of nitrates by nitrogen-fixing or recycling nitrogen compounds derived from decomposing organic matter is called nitrification, and is effected by a series of micro-organisms:

$$\text{Organic nitrogen} \xrightarrow{(1)} \underset{\text{ammonium}}{NH_4^+} \xLeftrightarrow{(2)} \underset{\text{nitrite}}{NO_2^-} \xLeftrightarrow{(3)} \underset{\text{nitrate}}{NO_3^-} \qquad (1)$$

Stage 1 is the result of decomposition by a range of different micro-organisms, and can proceed under aerobic or anaerobic conditions. Stages 2 and 3 are carried out by *Nitrosomonas* and *Nitrobacter*, respectively, both of which are active only under aerobic conditions.

Denitrification is the transformation of nitrate to gaseous nitrogen or nitrous oxide:

$$NO_3^- \rightarrow NO_2^- \rightarrow NO \rightarrow N_2O \rightarrow N_2 \qquad (2)$$

This transformation results from the activity of denitrifying bacteria, which are not limited by anaerobic conditions.

In well-drained soil, under aerobic conditions, nitrification recycles organic nitrogen:

$$\underset{\text{decomposition}}{\text{Organic nitrogen}} \rightarrow NH_4^+ \underset{\text{nitrification}}{\rightarrow} NO_3^- \underset{\substack{\text{uptake}\\\text{by plants}}}{\rightarrow} \text{Organic nitrogen} \qquad (3)$$

In waterlogged soil, under anaerobic conditions, decomposition will proceed only to the ammonia stage, and most nitrates present will be denitrified, although a small proportion will dissimilate into ammonia:

$$\underset{\text{decomposition}}{\text{Organic nitrogen}} \rightarrow NH_4^+ \underset{\text{dissimilation}}{\leftarrow} NO_3^- \underset{\text{denitrification}}{\rightarrow} N_2 \qquad (4)$$

Where waterlogged and aerated soils occur in close proximity, nitrifying bacteria can, ironically, contribute to denitrification.

Nitrification in the aerated soil removes ammonia, creating a diffusion gradient so that ammonium diffuses from waterlogged soil. This, in turn, is converted into nitrate but, because denitrification is removing nitrate from the waterlogged soil, a second diffusion gradient is created, causing nitrate to diffuse into the waterlogged soils, where it is broken down.

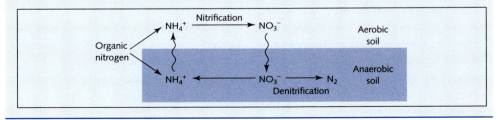

bathed by water with a relatively rapid turnover will have a constantly replenished nutrient supply and can be very nutrient-rich.

Toxins

In the absence of oxygen, reducing conditions are created in which toxic chemical species are generated, reduced forms of iron (Fe^{2+}) and manganese (Mn^{2+}), for example, being more soluble and therefore more chemically reactive, than their oxidized forms (Fe^{3+}, Mn^{4+}). Reduction of these metals, which will not occur until oxygen and nitrates have been completely exhausted, is carried out by anaerobic bacteria. Sulphates (SO_4^{2-}) too are reduced, in the total absence of oxygen or nitrates, to iron sulphide (FeS) or hydrogen sulphide (H_2S), again by anaerobic bacteria. In the absence of iron, toxic H_2S will be produced as a gas; its release when anaerobic sediments are physically disturbed gives off the distinctive sulphurous smell characteristic of wetland soils. Under extremely anaerobic conditions, methanogenic bacteria reduce carbon dioxide (CO_2) to methane (CH_4), another gas which is toxic at high concentrations.

8.2.3 Interaction with open water bodies

Water connections

Aquatic marginal wetlands cannot be considered in isolation from their parent water bodies. A fringe wetland will merge with the open water body which it fringes, and most large wetlands incorporate stretches of open water from which they cannot realistically be functionally separated. Any seasonally flooded wetland, for example, will have channels into which water drains, channels which in managed wetlands are straightened but otherwise equivalent to those which would occur naturally. Saltmarshes and mangals (mangrove swamps) contain networks of drainage channels, permanently flooded or into which sea will penetrate during every high tide (Figure 8.3); they

often experience salinity gradients equivalent to those in estuaries (Section 4.2). River floodplains may contain relatively large, permanent water bodies such as oxbow lakes, remnants of former river channels (Figure 7.1 and Box 7.1). These have no clear or permanent inflow, but are replenished by groundwater percolation and by occasional inundation during floods. Their equivalents in coastal wetlands are lagoons, generally physically separated from the adjacent sea and virtually tideless, but with some influx of sea water, particularly during flood tides.

Despite being formed or maintained without the requirement for open water bodies, mires will not exist independently of other aquatic systems. If a wetland has a fluctuating topography, some parts will inevitably fall below the water table, allowing open water bodies to form. A major difference between aquatic marginal and mire wetlands, in terms of their relationships to other water bodies, is that whereas aquatic marginal wetlands owe their existence to rivers, lakes or the sea, mires are reservoirs of water that actively contribute to the formation of other aquatic systems; even blanket bogs, maintained solely by rainwater, will almost inevitably act as a source for rivers, and commonly their surfaces are dotted with open pools. This distinction does, however, mask the role of flood wetlands in regulating the flow of rivers. Such wetlands act as reservoirs of excess floodwater, much of which will gradually drain back into the channel, dampening major fluctuations in discharge; isolated from its wetlands, a river's discharge may fluctuate rapidly in response to precipitation and, during dry periods, it may cease to flow altogether.

Nutrient and sediment exchange

Fringe wetlands, with their direct aquatic connections, exchange nutrients with the open water body, while flood wetlands rely upon the inputs of sediment and, of course, water during flood events, but in turn can be major providers

8

of organic detritus into river food webs. The lateral connections and movements of organisms between rivers and their floodplains are discussed in Section 3.6. Saltmarshes and mangals are often closely associated with estuaries, where their formation is intimately connected to the sedimentation process; they contribute much to the high productivity of so many estuaries and coastal seas (Section 4.4).

Much of the discussion in this chapter, particularly with respect to adaptations of wetland organisms, is equally applicable to the littoral zone of lakes; the distinction between a deeply flooded fringe wetland and a shallow lake is artificial.

8.3 Ecological effects of the abiotic environment

8.3.1 Anaerobic conditions

Anaerobic conditions in wetland soils require adaptations to overcome the problems associated with respiration and high concentrations of toxins. Relatively few plant species possess these adaptations, although many species typical of seasonally flooded wetlands avoid physiological stress by passing the flood period in a state of dormancy. Some wetland plants can respire anaerobically for short periods, but all will require oxygen eventually. Most species overcome the problems of low oxygen concentrations in the soil by development of aerenchyma (aerated tissues), creating interconnected air spaces within the root system, either within existing roots or, upon flooding, in new roots. The mangrove genus *Avicennia* develops roots which bend upwards, growing out of the soil to allow direct gaseous exchange with the atmosphere (**Figure 8.5**). Shoots, too, grow in some semi-aquatic species in response to flooding, to restore contact with the open air (Blom and Voesenek 1996). Oxygen trans-

ported to roots reacts with metal ions, transferring them into oxidized, and therefore less toxic, forms, but some wetland plants tolerate high concentrations of toxins by incorporating them into their tissues.

Animals, too, have to be adapted to the special conditions of wetlands. Seasonally flooded wetlands will support aquatic species when inundated and terrestrial species when dry, these animals either moving away, becoming dormant or even metamorphosing into a terrestrial life stage when conditions are unfavourable (Section 8.5). Fringe wetlands and other permanently flooded areas support a community reminiscent of lakes except that it tends to be more detritus- than phytoplankton-based. Stagnant water containing large quantities of decomposing organic matter reduces oxygen concentrations, but many invertebrates overcome this by breathing air, either rising to the surface to breathe or transporting and storing bubbles of air under water; others pierce aquatic macrophytes and breathe air in the aerenchymae. Animals living within the substrate have to overcome the same problems of oxygen shortage and toxicity faced by plant roots.

Floating vegetation mats

Anoxia in flooded wetlands is a contributory factor in the creation of floating mats of vegetation, formed when an interconnected mass of vegetation, generally dominated by a single species, breaks free from the underlying mineral substrate. Such mats float because their aerenchyma tissue makes them less dense than water and also, as they develop, because anaerobic decomposition within the mat generates trapped bubbles of methane, which will keep the mat afloat even if living tissue on the surface is removed.

Floating mats are especially prevalent in equatorial Africa, where they are known as 'sudd', from the Arabic for 'blockage' or 'obstacle' (Rzóska 1974), and have given their name to the Sudd wetland of Sudan (**Box 8.2**). They are normally dominated by papyrus (*Cyperus*

Figure 8.5 Mangroves of the genus *Avicennia* develop aerial roots, which grow vertically upward, allowing them access to atmospheric oxygen even during high tides: Poka, Ambon, Indonesia. Photo C. Frid.

Box 8.2   Ramsar sites

The Ramsar Convention of 1971, named after the Iranian town of Ramsar on the Caspian Sea, was the first modern global intergovernmental treaty on conservation and sustainable use of natural resources. Its official title—*The Convention on Wetlands of International Importance especially as Waterfowl Habitat*—demonstrates the original aim of the treaty, which was to conserve birds, but it is more normally known nowadays as the Convention on Wetlands, reflecting its wider remit to cover all aspects of wetland conservation and management.

The Ramsar Convention defines wetlands as:

areas of marsh, fen, peatland or water, whether natural or artificial, permanent or temporary, with water that is static or flowing, fresh, brackish or salt, including areas of marine water the depth of which at low tide does not exceed six metres,

and that further it 'may incorporate riparian and coastal zones adjacent to the wetlands, and islands or bodies of marine water deeper than six metres at low tide lying within the wetlands'. The definition of a wetland is therefore extremely broad, incorporating as it does a significant proportion of all the aquatic habitats (other than open oceans) covered in this book.

Countries that join the Convention are expected to designate at least one site as a wetland of international importance or Ramsar Site, and to encourage conservation and sustainable use of the wetland and its features of biological importance. The designation is not a guarantee that a site will be protected, but signatories are expected to establish and manage wetland nature reserves wherever possible.

8

Box 8.2   Continued

Criteria for determining whether a wetland should be listed are:

1 It contains a representative, rare, or unique example of a natural or near-natural wetland type found within the appropriate biogeographic region.

2 It supports vulnerable, endangered, or critically endangered species or threatened ecological communities.

3 It supports populations of plant and/or animal species important for maintaining the biological diversity of a particular biogeographic region.

4 It supports plant and/or animal species at a critical stage in their life cycles, or provides refuge during adverse conditions.

5 It regularly supports 20,000 or more water birds.

6 It regularly supports 1% of the individuals in a population of one species or subspecies of water bird.

7 It supports a significant proportion of indigenous fish subspecies, species or families, life-history stages, species interactions and/or populations that are representative of wetland benefits and/or values and thereby contributes to global biological diversity.

8 It is an important source of food for fishes, spawning ground, nursery and/or migration path on which fish stocks, either within the wetland or elsewhere, depend.

By November 2007, 157 countries had signed the treaty and between them had designated 1610 wetlands of international importance, covering over 145 million hectares. Recent designations include the 41,000 km^2 Inner Niger Delta (see Section 8.5) in February 2004 and the 57,000 km^2 Sudd in Sudan in June 2006. The most prolific designators to date are the United Kingdom (164 sites, including 14 in its overseas territories), Mexico (65 sites) Australia (64 sites), Finland, Netherlands and Spain (49 each) and Ireland (46 sites), although many of these are very small; 10 of the UK sites, for example are less than 20 ha in area.

papyrus), which can form extensive rafts of vegetation held together by interconnected rhizomes, with plants such as *Vossia* growing on the fringes (**Figure 8.6b**). These mats are created when floating swamp on the edge of a water body breaks free during stormy conditions or by a sudden rise in the water table.

8.3.2 Nutrients

Aquatic marginal and seasonally dry wetlands can be nutrient-rich (Section 8.4). However, permanently waterlogged wetlands with little input of flowing water, such as ombrotrophic mires, are subject to anaerobic denitrification (**Box 8.1**), as a consequence of which inorganic nitrogen is generally the limiting nutrient. Ironically, the very presence of well-adapted wetland species can contribute to denitrification: some of the oxygen transported into their roots will diffuse out into the surrounding substrate and create an aerobic environment in which nitrifying bacteria can convert ammonium to nitrate; this nitrate can then be broken down into gaseous nitrogen by denitrifying bacteria. It is the presence of aerobic and anaerobic conditions in very close proximity over a large surface area around the roots which enhances this process. In saltmarsh soils, too, anaerobic conditions promote denitrification

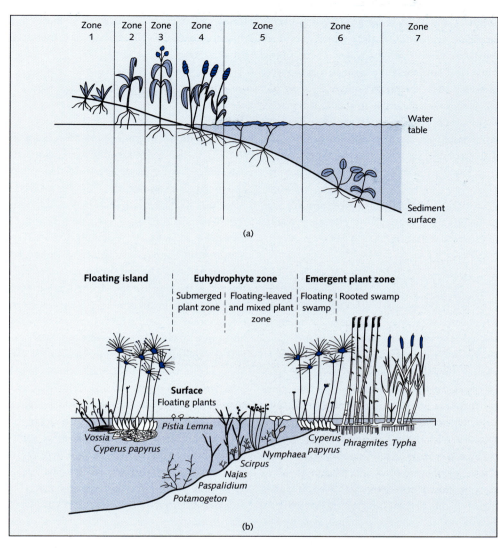

Figure 8.6 Vertical zonation in wetland plant assemblages. (a) A generalized representation of seven zones which occur at the land–water interface, and the life forms of plant species which occupy each, including the position of above-ground structures and roots in relation to the sediment surface and water table. See text for more details. (b) A typical vegetation zonation in tropical African fringe wetlands. Note the floating swamp and floating islands formed by *Cyperus papyrus*. From Denny (1995).

and biologically available nitrogen can be a limiting factor to growth.

To overcome the problems of nitrogen availability, wetland communities typically support plant species which can exploit alternative sources of nitrogen, including, in Europe, alder (*Alnus glutinosa*) and bog myrtle (*Myrica gale*) which maintain a symbiotic relationship with nitrogen-fixing bacteria in their root nodules.

Ombrotrophic mires support carnivorous plants such as sundew (*Drosera* spp.), which obtain nitrogen compounds from insects which they trap and digest.

Microflora and nutrient availability

Wetland macrophytes, even if submerged, gain much of their nutrient requirement from the sediment. They will, however, support a diverse epiphytic microflora of algae which gain their nutrients mainly from the water. There is little transfer of nutrients from algae to macrophyte, although some from macrophyte to algae but when they die, macrophyte stems, along with the microflora they support, collapse into the sediment. The nutrients absorbed from the water column by epiphytic algae are, in this way, transferred into the sediment and available for macrophytes. Therefore epiphytic microflora, despite shading macrophytes and reducing their photosynthetic efficiency, can be indirectly beneficial to their hosts.

8.3.3 Salinity

Plants inhabiting coastal wetlands have to withstand the further physiological stress of variable and often high salinity. Wetland plants employ several adaptations to reduce the detrimental impact of salt. Many mangrove trees physically exclude salt from entering their roots; others can partition salt within the cells, confining sodium and chloride ions to the cell vacuoles with osmotic balance being ensured by high concentrations of non-ionic solutes in the cell cytoplasm; some species deposit salt in bark or in senescent leaves or have salt glands on leaves that deposit salt crystals externally.

Response to salinity is complicated by the high requirement for nitrogen of species such as the saltmarsh grass *Spartina alterniflora*, possibly because nitrogen compounds are required to maintain osmotic balance under saline conditions. Plant productivity in estuarine salt marshes is normally higher in the low marsh than the high marsh, even when the same species are compared. This can be explained by the frequent influx of nutrient-rich water in low marshes, but also by the lower salinity of interstitial water around the roots, as a consequence of flushing by diluted estuarine water.

Ironically for a wetland plant, acquisition of water by mangroves is costly because of the problems of coping with salt. Where soil salinity is high, water loss is minimized by reducing transpiration rates through the stomata, but this also reduces CO_2 uptake and therefore growth rates.

8.3.4 Wetlands and water quality

Aquatic marginal wetland plants, being adapted to nutrient-rich environments, can respond to eutrophication by increasing their absorption and use of the extra nutrients. If nutrients are at an artificially high concentration, wetlands can remove the excess and, as such, are valuable in improving water quality. Some nutrients will be incorporated into plant tissue as enhanced growth, but more important for nitrogen compounds is the nitrification–denitrification process described in Section 8.3, in which local oxygenation of soil adjacent to plant roots facilitates conversion of excess nitrates into gaseous components which are then lost to the atmosphere, whereas phosphates are absorbed by microflora or retained in plant litter and sediments. Reedbeds are particularly effective because, not only can they absorb dissolved nutrients and enhance oxygen-dependent decomposition of organic matter, but the vegetation structure acts both as a net, causing particles in suspension to settle into the sediment, and as an attachment surface for microorganisms. They can, in this way, efficiently strip water of its nutrient load and improve its quality.

Removal of nutrients by soils in aquatic marginal wetlands can have a major influence on water quality. Pinay and Décamps (1988) estimated that a 30 m wide strip of riparian woodland along a tributary of the River Garonne in south-west France would be enough to remove all of the nitrate entering the groundwater from surrounding agricultural land. The same process can work in reverse: as floodwater from the River Rhine in eastern France infiltrates the soil, it is cleansed of excess nitrates and phosphates such that the water which enters the groundwater is significantly less polluted than the river water. This process is biologically controlled, in that the areas with

8

the most developed wetland vegetation are most efficient at purifying the water.

The ability of wetland vegetation to clean water is being exploited commercially, with beds of emergent reeds (*Phragmites*) or cattails (*Typha*) proving most successful. Water treatment wetlands can be created and maintained with relative ease; natural wetlands too can be effective, but will suffer reduction in species diversity, as nutrient enrichment favours the most competitive species at the expense of others (Section 8.5). Excess nutrient inputs may eventually destroy macrophytes in permanently inundated fringe wetlands, effectively destroying the wetland and creating an open water environment. This process normally proceeds according to the mechanisms described in Section 7.5, in which eutrophication of lakes leads eventually to domination by phytoplankton, but can be a consequence of more subtle changes. In the Norfolk Broads, in south-east England, there is a clear relationship between nitrogen loading of water and the loss of stands of *Phragmites australis*. As nitrogen concentration increases, *Phragmites* grows more vigorously, increasing the proportion of aboveground to below-ground growth. Most of the losses in the Norfolk Broads have been from a growth form known as 'hover', which consists of floating mats of *Phragmites*; these depend upon their root and rhizome mat for mechanical strength and become unstable if emergent shoot growth is too high, making them vulnerable to erosion.

8.4 Energy inputs

8.4.1 Primary production

Wetlands support some of the highest and some of the lowest productivity values recorded (Table 8.1; see also Section 4.4). At one extreme is reedswamp, whose productivity can be matched only by tropical rainforests and the most intensively managed agricultural systems, while at the other are ombrotrophic mires, some of whose productivities are on a par with those of tundra or desert ecosystems.

The most productive wetlands are fringe wetlands, bathed in a continually replenished source of nutrients to replace those scavenged by bacteria or incorporated into sediment. Single species stands, such as reedswamp or papyrus swamp, are dominated by species which, despite being perennials with the capacity for year-round growth, are not woody, so that little energy is expended on growing or maintaining non-photosynthetic structural features. Such wetland dominants are essentially giant leaves. Furthermore, living in an environment where water is plentiful, they have no need to limit transpiration rates. Seasonally inundated flood wetlands can also be highly productive, the flood period bringing in sediments and nutrients while the dry season allows aerobic decomposition of detritus and consequent nutrient release. In várzea floodplains

8

Table 8.1	**Net primary productivity (NPP: g m^{-2} yr^{-1}) of different freshwater wetland types**		
	Bog	**Fen**	**Aquatic marginal**
Polar (65–90°)	100–300	100–333	–
Boreal (55–65°)	300–700	400–700	500–1000
Temperate (30–55°)	400–800	400–1200	700–2000
Tropical (0–30°)	600–1200	–	1500–4000

Figures quoted show the normal range—examples of extremely high and extremely low productivities are given in the text. No attempt has been made to distinguish between different aquatic marginal wetland types, because of the range of definitions employed. For examples of productivity in coastal wetlands, see Table 4.1 from data in Aselmann and Crutzen (1989).

along the Amazon, the grass *Echinochloa polystachya*, which grows very rapidly and can achieve standing stocks approaching 10,000 g m^{-2}, absorbs large quantities of nutrients from the river. Dry season decomposition of this species, which grows at the lowest points on the floodplain, releases nutrients which are carried by rising floodwaters to higher levels so that while one generation is absorbing nutrients from the river the decomposition products of the previous generation ensure that floodwaters which penetrate further into the floodplain remain nutrient-enriched.

The least productive wetlands—ombrotrophic mires—are also dominated by non-woody species, but are limited by permanently low nutrient availability. Despite this limitation, they can have surprisingly high productivity values: Clymo and Hayward (1982) quote a typical value of 800 g m^{-2} yr^{-1} for *Sphagnum* growing in pools in British bogs, somewhat higher than the productivity of temperate grassland, although productivity on hummocks (Box 8.3) is typically less than one quarter of this value.

High net productivity can be maintained indefinitely in fringe wetlands if decomposition rates or removal of detritus by the parent water body match production, suppressing accumulation of organic matter. Blanket and raised mires accumulate organic matter, as a result of extremely low rates of decomposition in the cool temperate environments in which they usually form, but by raising the water table as they develop (Box 8.3), continued domination by *Sphagnum* is assured. In other wetland environments, such as rheotrophic fens, accumulation of organic matter can lead to relative lowering of the water table and an increased proportion of species adapted to drier environments. Flood wetlands, too, experience reduced inundation stress during the dry season, allowing more terrestrially adapted species to persist. In such situations, total biomass may be higher than in fringe wetlands, but net productivity is lower.

Box 8.3 *Sphagnum* control of water tables

Sphagnum mosses lack roots, whereas most of their potential competitors are rooted vascular plants, so *Sphagnum* strategies for reducing competition involve creating adverse conditions below ground, to suppress root growth. Waterlogging clearly impedes non-wetland species, but *Sphagnum* also increases the acidity of its environment and reduces nutrient availability, its leaves intercepting atmospheric nutrients before they can penetrate the surface layers, and through slow mineralization as a result of reduced decay rates. Finally, its poor heat conducting properties ensure that the below-ground parts of the bog remain cool. Suppression of vascular plants increases light availability and reduces water loss through transpiration, both of which maintain optimum conditions for *Sphagnum* itself.

Sphagnum shoot growth can exceed 10 cm yr^{-1} in pools, but is typically 1–5 cm yr^{-1} in hummocks. Vascular plants growing with *Sphagnum* in peat-forming systems must, therefore, match this growth to avoid being smothered. The growth form of sundew (*Drosera rotundifolia*), for example, comprises a vertical stem and a series of leaf rosettes, illustrated in the figure to the right; stem growth keeps pace with that of *Sphagnum*,

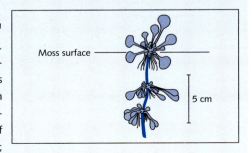

Moss surface

5 cm

From Svensson (1995).

Box 8.3   Continued

but at the expense of successive leaf rosettes, which are abandoned within the *Sphagnum* moss as they become smothered. By adopting strategies such as this, vascular plants can maintain their presence in *Sphagnum* bogs. In turn, they have little detrimental effect upon *Sphagnum* itself, although its growth will be suppressed by shading and particularly by accumulation of above-ground litter. The competitive advantage held by *Sphagnum* does, however, require maintenance of the high water table. Drought conditions will benefit vascular plants, which have effective root systems for gain-

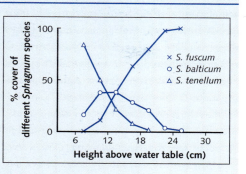

From Rydin (1993).

ing water. Paradoxically, those with shallow roots, including *Drosera*, rely on the water-conducting capacity of *Sphagnum* for their water supply, so will also suffer under drought conditions.

A *Sphagnum* bog consists of raised hummocks, above the water table and often relatively dry, hollows which are permanently waterlogged or even flooded, and flat lawns. Although up to ten different species of *Sphagnum* may coexist in a single mire, most are confined to a narrow vertical range and can be identified as 'hummock species' (e.g. *S. rubellum* and *S. fuscum*) and hollow species (e.g. *S. balticum* and *S. tenellum*), with a range of species from both groups occupying lawns. Within this division, there is often further, more subtle zonation, such as that shown in the figure above, in which *S. fuscum* occupies the upper part of hummocks and *S. tenellum* the lower part.

Hummock species take advantage of their high capillarity, greater tolerance of low nutrient levels and more stable dead tissue to grow above the open water level and form hummocks, the upper limit for each species being set by its physiological tolerance of an increasingly drier environment and reduced nutrient levels. The determinants of the lower limits to each species are, however, much less clear; transplantation experiments have demonstrated that it is not simply a case of hollow species being more competitive, as hummock species, when transplanted to hollows, have persisted and even thrived. How so many species can coexist when several apparently occupy each niche without any apparent habitat differentiation is also a mystery.

The hummock–pool sequence is very effective at reducing runoff and therefore retaining the water table at a level advantageous to *Sphagnum*. The hummocks, however, are more oxygenated and therefore suitable for vascular plant species, so they support the growth of bog shrubs, such as heather (*Calluna vulgaris*) and even trees. Ironically, the roots of vascular plants form a firm matrix which supports the spongy *Sphagnum*; without this matrix, there is a danger that hummocks would collapse. The *Sphagnum*, therefore, relies upon vascular plants to maintain the high water table which gives it the competitive edge over these same plants.

8

8.4.2 Secondary production

Grazing

Despite the high productivity of wetland plants, relatively little is consumed by herbivores.

Most macrophytes, and particularly emergent species, have tough cell walls and high lignin content, making ingestion difficult. True grazers on macrophytes are mainly found among mammals and birds. Barasingha (*Cervus duvaucelii*), the endangered swamp deer of south-east Asia, relies heavily upon sub-merged macrophytes as a source of sodium and will actively seek otherwise nutritionally poor species from ponds and ditches; moose (*Alces alces*), which occurs widely in North America and northern Eurasia, is another deer species which habitually feeds upon aquatic plants, apparently choosing sodium-rich species. Perhaps the most effective mammalian grazers, however, are the sirenians, mainly marine but including the American manatee (*Trichechus manatus*) which moves freely between marine, estuarine and freshwater habitats, and the only exclusively freshwater species, the Amazonian manatee (*Trichechus inunguis*). This latter species, which weighs up to 450 kg, feeds upon aquatic and semi-aquatic macrophytes, con-suming about 8% of its body weight per day. The movements of this species are determined by the hydrological cycle of the Amazon. As the rivers rise, between December and June, mana-tees spread into the várzea and igapó areas to feed on the lush new growth of macrophytes; when river levels drop, manatees migrate into perennial lakes where, if macrophytes are in short supply, they will eat detritus off the lake bed, or simply not feed at all.

Among invertebrates, there are efficient grazers such as apple snails (Ampullaridae) in tropical South America. Among species that are norm-ally considered to be detritivores, herbivory often occurs, particularly when alternative food sources are scarce. Crayfish are omni-vorous, adults eating mostly vegetation and detritus; they prefer fully submerged species to emergent forms and can affect macrophyte numbers both through direct consumption and by cutting plant stems near the base.

Exploitation of detritus

Conditioning is important in improving palat-ability of detritus derived from autochthonous wetland plants, just as it is with respect to allochthonous inputs to other aquatic systems (see Sections 3.4 and 4.4 for discussion of con-ditioning in rivers and estuaries, respectively). The high productivity of many wetlands gener-ates large volumes of detritus, this standing stock being enhanced, in the case of fringing and flood wetlands, by detritus from external sources washed in from the adjacent water body; wetland plants act as traps for algae, detached pieces of macrophytes and detritus originating from the pelagic zone.

Invertebrate detritivores often rely upon macro-phytes as structural habitat features, benefiting from their high surface area. In Tivoli South Bay, a small wetland on the Hudson River in New York, the detritivorous chironomid *Cri-cetopus* sp. accounts for 73% of chironomid larvae in early summer. Although a consumer of fine particulate detritus, it uses the dominant macrophyte, water chestnut (*Trapa natans*) as an attachment structure, reaching densities of 5000 m^{-3} on the underside of its leaves; num-bers of chironomids decline, however, later in the year, possibly as a result of heavy predation but also because the water chestnut plants become more emergent, reducing underwater surface area available for colonization.

Export of production

Fringe wetlands act not only as sinks for allochthonous detritus and nutrients, but also as sources of energy for adjacent open water bodies. The role of saltmarshes in providing energy to estuaries is considered in Section 4.4.

The impact of saltmarshes on productivity of adjacent open coastal waters is not simply due to export of detritus, as inputs can be inter-mittent and often very low; neither is it a consequence of nutrient export, as these are efficiently recycled within the wetland system. In Florida, Stevens *et al.* (2006) demonstrated that movement of nekton was a major mechan-ism by which energy transfer occurred; many fish feed in saltmarshes before moving out into the open water, and their contribution in this way to energy transfer was estimated at 1–2% of total saltmarsh primary production.

8

8.5 Community structure

■■■■■■■■■■■■■■■■■■■■■■■■■■■

The soil conditions, the risk of inundation and, in flood wetlands, the seasonality of the inundation cycle, all make a wetland an extreme environment for rooted plants. The precise physiological tolerances of most wetland plants ensure that community structure is intimately related to the water table. Conditions may be so extreme that no species can be optimally adapted to more than a narrow range of requirements and will be outcompeted elsewhere.

8.5.1 Zonation

Plant zonation

Vertical zonation is commonly seen in wetlands, where a change in elevation of only a few centimetres will create a range of different conditions, each requiring different strategies and therefore supporting different species (Figure 8.6). A series of zones, related to the height of the water table with respect to the sediment surface, can be recognized at the land–water interface, representing different sets of conditions to which their inhabitants must be tolerant. These zones are illustrated in Figure 8.6a. Zone 1 represents the fully terrestrial environment, with freely draining, aerated soils. Plants living in zone 2 have roots which are, for the most part, in aerated soil, but may have to overcome problems of anoxia and toxins at depth. Plants living in zones 3 to 6 must be tolerant of permanently waterlogged sediments; in zone 3, their above-ground parts are above the water and, in zone 4, the water is shallow enough to allow emergent species to grow. Zones 5 and 6 are too deep for emergent species, and their flora is fully aquatic (euhydrophytes), while zone 7 represents those parts of the water body beyond the littoral zone which are too deep for

light to penetrate to the bed, and so are devoid of attached macrophytes. This change in plant community composition across an environmental gradient is referred to as a coenocline; such zonation is particularly clear in fringe wetlands, such as that illustrated in Figure 8.6b, but can be found in most wetland types. Zonation in *Sphagnum* bogs is considered in Box 8.3.

Zonation of species analogous to that of marine algae occurs in coastal wetlands, where water level fluctuations are predictable yet frequent. Mangroves occur in tropical coastal zones from mean sea level to the highest spring tide level, within which are clear zones, each with its dominant species. Species-poor Central American mangals are dominated by red mangrove (*Rhizophora mangle*) on the seaward side, where tidal inundation is daily, black mangrove (*Avicennia tomentosa*) where tidal inundation is less frequent, and white mangrove (*Laguncularia racemosa*) at levels flooded only by spring tides. Indomalayan mangals are much more speciose, but similar zonations occur. Saltmarsh can generally be divided into three vertical zones. Low marsh, or pioneer zone, supports very few macrophyte species and can contain large areas of bare mud, although in eastern North America this zone may be dominated by the grass *Spartina alterniflora* (Section 4.4) and, increasingly in the British Isles, by *S. anglica*, a hybrid of the alien *S. alterniflora* and the native *S. maritima*. Above the MHW (Figure 5.1) is middle marsh, or turf, extensively vegetated with halophytic grasses, while above MHWS is high marsh, supporting halophytes and some freshwater species, such as common reed (*Phragmites australis*), which will persist so long as freshwater inputs are great enough to flush out occasional saltwater incursions.

Animal zonation

Zonation has been recorded among invertebrates associated with vegetation in permanently

8

flooded wetlands. In coastal wetlands, reed-swamp dominated by *Phragmites australis* may extend across a salinity gradient from fully fresh water to strongly brackish; in a wetland of this type in South Wales, the aquatic invertebrate assemblage changed markedly from one dominated by typical freshwater species (aquatic beetles, pond snails, hoglouse, etc.) to one supporting species more commonly encountered in estuaries, such as ragworm (*Hediste diversicolor*) and the estuarine shrimp *Gammarus duebeni* (Arnold and Ormerod 1997).

Changing physical conditions, too, can create spatial zonation patterns among animal species. The North American Great Lakes are fringed by numerous wetlands, which are exposed to fluctuating water levels at temporal scales ranging from hours, caused by seiches, to decades, caused by long term changes in precipitation; superimposed on these changes, which penetrate deep into fringing wetlands, are the effects of wave action, which are mainly confined to their fringes. Based on a series of studies from coastal marshes on Lake Huron, Burton *et al.* (2002) developed a conceptual model to explain the patterns observed. They suggested that exposed littoral areas are unable to accumulate organic sediment, as it is dispersed deeper into the wetland or out into the open lake, and so the dominant invertebrates are filter-feeders and grazers, the latter feeding on the epilithic algae that grow abundantly in this zone. The inland areas, in contrast, accumulate high biomass of organic sediments, which settles in the relatively still conditions, so collector-gatherers dominate, along with visual predators taking advantage of the clear water. Accumulation of organic matter and low replenishment of water result in low oxygen concentrations in the inland areas, so fish diversity is low, whereas in the littoral area oxygen levels, and therefore fish diversity, are high.

Many wetland species require the mosaic of habitats that the natural cycle of wetland change creates. For example, the European bittern (*Botaurus stellaris*), a cryptic heron found almost exclusively in reedbeds, requires both large areas of reed in order to hide and nest, and areas of open water. This is because it feeds mainly on fish, which are most common in the open water and immediately adjacent reed bed. The bittern uses reed stems as a fishing platform, preying on fish that enter the vegetated area from the open water. In England Self (2005) demonstrated that bittern territories required a complex mix of open water and reedbed, with 170–970 m of edge per hectare.

8.5.2 Competitive dominance

Within each zone in productive systems, highly competitive species can dominate to the extent that virtual monocultures are created, hence the use of terms such as reedswamp to describe domination by *Phragmites australis*, and papyrus swamp where *Cyperus papyrus* is the dominant species. The height and shoot density of these species makes them competitively superior to potential colonists, but they can manipulate the environment in more subtle ways. This is particularly true of *P. australis*, which physically impedes colonization by seeds of other species and then reduces their survival by lowering soil oxygen concentration and raising sedimentation rates. It also deposits large quantities of litter, which both shades seedlings and in tidal areas can be suspended by the rising tide and physically break fragile seedlings through its movement (Michinton *et al.* 2006). In coastal eastern North America, *P. australis* was formerly confined to the terrestrial edge of freshwater and brackish wetlands but is now causing particular concern because it is expanding and eliminating diverse assemblages of plants elsewhere, often very rapidly. The sudden expansion has been attributed to the recent invasion of a non-native genotype from Europe that is tolerant of a broader range of environmental conditions and produces more seeds than the native form (Saltonstall 2003).

Monocultures of species such as *P. australis* generally only occur in permanently or frequently inundated areas. As the elevation rises,

8

there is a transition to plants such as rushes (*Juncus* spp.) in less flooded zones, or even to fully terrestrial species. In these transitional zones, competitive interactions among species can be complex, maintaining a higher diversity of plant species (Lenssen *et al.* 2004).

In mangroves, the differential trade-off between growth rate and salt tolerance (Section 8.3) manifests itself in competitive interactions among species. For example, two Australian species, *Sonneratia lanceolata* and *S. alba*, have adopted different strategies: both show optimal growth at a salinity of 2 but, whereas *S. alba* can tolerate salinities up to 37, *S. lanceolata* can only grow in salinities up to 20. The growth rate of *S. alba* under optimal conditions is only half that of *S. lanceolata*, so the latter dominates lower salinity zones, while the former is able to persist in high salinity zones, beyond the salt tolerance of *S. lanceolata*.

Terrestrialization

Adaptations to permanently waterlogged conditions can lead to problems living in well-drained soils. Some wetland species have little control over transpiration rates from their leaves and will dry out rapidly in the absence of a constant supply of water, while others, growing shallow roots to avoid anoxia, have little capacity for extracting water from dry soils, but wetlands support many species adapted to waterlogging and low nutrient levels, which are physiologically able to survive or even flourish in well-drained soil. Most of these species are, however, confined to wetlands by their poor competitive abilities elsewhere, as adaptation to waterlogging leads to strong specialization and therefore a competitive disadvantage against terrestrial species if the environment dries up. This is illustrated by changes wrought to the nutrient-poor Pinelands wetlands in the coastal region of New Jersey, USA, parts of which have undergone drying and nutrient enrichment as a consequence of nearby urban development. Changes to water quantity and quality have little effect upon the overall phys-ical appearance of the habitat, or on the relative abundance of the various life forms of plants: trees, shrubs and herbaceous vegetation. What does occur, however, is a change in species composition, as a result of two processes. Lowering the water table allows the soil to dry, removing the constraints of toxicity and low oxygen concentrations which had previously impeded invasion by terrestrial species, while nutrient enrichment allows them, once established, to grow rapidly. Wetland plants, adapted to a normally nutrient-poor environment, are low in stature and slow-growing, and in the presence of nutrients are outcompeted by the more vigorous terrestrial species, which have taller stems, broader leaves and more rapid growth rates. Ehrenfeld and Schneider (1991) found that the trend following urbanization was towards increased species richness, but that this was actually hiding a loss of wetland species and their replacement with terrestrial plants, rather than a simple mixing of the two sets of species assemblages.

8.5.3 Seasonal changes in wetlands

Seasonality is a feature of many wetlands, resulting in seasonal patterns of abundance among wetland organisms. This is most pronounced in the Aquatic Terrestrial Transition Zone (ATTZ) (Box 3.5), a flood wetland which can alternate between fully aquatic and fully terrestrial environments, and in which both terrestrial and aquatic assemblages may occupy the same site at different times, the community changing rapidly in response to water levels (Junk 2005).

Most terrestrial animals migrate to non-flooded areas during wet periods, but other flood responses, commonly observed in tropical ATTZs, include migration along the advancing flood line or, in the flooded forests of Amazonia, vertically up trees, or dormant stages as eggs or in retreats. Some terrestrial invertebrates, such as the Amazonian millipede *Myrmecodesmus adisi*, are even able to remain

8

active for several weeks while completely submerged (Adis and Junk 2002). In temperate zones, such adaptations are rare as flood periods tend to be relatively short and less predictable than in the tropics, and most invertebrates are opportunistic, simply recolonizing from adjacent permanent habitat (Bedford and Powell 2005).

The vertebrate cycle

Spatial variations in topography enhance diversity, as do temporal changes in water level, particularly in seasonal floodplains where different stages in the flood–drainage cycle favour different groups of species. The movement in and out of floodplain wetlands by river fish has been considered in Section 3.6, but similar

migrations occur among terrestrial and semi-aquatic groups. The vertebrate cycle of temperate wetlands illustrates this well. During high water levels in the winter, the wetland is dominated by aquatic birds, most of which are wintering species which migrate to breeding grounds in the spring. As the water level recedes, aquatic species are confined to ever smaller areas but terrestrial grazers and, later, fruit- and seed-eating species move in, exploiting the high productivity. Equivalent cycles occur in tropical wetlands: many of the largest African wetlands are in semi-arid zones, and mass migration of grazing mammals—and their predators—into wetland areas during the dry season is common (Box 8.4). When water levels rise once more, aquatic species return. In contrast, terrestrial birds may enter reed beds in

---

Box 8.4   The Okavango Delta

The Okavango Delta in Botswana is one of the most famous and least disturbed of the World's large wetlands. It is an alluvial fan with a very low gradient (1:3300), fed by the Okavango River which is itself derived from the Quito and Cubango Rivers, flowing from central Angola. Immediately upstream of the delta is an incised corridor known as the Panhandle, through which the Okavango River flows. The Okavango has a clear flooding cycle, caused by rainfall in the catchment during December–March. This creates floodwater which reaches the northern end of the Panhandle in April and then takes 4–5 months to traverse the fan. The system contains about 3000 km^2 of permanent wetland, which expends during the flood season to an average of 10,000 km^2, although the actual area varies from 6000 to 14,000 km^2. Almost all of the floodwater which enters the Okavango is lost through evapotranspiration or groundwater recharge, only 3% discharging out of its southern end. The Okavango Delta is famous for its wildlife. It supports the largest populations in the world of two semi-aquatic antelopes: sitatunga (*Tragelaphus spekei*) and red lechwe (*Kobus leche*). It also supports large numbers of migratory mammals, including elephant (*Loxodonta africana*), wildebeest (*Connochaetes gnu*) and Burchell's zebra (*Equus burchelli*), along with a large population of the endangered African wild dog (*Lycaon pictus*). There are also 650 bird species and 68 fish species. Although very few of its known species are endemic, the sheer diversity makes the Okavango an important area for conservation.

The organisms living within the Okavango Delta are important determinants of its structure and function, as the following examples demonstrate.

Tree islands

Trees can only establish in areas of relatively high elevation. These are often created by termites, which in exceptionally dry years establish colonies in areas that would normally be seasonally flooded; eventually, floods in subsequent years will destroy the colony, but its mound remains and is colonized by

---

8

Box 8.4 **Continued**

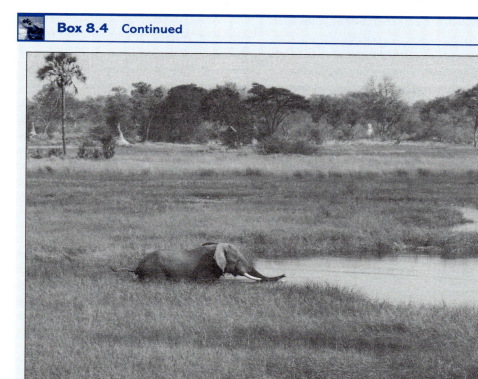

Photo copyright A. Thomas.

trees. The trees transpire large volumes of water out of the ground, resulting in deposits of calcite whose deposition enlarges the islands. Eventually, over a period of 100–200 years, the groundwater beneath the islands becomes too saline for trees. The trees therefore destroy their habitat through transpiration, but by concentrating the salt in relatively small areas, they probably contribute to maintaining a freshwater wetland in a semi-arid area, where annual evaporation is four times precipitation.

8

Hippo trails

Hippopotamus (*Hippopotamus amphibius*) spend the day in the deep water of permanent wetland areas, and then move out at night to grazing grounds, known as 'hippo lawns', which are usually situated on the tree islands. As the hippo lawns can be several km from deep water, this nightly movement creates a series of trails, up to 2 m wide and 1 m deep, which are kept free of vegetation by the movement of the animals. Main trails follow deep channels, and act as conduits for free movement of water; side trails branch off to the hippo lawns. As the islands salinify and the grazing is lost, the side trails will be abandoned and the trail, heavily fertilized by defecation, becomes a ribbon of luxuriant vegetation. The major distributory channels are bordered by peat, which hippo trails breach, allowing water and sediment to flow into and maintain back swamps. Eventually the natural processes of sedimentation infill the main channel, and fires will burn away the shallow peat layer, leaving the sediment-filled former hippo trails—initially cut into the peat—as ridges of sand up to 1 m higher than the surrounding land. These ridges may later become the foundation for new islands to form as the area re-floods, islands to which hippos will create trails in order the exploit new hippo lawns.

large numbers during migration or in search of food during winter. Floodplain wetlands are not, however, simply the haunt of river or terrestrial species which move in when conditions are appropriate. There are also true wetland species, adapted to cope with the seasonally changing conditions. Examples are amphibians, such as frogs, which require pools of water during the spring for larval development but become more terrestrial as adults, seeking out open water once more in the autumn as a site for hibernation. An alternative strategy, exemplified by the lungfish of tropical Africa and South America, is to spend the dry season in a dormant state, known as aestivation, buried in the ground awaiting the return of the floods. Lungfish are also able to exploit the permanently anoxic waters of the interior of large fringing wetlands (Chapman *et al.* 1996).

Wetlands of the south-eastern USA flood during spring and early summer, leading to a contrasting cycle in which detritus-based food chains predominate during the growing season, when primary production is high, while herbivore pathways dominate over winter, when primary production is low. This is because herbivorous mammals such as deer and rodents are driven away by rising water levels, leaving only specialist semi-aquatic species such as muskrat (*Ondatra zibethica*); herbivorous

birds either migrate to northern breeding grounds or, if resident, shift their diet to take advantage of large numbers of emerging insects. During the summer, therefore, the flooded wetlands are dominated by fish and amphibians, along with aquatic invertebrates, most of which are detritivores or predators, and it is only during the autumn, when water levels recede and, coincidentally, primary production declines, that herbivorous mammals can invade.

The human exploitation cycle

The vertebrate cycle has its parallels in human exploitation of wetlands. The Inner Niger Delta in Mali supports 2–3 million head of livestock which move, along with their human guardians, in an annual cycle dictated by floodwater (Figure 8.7). During the rainy season in August to October, the delta is flooded and livestock are grazed on the surrounding higher ground. In October and November they gather around the edge of the delta, ready to move into the newly exposed pasture as the floods recede. At first, they range widely but, as the dry season progresses, they aggregate around permanent water holes and channels, before moving out to higher ground once more as floods return in July. In a further analogy to the vertebrate

8

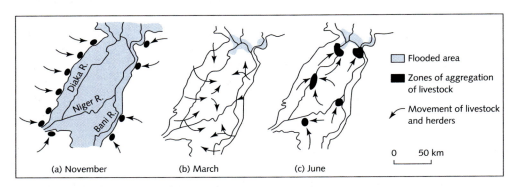

| (a) November | (b) March | (c) June |

Labels within figure: Diaka R., Niger R., Bani R.

Legend:
Flooded area
Zones of aggregation of livestock
Movement of livestock and herders
0   50 km

Figure 8.7 The human exploitation cycle: pastoral migration in the Inner Niger Delta. (a) during October and November, herdsmen assemble their cattle along the edges of the flood plain; (b) by March, herds have dispersed across the flood plain, following the receding water; (c) in June, herds reassemble around remaining water holes, ready to move to higher ground as floodwaters rise. After Gallais (1984).

cycle, as the livestock and human pastoralists move out, the fish—and the people who fish for them—return.

Sedentary agriculturalists, too, can exploit the predictability of the seasonal flood. Seasonal floodplains receive an annual input of generally nutrient-rich sediment and, when the soil aerates, can be very fertile. This fertility can be exploited by growing crops during the dry season. Managing the floodwater using drainage channels and sluices in levées, to hasten its removal, is a long established and effective system, and has been practised along the Lower River Nile in Egypt, for example, for thousands of years. In England, water meadows are a consequence of this technique. The flooding ensures that they are naturally fertilized every year, so they can be very productive and heavily grazed over summer. The result is fertile, botanically rich meadows, with waterlogged drainage ditches supporting a great variety of wildlife, and the value of water meadows to conservation in areas heavily modified by human activity is well understood.

8.5.4 Species diversity

Species diversity in wetlands is determined to a large extent by succession and seasonal change, both of which allow a wide range of species assemblages to coexist in close proximity. Some wetland types can be species-poor, good examples being reedswamp. A feature of wetland areas, however, is that so often several different habitat types, each with its representative species, are mixed together in a small area. The precise adaptations of different wetland plants to specific degrees of waterlogging contribute to this high species diversity, with very small changes in topography across a small area allowing different species assemblages to coexist (Figure 8.6). The river in Figure 8.1a, for example, has an open water habitat, a fringing wetland, a normally dry levee and a flood wetland, all within a few metres of each other.

Just as the water table produces spatial patterns in wetland biota, so any fluctuations will create temporal patterns. The seasonal changes described above illustrate this well: during any one season, vertebrate species diversity may be low, but when all seasons are considered together, diversity is enhanced by seasonal turnover of species.

8.5.5 Ecological functioning

As a wetland is defined by the vegetation it supports, the role of living organisms in determining function is greater than in any other aquatic habitat. This influence is further increased by the role of soil microorganisms in controlling nutrient availability and soil toxicity (Section 8.2). Sediment trapping by plant stems, plus aggregation of detritus, both contribute to the process whereby a wetland will naturally infill and progress towards a more terrestrial stage (Section 8.6), while establishment of trees can reduce water tables and accelerate terrestrialization due to increased evapotranspiration rates. Perhaps the most striking example of an organism influencing its environment, however, is that of *Sphagnum* mosses controlling and enhancing the water table to their own competitive advantage (Box 8.3).

Influence of animals

Large wetland animals modify their local environment to their own benefit, often indirectly influencing other organisms. North American beaver (*Castor canadensis*) can create new wetlands by damming rivers (Figure 3.14); prior to European settlement it has been estimated that, due to the influence of around 40 million beavers, 207,000 km^2 of the upper Mississippi and Missouri river basins were flooded beaver ponds. Also in the United States, burrowing by muskrat (*Ondatra zibethica*) changes flow patterns in wetlands, while 'eat outs', in which they completely denude patches of vegetation through grazing, can have important influences on local hydrology, sedimentation and of course biomass. Alligators (*Alligator mississippiensis*) excavate 'gator holes', which act as a

8

permanently flooded habitat for other organisms such as fish and turtles during the dry season. These influences are locally important in determining the structure and function of areas of wetland, but in some cases animals have been implicated in more fundamental restructuring of wetland areas, as in the case of hippopotamus in African wetlands (**Box 8.4**).

Tree islands

Very large wetlands may be characterized by tree islands, relatively high elevation areas that are therefore able to support trees. These islands are often created by biological activity (**Box 8.4**) and the trees, in turn, influence the wetland environment. In the Florida Everglades, tree islands are important in maintaining the generally oligotrophic nature of the water, as they concentrate phosphates, which can be up to 100 times more concentrated in the islands than in surrounding areas (Wetzel *et al.* 2005). This is partially a consequence of the trees themselves, as their evapotranspiration draws water in, which then deposits its nutrients, and partially because the trees act as roosts for birds, whose guano concentrates nutrients. Wading birds in the Everglades disperse widely to feed but roost and nest aggregated into colonies in trees, creating areas where excretion and mortality are concentrated. Harris (1988) has calculated that, during the late nineteenth century, there were an estimated 2.5 million wading birds in the Florida Everglades which, assuming 50% of excretion and mortality occurred at roosting sites, would have transferred 400 t yr^{-1} of ash (the inorganic nutrient component of excretory products and corpses) from the wider wetland to relatively small areas around roosting sites. As the number of birds has been reduced by habitat loss and degradation and, until early in the twentieth century, hunting for plumes, so the extent of this redistribution of nutrients will have declined significantly.

The influence of tree islands on salinity in the Okavango Delta is detailed in **Box 8.4**.

8.6 Succession and change

Wetlands are very dynamic systems, in which patterns of succession and seasonal cyclical change are more marked than in any other aquatic habitat type.

8.6.1 Origin of aquatic marginal wetlands

The creation of freshwater flood wetlands is enhanced by the sediment-carrying properties of rivers. When a river floods and spreads across its floodplain, the reduced water velocity lowers its sediment-carrying abilities and it deposits material that it is carrying. Coarse materials, requiring the most energy to keep in suspension, tend to be dropped first, close to the edge of the channel, creating a barrier, or levée, which effectively raises the height of the channel rim. The effects of a levée are twofold: it reduces the number of times a river will spill over onto its floodplain, increasing the seasonal nature of flooding, but at the same time it impedes drainage, so that the wetted area will remain waterlogged for a long period after the flood has subsided.

8.6.2 Development of mires

Fens are created in areas where open water accumulates, including depressions in bedrock or glacial drift. A fen may develop directly as such, but typically is the result of infilling a lake or even a semi-enclosed arm of the sea, evolving, therefore, from an aquatic marginal wetland. The process of vegetation growth will lead to infilling of the aquatic component of the wetland and gradual terrestrialization through a series of successional stages, each supporting distinctive plant assemblages.

8

Fen development from a lake

The various stages of the infilling of a lake, known as hydroseres, are illustrated in **Figures 8.8** and **8.9**. Infilling proceeds from the edge of the lake **(Figure 8.8a)**; submerged macrophytes grow in the littoral zone, producing detritus and facilitating sedimentation by trapping and retaining silt within their roots and rhizomes. When the water is shallow enough, emergent species such as *Phragmites*, which can establish in water up to 1 m deep, will invade, often forming large stands **(Figures 8.8b, 8.9a)**. These are very productive, generating large quantities of detritus which accelerate deposition even further until, as sedimentation rises above the water table, species typical of fens, such as sedges, begin to encroach **(Figure 8.8c)**.

At this stage, the sediment, although water-logged, is not generally immersed **(Figure 8.9b)**. Sedge peat will begin to accumulate, marking the transition of the wetland from a fringe wetland to a fen. Once the fen has developed, wetland trees such as willow and alder will establish, increasing detritus inputs and, at the same time, lowering the water table through elevated rates of transpiration. The former wetland, therefore, can become a fully terrestrial environment **(Figure 8.8d)**.

This sequence of hydroseres, in which each facilitates development of the next by modifying sediment levels, would appear to be the most likely course that wetland succession should take, but, in reality, the successional process is much more complex. Walker (1970) found a range of different sequences from a series of British peat cores, demonstrating that the pattern illustrated in **Figure 8.8** is by no means the same in different sites. It may be predicted that the inevitable conclusion of wetland succession is fully terrestrial vegetation, but this scenario, though widely quoted, is not borne out by the evidence. The initial stage, in which reedswamp develops from submerged or aquatic macrophytes, occurs commonly, but reedswamp, in turn, can be replaced by open sedge fen, carr woodland (formed by wetland-tolerant trees growing directly on peat) or *Sphagnum* bog. A series of successional pathways can be identified, but almost all culminate in *Sphagnum* bog **(Figure 8.10)**, to the extent that Walker concluded that the true climax of hydroseres in the British Isles is not terrestrial woodland but *Sphagnum* bog.

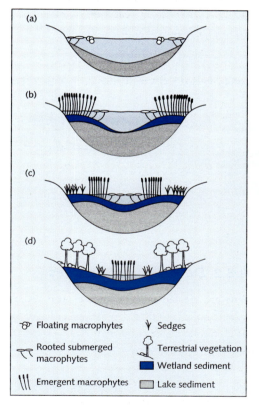

8

Figure 8.8 Stages in the development of a mire from a lake. See text for more details. (a) Open water with floating macrophytes; (b) encroachment by emergent vegetation; (c) infilling with *Carex* fen; (d) encroachment by terrestrial vegetation.

Legend:
- Floating macrophytes
- Rooted submerged macrophytes
- Emergent macrophytes
- Sedges
- Terrestrial vegetation
- Wetland sediment
- Lake sediment

Bog development

Bogs will form where precipitation exceeds evaporation and drainage is very slow. Establishment of *Sphagnum* further impedes drainage; *Sphagnum* lacks internal water conducting tissues, so draws water by external capillary action through spaces formed as its branches hang against its stem. *Sphagnum* grows in sponge-like aggregations which hold

(a)

8

(b)

Figure 8.9 Stages in the creation of a fen from a shallow lake. (a) A shallow lake, showing encroachment by emergent macrophytes: Decoy Lake, Orielton, Wales; (b) a fen on the site of an infilled lake: Tullanavert, Northern Ireland. Photos M. Dobson.

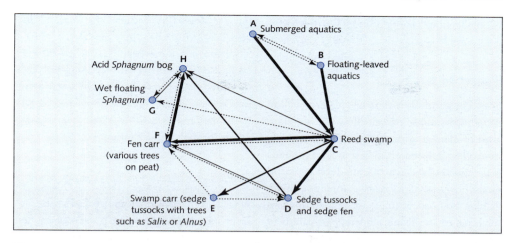

Figure 8.10 Successional stages which have occurred in lake hydroseres in the British Isles, as determined by analysis of peat cores. Arrows indicate direction of change, the width of each arrow being proportional to the number of times this change has been identified. In most cases, the sequence terminates with acid *Sphagnum* bog. Modifed from Chapman and Reiss (1992).

large volumes of water, such that up to 98% of a living *Sphagnum* carpet can be water. It has no roots and grows from the top, new growth continually shading and killing older parts. The dead tissues collapse and become compacted into a water-saturated, poorly permeable mass, the catotelm, covered by the living layer, the acrotelm, up to 40 cm thick and very permeable. As the moss continues to grow, the bog adds more organic matter to the catotelm, allowing it to rise above the surrounding land and raising the water table with it. Such raised bogs typically develop over the top of fens (but see **Box 8.3**), a rheotrophic mire being replaced, through the action of *Sphagnum*, by an ombrotrophic mire. The sequence is illustrated in **Figure 8.11**, starting with encroachment of fen vegetation in a rapidly infilling basin (**Figure 8.11a**). Eventually the fen completely dominates the basin (**Figure 8.11b**); the water table is high, allowing patches of *Sphagnum* to establish. These patches expand and coalesce and *Sphagnum* expands across the fen (**Figure 8.11c**). As the *Sphagnum* continues to grow, it develops a catotelm, which remains saturated, raising the water table. This allows the *Sphagnum* to expand upwards, creating a

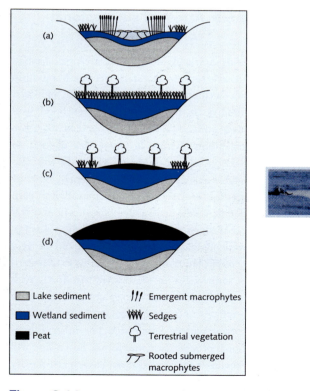

▢ Lake sediment	𝄇𝄇𝄇 Emergent macrophytes
▪ Wetland sediment	𝖶𝖶 Sedges
▪ Peat	♀ Terrestrial vegetation
	⟙⟙ Rooted submerged macrophytes

Figure 8.11 Development of a raised bog on top of a fen. See text for more details. (a) Fen encroachment and sedimentation; (b) a mature fen; (c) *Sphagnum* encroachment across the fen; (d) a raised *Sphagnum* peat bog.

8

raised bog of *Sphagnum* peat on top of fen peat (**Figure 8.11d**).

Blanket bogs

Cool, constantly wet climates, in combination with impermeable rocks, allow the development of blanket bogs. Again, *Sphagnum* is the controlling influence, but these are not confined to previously flooded areas, but develop on and spread over flat or gently sloping ground. Blanket bogs are features of temperate zones subjected to an oceanic climate, particularly in north-west Europe, and are typical of much of the north and west of the British Isles, where they can extend over large areas. The Flow Country of northern Scotland, for example, is a more or less continuous expanse covering nearly 4000 km^2.

8.6.3 Arrested successional communities

Fringe wetlands along rivers rarely pass beyond an early successional stage, because floods flush out accumulated silt and detritus. Similarly, floodwaters moving rapidly over flood wetlands shift sediment around and break down vegetation. Rivers in areas flat enough to develop aquatic marginal wetlands will also tend to change their course frequently, either by gradual erosion or catastrophically in flash floods, unless artificially constrained, and will cut through wetland areas. For these reasons, wetlands along river floodplains can be very heterogeneous, but rarely progress to terrestrial or mire stages. If floodwaters rise slowly with little kinetic energy, then physical changes are less likely to occur; encroachment by terrestrial species will be suppressed but other successional processes are unaffected. Succession will remain in an arrested state, however, if the dry season allows the mineralization of organic layers that have built up during the flood. If a river's ability to flood is curtailed, by damming or embanking, then terrestrial vegetation will

quickly dominate, because the wetland has been cut off from its supply of water.

Fringing marine wetlands may go through cyclical changes in which succession proceeds to a given stage and then is reset by storm activity. Dry land stages will only be reached rapidly if the wetland is actively expanding, for example through rapid siltation.

Cyclical succession in Sphagnum bogs

A well-established theory associated with ombrotrophic *Sphagnum* mires is the regeneration cycle, a cyclical succession which occurs in the hummock–hollow complex of these environments. The regeneration cycle hypothesis can be summarized as follows: as hummocks develop, they rise above the water table until they are dry enough to support terrestrial species (such as *Calluna*) which produce very little peat, so the hummock ceases to rise and may begin to erode. Within hollows, meanwhile, constant waterlogging ensures deposition of large amounts of peat, so that the hollows infill. Eventually, the hollows will reach the level of former hummocks, which become waterlogged and, themselves, converted into hollows, the former hollows, which continue to grow, become hummocks. In due course, these new hummocks will dry out and cease to grow, while the new hollows will infill with peat. By constant repetition of this cycle, a single point will, alternately, be occupied by hummock and hollow. This hypothesis is intuitively very appealing, as it describes an apparently logical progression of bog development, and has been widely presented as an established truth. Unfortunately, as Backéus (1990) points out, there is no evidence for its occurrence and, in all probability, the regeneration cycle does not occur.

8.6.4 Human influence on succession

The raised bog illustrated in **Figure 8.11** may be an end point in itself but, if precipitation is

very high and evaporation very low, then expansion into a blanket bog extending over the surrounding land is a possible outcome. Blanket bogs expanded over much of Britain, Ireland and Norway around 5000 BP, at a time of climatic deterioration into cooler, wetter conditions. This, in itself, may have been enough to trigger their development, but the role of human activity in this bog expansion is also implicated. Areas that are climatically susceptible to blanket bog expansion may remain dry if there is adequate tree cover; removing trees reduces transpiration losses and interception of rainfall, leading to increased groundwater levels and waterlogging, which favours development of *Sphagnum*. The expansion of blanket bogs in north-west Europe, most of which cover formerly forested areas, was probably enhanced by human activity, cutting and burning tree cover, particularly in areas marginal for blanket mire formation but where trees were under climatic stress.

Attempts at conservation of wetlands can activate successional processes. At the head of the Bay of Fundy in New Brunswick is a series of wetlands which have been contained within embankments, stabilizing water levels, in order to create open water habitats for waterfowl. Whereas these wetlands would formerly have dried out for short periods during the summer, allowing decomposition of organic matter and remineralization of nutrients, this can no longer occur where embankments impede drainage. In one such impoundment, Hog Lake, raised water levels since embankments were constructed in 1973 have led to domination by *Typha* × *glauca*, which forms floating mats consisting almost exclusively of this single species. Where these mats form, they insulate the water beneath from summer warming, reducing decomposition rates and nutrient release even further. This, is turn, leads to lower *Typha* vigour and invasion of mats by *Sphagnum*. It is likely that, in due course, *Sphagnum*, which acidifies the water and further suppresses *Typha*, will dominate.

8.7 Wetland extent and loss

8.7.1 Extent of wetlands

Calculation of the worldwide extent of wetlands is difficult because they are very scattered and definitions are often hard to apply. Wetlands do not form a discrete biome, so their extent cannot be estimated by delineating an appropriate climatic type. A reasonable estimate would be that wetlands cover about 6% of the Earth's land surface, or around 8.5 million km^2, of which coastal wetlands account for about one quarter of the total area. Freshwater wetlands are, however, very unevenly distributed. Their greatest extent is in two climatic zones—the boreal and tundra of the northern hemisphere, and more fragmented, but still sizeable, patches in equatorial regions (Figure 8.12a). Aselmann

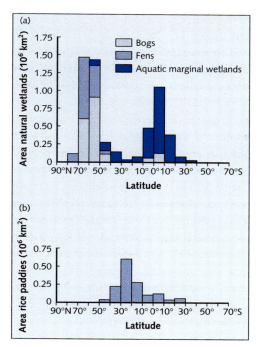

Figure 8.12 Distribution of wetlands by latitude. (a) Natural wetlands; (b) rice paddies. From Aselmann and Crutzen (1989).

8

Figure 8.13 Rice paddies, artificial wetlands managed for food production: Tamil Nadu, India. Photo M. Dobson.

and Crutzen (1989) estimated that approximately half of all natural freshwater wetlands are in an almost continuous expanse of mire across Canada, Alaska and Russia, and a further quarter is aquatic marginal vegetation associated with rivers and floodplains in the Amazon region of South America, these areas accounting for the two peaks of distribution in Figure 8.12a. Outside these regions, wetlands are generally small and scattered. There are exceptions, such as the Florida Everglades in the USA and the Okavango Delta (Box 8.4) in Botswana, but more typical are those on narrow river floodplains, or small, discrete wetlands in glacial hollows, such as the prairie pothole region of the north-central USA or the border region of Northern Ireland.

To the area covered by natural wetlands must be added cultivated rice paddies (Figure 8.13), which cover a further 1.3 million km², almost 90% of which is in South-east Asia. The distribution of natural wetlands shows a marked trough in northern subtropical latitudes (Figure 8.12a); to a large extent this is expected, because this marks an arid climatic zone, but it is also the latitudinal region with the greatest concentration of rice paddies (Figure 8.12b), many of which are former natural wetland areas which have been converted to cultivation.

Coastal wetlands, in contrast to freshwater wetlands, show a relatively even distribution around the world's coasts, although salt-marshes are confined to temperate regions and mangals to the tropics, with very few areas of overlap. This distribution is related to climate: mangroves, some species of which grow in excess of 10 m high, easily outcompete salt-marsh plants by overshading, but are very sensitive to frost. In Florida, one of the few places where both wetland types coexist, black mangrove (*Avicennia nitida*) is normally dominant, but is occasionally killed by frost, allowing salt-marsh species to flourish until it recovers.

8.7.2 Wetland loss

Calculation of the extent of wetlands is further compounded by the scale of their loss, particularly in Europe, North America and South-east Asia. One of the most infamous of the world's lost wetlands is the East Anglian Fenland of southern England, reduced from 3380 km^2 during the seventeenth century to around 10 km^2 today (**Figure 8.14**). The United States has had its area of wetland cover reduced from maybe 900,000 km^2 in the sixteenth century to around 400,000 km^2 today, but losses have been unevenly distributed, California and Ohio having lost 91% and 90%, respectively, of their original wetland cover. Such losses occur through active destruction, particularly to create agricultural land, but also indirectly by lowering the water table. The Coto Donãna in southern Spain is one of Europe's largest wetlands, and is protected as a National Park and a series of Natural Areas acting as buffers to development, but extraction of groundwater in the surrounding region has reduced its supply to the extent that its persistence as a wetland is under severe threat, despite its designation in 1983 as a Ramsar Site (see **Box 8.2**).

The loss of some coastal wetlands, in contrast, has been blamed on rising water levels. South-east England originally supported 40,000 ha of saltmarsh, of which less than 4500 ha now remain. Most of the loss was due to land reclamation over several centuries, but since the middle of the twentieth century, losses due to erosion have become severe, and now an estimated 40 ha per year is lost in this way. The standard explanation for this loss is that it is caused by coastal squeeze. Sea level rise would normally be countered by the saltmarsh migrating inland, but if the saltmarsh is restricted by a sea wall, it cannot do this and therefore simply erodes without being able to develop replacement marsh inland. In the south-east of England, relative sea level is rising due to isostatic land subsidence, and saltmarsh loss is extensive; elsewhere in the United Kingdom, where relative sea level is not rising, saltmarsh losses are minor.

There are two problems with this theory. First, relative sea level has been rising in many parts of the world over the past 10,000 years, at rates similar to those of today, and yet saltmarshes developed in these areas. The reason is that, if the supply of sediment is sufficient, saltmarsh will migrate vertically rather than horizontally in response to sea level rise. Saltmarshes can also extend seaward, even under conditions of sea level rise, as the many former seaports in south-east England, now several kilometres inland, testify. Second, coastal squeeze would be expected to result in loss first of high marsh species, then mid marsh species and finally low marsh pioneer zone species. However, the greatest losses have been of pioneer zone species.

An alternative theory is that recent losses are due to bioturbation and herbivory by invertebrate infauna (Hughes and Paramor 2004). Excluding invertebrates such as ragworm (*Hediste diversicolor*) from sections of intertidal sediment allows saltmarsh species to establish. There may be two alternative stable states (see Section 7.5), one in which ragworm disturbance

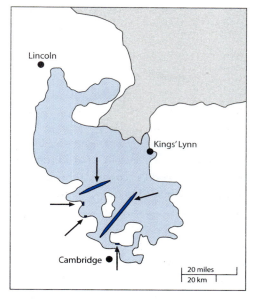

Figure 8.14 The decline in areas covered by wetland in the Fenland area of England between 1600 and 1900. The pale blue shading denotes the original extent of wetland. The arrows point to the five remaining wetland areas in this region.

of sediment suppresses plant establishment, the other in which plant roots suppress ragworm burrowing activities.

Ragworm has apparently increased in abundance in the region, supplanting bivalves as the main infaunal species. Its increase is possibly in response to increased organic matter loads caused by sewage input, but ironically it may be a consequence of widespread planting of the saltmarsh grass *Spartina anglica* in the early twentieth century. This increased the supply of coarse detritus on which ragworm can feed but other detritivores, such as bivalves, are unable to consume; now, with *Spartina* dying back once more, ragworm is benefiting again from increased food supply.

8.7.3 Wetland degradation

Wetlands can become badly degraded by inappropriate activities in their catchments. They are often sinks for excess nutrients and pollutants and, generally perceived as 'wasteland', small wetlands can become centres for dumping of waste products.

Even large wetlands can be degraded. The Florida Everglades system still covers most of its original extent, and is protected as the Everglades National Park and a series of Water Conservation Areas (Figure 8.15). However, it has suffered serious degradation, particularly as a result of altered hydrology. The Everglades is fed by the Kissimmee River, which flows south from its headwaters near Orlando into Lake Okeechobee. Before human intervention, this lake then overflowed along its entire southern margin, creating a 48 km wide channel that flowed an estimated 34 m per day south through the Everglades and into Florida Bay, a distance of around 150 km. As it flowed southwards, the 'river of grass' created a wetland dominated by sawgrass (*Cladium jamaicense*) immediately

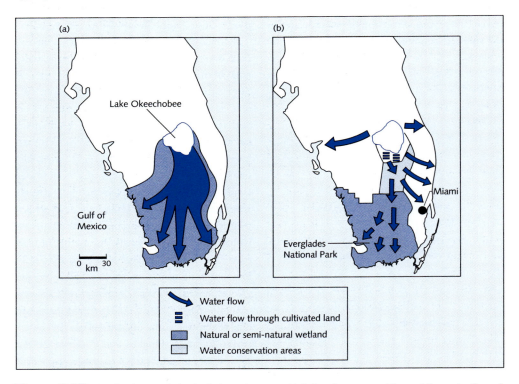

Figure 8.15 The Florida Everglades hydrological system. (a) Before European settlement the water flowed south from Lake Okeechobee, creating a shallow river many km wide; (b) today, the water is mainly deflected east and west and channelled through cultivated land (the Everglades Agricultural Area). Arrows show direction of flow of water.

south of the lake, and then a ridge and slough landscape, the sawgrass-dominated ridges standing up to 1 m higher than the water-filled sloughs. This landscape, maintained by the flow, was important in maintaining the bio-diversity of the wetland, as the sloughs remained water-filled in all but the driest years, providing a refuge for aquatic organisms, while the ridges provided refuges for those organisms requiring shallow water or dry land, and many were col-onized by trees, creating the characteristic tree islands of the system (Section 8.5).

In the mid twentieth century, four major alterations were made to this system. A levee was constructed around Lake Okeechobee, to stop the southward flow of water; the sawgrass plain immediately south of the lake was drained to create the Everglades Agricultural Area (EAA); the water from the lake was diverted west to the Gulf of Mexico and east to the Atlantic Ocean; and a series of barriers was built perpendicular to the flow in the wet-lands south of the EAA. This series of changes has had many consequences. Nutrient levels are rising as a consequence of inputs from the EAA; the ridge and slough system is degrading, with loss of tree islands, which further exacer-bates eutrophication; the dammed river bed has become a series of hydrologically separate systems, which drain out of their northern (high elevation) areas and into their southern (low elevation) areas, so that the southern parts flood too deeply while the northern parts are often dry and at risk from fire. Finally, within the Everglades National Park itself, the reduc-tion in freshwater inputs means that sea water and mangroves are encroaching, at a rate of 4 m per year.

In 2000, the Comprehensive Everglades Restoration Plan (CERP) was initiated, to try to remedy these problems. Its remit was simple: to allow water to flow south as it originally had done, rather than east–west. In practice, this requires piping water through the EAA, and then ensuring that the restored flow does not flood cities that have developed along its mar-gin, including Miami and Fort Lauderdale. In

view of the number of conflicting interests in the region, the success of the project is currently in doubt.

8.7.4 Wetland conservation

Natural successional processes lead to the loss of many wetlands to terrestrial environments, but these are countered by creation of new wetlands, particularly along coasts and river floodplains. The process of destruction has, however, been accelerated throughout the world by drainage and reclamation; at the same time, natural wetland creation processes are being impeded. As more rivers are impounded and coasts walled, so fewer new wetlands will be created to replace the losses.

To maintain their conservation value, there-fore, wetlands isolated from the natural dynamics of creation and destruction need to be heavily managed. In western Europe, where most remaining wetlands are small (less than 200 ha), sluices are commonly used in pro-tected areas to regulate water levels and to restore an approximation of natural flood cycles, while encroachment of terrestrial veget-ation is stemmed by clearance. Early succes-sional stages of reedbed development can be maintained by commercial harvesting of reeds, and in England the highest densities of bittern occur where up to 30% of the reed bed is har-vested in this way per year.

The first stage in drainage of a wetland is nor-mally creation of open water drainage chan-nels. However, these can then support a high diversity of organisms, often of conservation importance. Ironically, therefore, the open water channels are generally maintained as such in nature reserves, but flow is managed by sluices to maintain high water levels in the sur-rounding wetland area (Figure 8.16).

8.7.5 The value of wetlands

Quite apart from their intrinsic value as high diversity environments, wetlands provide much of benefit to human societies. Their potential

(a)

8

Figure 8.16 Examples of drainage ditches in wetlands. Ditches are dug to control water, but often develop into valuable conservation features, with a high diversity of species. (a) Inhaca Island, Mozambique; (b) Wicken Fen, southern England. Photos M. Dobson.

(b)

Box 8.5 Schwingmoors

Succession in kettle hole lakes may proceed directly to *Sphagnum*-dominated bog. Floating macro-phytes which grow outwards from the bank, such as leatherleaf (*Chamaedaphne calyculata*) in north-eastern North America, are important in this succession as they provide a substrate upon which *Sphagnum* and various sedges can become established. The peat produced from the partial decom-position of these species is slightly buoyant, creating a mat of peat which floats, albeit mainly below the surface. Organic debris which erodes from the mat, along with planktonic and inorganic debris from the pelagic zone, form a distinct layer of debris peat. Thus kettle hole bogs contain two horizontal lay-ers, the debris peat and the mat peat layer, with remains of leatherleaf in between. The mat peat can be further subdivided into three vertical zones. The outer edge is a zone of thickening of the floating mat, as more organic detritus is deposited, pushing the mat deeper into the water. When it becomes so deep that it comes into contact with the debris peat, further thickening cannot occur, so addition of new peat at the surface results in compaction. Closest to the shore is peat at its maximum density; no further compaction can occur and a zone of equilibrium is produced, in which organic matter addition is matched by decomposition at the surface.

Peat encroachment will eventually cover the open water and may overtop the lake centre, leaving a lens of water trapped within the centre of the fen, creating a quaking bog or schwingmoor (illustrated below), so-called because the lens of water allows it to move underfoot. When the open water is com-pletely covered, then the water table is able to rise and *Sphagnum* growth extends upwards, creating a raised bog.

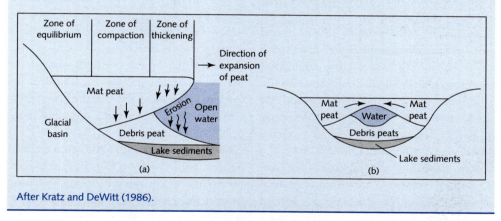

After Kratz and DeWitt (1986).

role in improving water quality has been considered in Section 8.3, and further benefits are outlined below.

Hydrological benefits

Wetlands associated with rivers reduce storm flows by absorbing excess surface water and releasing it slowly. In this way, flooding is local-ized in an environment adapted to high water tables, and the potentially damaging effects of

floods further downstream are reduced. Removal of wetlands into which excess water can drain, therefore, increases the frequency and magnitude of floods. **Table 8.2** shows the effect of increased urbanization and expansion of agricultural land around Baton Rouge, Louisiana, as wetlands and wet forest have been lost. The total volume of water passing along the Comite River, which drains this area, had not changed, but, as wetland sinks were lost, it became more prone to spates, its water

Table 8.2 Change in peak river discharge and number of floods along the Comite River, Louisiana, following wetland loss around Baton Rouge

	Percentage change in peak river discharge (1951–1960 vs 1961–1970)	Number of annual floods		
		1951–1960	1961–1970	1971–1978
Upstream of Baton Rouge	−2	2.4	2.6	2.6
Downstream of Baton Rouge	+23	3.8	3.9	5.3

After Stone *et al.* (1980).

passing through in flushes rather than being stored and released gradually. In the early 1970s, a series of options was considered for providing flood protection along the lower Charles River, near Boston, Massachusetts: construction of a reservoir capable of holding 68 million m^3 of water; extensive embankment of the river or of sensitive areas in the catchment; or protection of 3440 ha of wetlands. The purchase and protection of wetlands was eventually concluded to be an adequate flood control measure, and significantly less expensive than the other options.

Saltmarshes play a valuable role in coastal protection, dissipating the energy from storm surges and thereby reducing the requirement for expensive artificial flood defence structures. Saltmarshes are accreting environments, the vegetation slowing water flow over the marsh so that the sediment burden of the incoming tidal waters is deposited. This often leads to a marked change in level at the edge of the vegetated marsh compared to the adjacent tidal flats, this 'micro-cliff' and the extensive areas of gently sloping marsh acting to dissipate wave action and so prevent erosion of the coast at the high water mark. Concern has been expressed that rising sea levels will cause submergence of the marsh and so reduce its ability to provide coastal defence (Doody 1992), although this is unlikely unless sea level rise is particularly rapid. Establishment of new saltmarshes is being actively encouraged by flooding coastal meadows, in order to provide a coastal defence function. In the past, *Spartina anglica* was widely planted to enhance coastal defence in

temperate areas throughout the world, as it is a rapidly growing species that develops dense root systems, increasing silt deposition and binding sediments. Unfortunately, its invasive nature and rapid spread soon led to competition with native species and to losses of valuable tidal mudflats, and it has now become the subject of control attempts.

In Louisiana, more than 4000 km^2 of coastal wetlands have been lost to erosion since 1950, a consequence of sediment starvation following flood control works on the Mississippi River. Coastal wetlands are effective at absorbing the energy from storms, reducing the height of storm water by an estimated 5 cm per km. Therefore, the loss of wetlands around the Mississippi Delta has been implicated in the damage and loss of life experienced in New Orleans as a consequence of Hurricane Katrina in 2005 (Stokstad 2005).

Wetlands may also be important in recharging aquifers. In southern Florida, for example, the Everglades wetland system overlies a large expanse of porous limestone, the Biscayne aquifer, into which much of the water seeps. The area is very low-lying and channelization and drainage during the twentieth century have disrupted this recharge, allowing intrusion of saline water from the sea, a process compounded by heavy extraction from the aquifer for domestic supply.

Natural resources in wetlands

The Inner Niger Delta (Section 8.5, Figure 8.7) is but one of many African wetlands which support large numbers of people, dependent upon stock raising on the productive pasture during the dry season and fishing during the flood period. Floodplains and fringe wetland margins provide over one third of the entire freshwater fish catch in Africa and coastal wetlands are extremely important worldwide as spawning grounds and nursery areas for commercially important marine fish and crustaceans. To these benefits must be added the value of wetlands as sources of wood, reed thatch, peat and many other harvestable products.

Wetlands as refugia

The conservation value of wetlands lies in their intrinsic fauna and flora, and also in their role as refugia for species that have been persecuted in more accessible environments. Wetlands around Lake Victoria in East Africa are dominated by papyrus swamps which may represent an important refugium for native fish threatened by the introduced Nile perch (*Lates niloticus*), a situation described further in Section 7.7. *Lates* is inhibited by low oxygen levels and the structural complexity of the fringes of these wetlands. In contrast, many of the threatened native fish species, having evolved on the fringes of Lake Victoria or in isolated satellite lakes, are naturally predisposed to the oxygen concentrations and can take advantage of the structural diversity of the wetland fringe. Losses of native fish species in Lake Victoria are likely to be very high, but some of those believed to be extinct may persist in adjacent wetlands, either as remnants of formerly more widespread populations or species which have shifted their distribution away from the open water in response to heavy predation pressure (Chapman *et al.* 1996).

Bulleted summary

- Wetlands are areas of land defined by a water table which is permanently or frequently high, leading to waterlogged or flooded soils. Aquatic marginal wetlands are created by rivers, lakes and the sea. Mires are fed by groundwater, overland run-off or precipitation.

- Wetlands are characterized by anoxic soils, which often contain toxic chemicals and support bacterial activity which removes nitrates. Aquatic marginal wetlands have nutrient supplies replenished from the parent water body and, if seasonally dry, through decomposition, whereas permanently flooded mires may be nutrient-poor.

- Wetland plants generally have very specific tolerances to environmental conditions, leading to marked vertical zonation over very small elevational ranges. Their precise requirements ensure that they are very sensitive to fluctuations in water level.

- Seasonally flooded wetlands ('flood wetlands') and those on the edge of open water bodies ('fringe wetlands') are amongst the most productive environments on Earth, whereas rain-fed mires have very low productivity, but even slower decomposition in the waterlogged conditions leads to a build up of partially decomposed plant remains—peat.

- Many wetlands are grazed by large aquatic or semi-aquatic mammals and birds. Most invertebrate consumption is of detritus.

- Wetland diversity is enhanced by seasonal changes, particularly in those which are seasonally flooded and therefore support terrestrial and aquatic organisms at different times of the year.

8

Human exploitation of wetlands is often seasonal, including grazing livestock on rich pasture during the dry season and fishing during the wet season.

- Many processes in wetlands, including their formation and development, are mediated by the organisms inhabiting them. *Sphagnum* moss can control the water table to its own advantage, creating mires which often cover large areas. Few vascular plants possess the adaptations required to persist in *Sphagnum* mires.

- Groundwater-fed mires normally develop from infilling a lake or an arm of the sea. There is a clear successional sequence, the end point of which is often *Sphagnum* mire.

- Wetlands absorb excess nutrients and floodwater, as well as supplying natural resources. Losses, particularly to agriculture, have been extensive, and degradation of many wetlands is severe.

Further reading

Adams P. (1990) *Saltmarsh ecology*. Cambridge, Cambridge University Press.

Chapman V.J. (ed.) (1977) *Ecosystems of the world 1. Wet coastal ecosystems*. Amsterdam, Elsevier.

Gore A.J.P. (ed.) (1983) *Ecosystems of the world 4. Mires: Swamp, bog, fen and moor*. Amsterdam, Elsevier.

Hogarth P.J. (1999) *The biology of mangroves*. Oxford, Oxford University Press.

Hughes J.M.R. and Heathwaite A.L. (eds) (1995) *Hydrology and hydrochemistry of British wetlands*. Chichester, John Wiley and Sons Ltd.

Mitsch W.J. and Gosselink J.G. (2000) *Wetlands*, 3rd Edn. New York, John Wiley and Sons, Inc.

Whigham D.F., Dyjová D. and Hejný S. (eds) (1995) *Wetlands of the world I: Inventory, ecology and management*. Dordrecht, Kluwer Academic Publishers.

Assessment questions

1 Why do plants in many different wetland types need strategies to cope with low soil nitrate concentrations?

2 What is the role of seasonally fluctuating water levels in determining wetland species diversity?

3 What are the main threats to wetlands?

4 How do large mammals influence wetland extent and structure?

5 Which processes facilitate development of a bog from an open water body?

9

The Aquatic System

9.1 Introduction

At the beginning of this book we emphasized the similarities among aquatic environments, but then we proceeded to present aquatic habitats as separate blocks. To a large extent this was for convenience and, to some degree practical, as it differentiates habitats according to the ways in which they have historically been studied. It also, however, reflects natural divisions in energy flow and ecological processes. Aquatic habitats do separate naturally into discrete elements, but there are two points to bear in mind. First, aquatic habitats are not isolated from each other; rather, there are strong links and interconnections, considered in Section 9.2. Second, such divisions as can be identified among aquatic systems are more complex than the simple division between marine and freshwater environments. From an ecological perspective, more pertinent divisions can often be identified between pelagic and benthic communities, or between those with strong terrestrial links (rivers, estuaries, coastal seas, lakes, wetlands) and those with little terrestrial influence (deep oceans).

Within each habitat chapter, there is normally some homogeneity in form and function, although the extremes—large rivers versus small headwater streams or fringing reedbeds versus blanket bogs, for example—can be very different. The one exception to this is the lake chapter (Chapter 7). There are saline lakes, whose extremes of water chemistry—and particularly the fluctuations in chemical status to which they are subject—make them ecologically very different from either marine systems or freshwater lakes, and they are presented together only through a common feature of physical structure.

Despite the clear differences in physicochemistry, and in composition of the associated biota, common features can be identified among different aquatic systems, cutting across the marine–freshwater division. These parallel processes are considered in Section 9.3. Throughout the habitat chapters, the way in which the biota and the physical environment interact to produce an ecologically functioning system has been presented.

This can manifest itself in different ways, dependent upon the habitat and the organisms within it, but common patterns are considered in Section 9.4.

Increasingly, economic terminology is being applied to the biosphere. An example of this is the use of terms such as 'ecological goods and services'. This is often 'translated' by biologists such that the 'goods' equates to the living components of the system—the biota—and the 'services' are the biologically mediated processes occurring in the ecosystem. Humans therefore exploit both the goods, such as harvesting fish, and the services, including using the environment to assimilate our waste products. Concern about maintaining ecological goods and services often involves consideration of conservation of biological diversity, the range and abundance of the species present and concern about the delivery of ecosystem functions. Therefore, Section 9.5 extends ecological functioning of aquatic ecosystems to a consideration of the goods and services relevant to human systems. Through the habitat chapters, human influences have been introduced only insofar as they help to illustrate or understand natural processes in aquatic systems. From a human perspective, however, the attributes of aquatic systems with which we interact are crucially important, and our negative influences upon these attributes and processes may have serious repercussions. An integrated consideration of human influences upon aquatic systems is therefore pertinent.

9.2 Interconnections

Two processes link all the elements of the aquatic environment—the movement of water around the hydrological cycle and the flux of organic matter, particularly in the form of detritus. These links blur the distinctions between the various habitats and provide a major contrast between the energetics of terrestrial and aquatic ecosystems.

9.2.1 Links between aquatic habitats

The vast majority of the world's aquatic systems are strongly interconnected to other such systems. This is particularly clear in inland waters, where rivers, lakes and wetlands form a continuum of habitat types, often extending into estuarine and coastal systems. Wetlands are rarely separable from other water bodies, either being hydrologically fed and maintained by rivers or themselves being the source of open water systems. Wetlands, in turn, are important in providing energy and nutrients to rivers and lakes. Low-order rivers rely upon the terrestrial catchment for energy: higher-order rivers depend on export from upstream but also on lateral inputs from their floodplains, as emphasized by the River Continuum concept and the Flood Pulse concept respectively.

There are several spectacularly large and deep lakes in the world, but most are relatively shallow, filling low gradient basins which allow the development of extensive fringing wetlands and indeed the sedimentation and growth associated with wetlands serves to increase the area they cover. Large lakes may depend upon pelagic primary production, but most lakes are small and/or shallow enough for macrophytes to be the dominant source of production, particularly those with fringe wetlands. The strong interaction between a lake and its wetland fringe is emphasized by the successional process in which wetlands gradually encroach upon and infill lakes. Most lakes are connected to river systems, acting both as a sink for inputs from upstream and a source for outputs. Turbulent rivers have little plankton, but lake outlet rivers may contain high densities, constantly being exported from the lake pelagic and, in turn, supporting high concentrations of filter feeders taking advantage of this food resource (Figure 9.1).

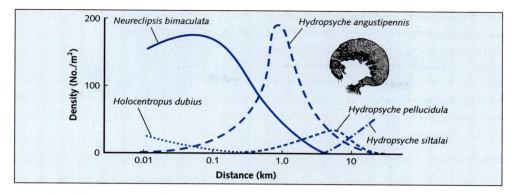

Figure 9.1 Downstream succession of five species of net-spinning caddis larvae in a lake-outlet stream in Sweden. From Giller and Malmqvist (1998).

The freshwater system spills out into coastal areas, where inputs from upstream, in combination with coastal wetlands, provide major fluxes of detritus which are exploited directly or provide nutrients to stimulate autochthonous primary production. The only aquatic system which is completely self-sufficient, albeit the one which accounts for the vast majority of habitable aquatic volume, is the open ocean. Productivity in this open ocean is, however, so low that even relatively small amounts of terrestrially or coastally derived plant detritus or nutrients can have an important effect upon local productivity. Within the oceans, there is a vertical connection between the upper illuminated layer and the dark layer beneath: the deep ocean relies upon export of excess production from the photic zone which, in turn, can become starved of nutrients in the absence of upwelling from depths.

Whereas the open ocean apparently does not require coastal, freshwater or terrestrial environments in order to function, these are dependent upon oceans as determinants of climate and sources of water. It is also interesting to speculate upon the ecological nature of oceans in the absence of the land masses which support coastal and freshwater systems; deprived of the major source of upwelling, bringing nutrients from the deep ocean, there would be an equivalent reduction in downwelling, with consequent impacts upon oxygenation at depth, and

so even the deep ocean would be ecologically very different to that which exists today.

9.2.2 Dispersal in aquatic organisms

Ephemeral habitats such as tree hole pools are apparently independent of all other water bodies but they are connected, at least to each other, in that their inhabitants require adaptations for dispersal and therefore provide a biological link. Dispersal ability is an important criterion for inhabitants of several different types of aquatic systems, and is positively correlated with the lifespan of the habitat. Estuaries, hydrothermal vents, lakes and ponds are all examples of systems whose inhabitants have generally high powers of dispersal and rapid colonization rates, in response to the relatively short existence of their habitat units. The few lakes currently in existence which have been present for more than a few hundred thousand years often have high endemicity, in contrast to the extremely low numbers or complete absence of endemics in most lakes, and it is probable that the same would occur in any estuary or hydrothermal vent which could persist for an evolutionarily meaningful timescale. In contrast, oceanic species are less well adapted to disperse between habitats, although perfectly able to redistribute themselves within a given habitat. The ocean is effectively a single

9

unit, with much opportunity for faunal exchange. Rivers are geologically long-lived features, but the variable physical environment tends to favour adaptable species which, although lacking the dispersal abilities of many lake species, allow enough dispersal to keep endemism low.

Segregation of life stages

The dispersal ability of many aquatic species is enhanced by spatial segregation of adult and juvenile phases within a species, one of the principal differences between aquatic and terrestrial communities. Such disjunctures in the life history have a number of ecological consequences. The spatial separation of the adults and larvae means they exploit different resources and so are not usually in competition with each other. The timing of breeding may allow synchronization with a seasonally abundant food supply, such as the spring phytoplankton bloom or the pulse of allochthonous detrital inputs associated with leaf fall, while the mobile stage allows colonization of new areas. In a system with a finite life, or in which there are continually renewed resource patches as a result of disturbance, the advantages to seeking a new patch are obvious.

Most aquatic populations are therefore highly fragmented, with a sessile or low mobility stage occupying one habitat and more mobile stage exploiting another habitat and set of resources and dispersing in search of new habitats. Populations of the sessile stage can be regarded as a metapopulation of the species, with the dispersal stages providing the route connecting them.

Migration

Migration is a periodic movement to or from a specific breeding area. Vertebrates, in particular, can migrate over long distances, well-known examples including movement of eels from fresh waters to the sea to breed, movement in the opposite direction of salmon, and calving migrations of whales.

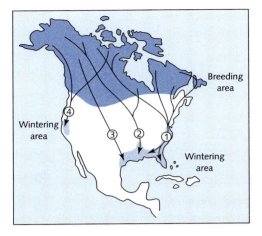

Figure 9.2 Waterfowl which nest in northern North America follow four main flyways to their wintering grounds in the south of the continent: (1) Atlantic coast flyway; (2) Mississippi flyway; (3) Central flyway; (4) Pacific flyway. Note that many of the wintering grounds are coastal, while breeding grounds are fresh water.

Wetland birds provide a connection, through migration, between wetland systems, often connecting freshwater and coastal wetlands. In North America, this has been intensively studied with respect to wildfowl, which are a valuable resource for sport hunting. In general, North American wildfowl breed in the north, in the prairie wetlands of the northern USA and southern Canada, but winter in the south, the breeding and wintering areas being linked by four major flyways (**Figure 9.2**). Not only does this process connect the larger wetland areas of north and south, but along each flyway, any other wetland, however small, takes on a crucial role as a stopover. In the eastern Hemisphere, the situation is probably even more complex, but not so well studied because many more countries are involved. Again, the major breeding grounds are in the tundra areas: Siberia, northern Scandinavia, Iceland, Greenland and even eastern Canada. Wintering grounds are many, but fall into four main regions: the many estuaries and mudflats of coastal northwest Europe, and the fewer, but individually

9

larger, wetlands in the Mediterranean, the Middle East and West Africa.

Human dispersal processes

The dispersal ability of many aquatic species is enhanced by transport of organisms beyond their native range by human activity. This process, reviewed in Frid and Dobson (2002), is often deliberate, particularly in the case of commercially important species, but accidental dispersal is also widespread.

Translocation in ballast water is perhaps the most well-known mechanism by which aquatic species are accidentally dispersed. Unladen cargo vessels carry ballast to ensure that they lie deep in the water, thus improving their stability; this is later jettisoned and replaced by cargo. Therefore, ballast is loaded at the arrival destination for the cargo and removed at the departure point for cargo. Most modern ocean-going vessels use water as ballast, transporting water and its inhabitants over large—often intercontinental—distances. A consequence of this process is that coastal and estuarine organisms have been widely dispersed in ballast water, often with serious consequences for the native fauna and flora. Some of these species— tolerant of both fresh and brackish water— have invaded fresh waters in this way, of which the prime example is the zebra mussel (*Dreissena polymorpha*), originally part of the Ponto-Caspian fauna of south-east Europe and the Caspian Sea region, but now established through Europe and, via the Great Lakes of North America, into the Mississippi River system. Having reached North America in ballast, it is now spreading naturally within the water bodies to which it was transported and, by attachment to boats that are then transported by road, into new river catchments **(Figure 9.3)**.

Transport by boat, whether as ballast or attached to the outer hull, is a passive transfer mechanism, but many aquatic species are extending their distributions by using artificial water bodies, including canals and interbasin transfer schemes. In central Europe, the con-

nection of the Rhine and Danube systems has allowed the movement of species into new catchments, and many crustaceans native to the Ponto-Caspian fauna are now extending their ranges into northern Europe via the Rhine and, through other canals, into adjacent catchments. This process is mainly confined to freshwater species, but includes also the Lessepsian fauna moving from the Red Sea to the Mediterranean via the Suez Canal.

9.2.3 Energy transfers

A common feature among aquatic systems is the importance of energy inputs which do not derive directly from photosynthesis. Autochthonous primary production is important only in shallow fresh waters and the upper mixed layer of oceans, while allochthonous inputs dominate all other aquatic systems.

The dependence of the aquatic system upon detritus is enhanced by transport between its various components, most obviously in the sinking of POM from the surface to the deep ocean or the hypolimnion of lakes, and in downstream transport along rivers and out into the estuary or open sea. In both cases, detritus undergoes modification as it progresses, much of this through the activity of intermediate consumers (including conditioning), and its nutritional value for macroconsumers is determined by its residence time in the aquatic system.

Detrital transport in low-order rivers is mainly from benthos to benthos, with the pelagic phase simply being downstream flushing (although microorganisms living on or in the detrital matter do not distinguish between these states; once established, they will persist whether the detrital particle is on the bed or in the water column). Only in larger rivers does autochthonous planktonic production become important, although even here benthos are heavily dependent upon POM transported from upstream. In shallow seas, too, the marine littoral fauna is dominated by sessile benthos dependent upon POM brought in by currents. In lentic and oceanic systems, transport is generally from

9

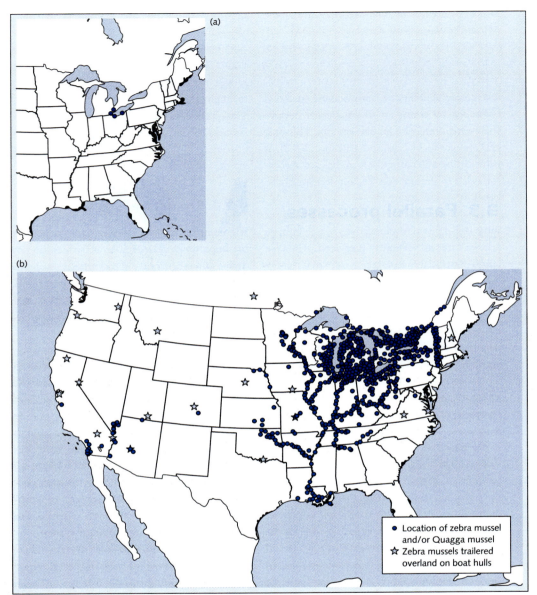

Figure 9.3 Distribution of zebra mussels and quagga mussels (*Dreissena* spp.) in North America (a) in 1988, soon after first being recorded in the Great Lakes region, and (b) in 2008. Mussels have spread along river systems connected either naturally or via canals to the Great Lakes, and have been carried between catchments on the hulls of boats transported by road. Data derived from the USGS.

pelagic production, via deep pelagic consumption, to benthos, where it is finally mineralized. Resuspension does, however, occur and, although much of this will sink to the bed, it is also utilized by pelagic consumers. Consumption by benthic detritivores can generate DOM

which provides resources, via the microbial loop, for plankton, while grazing or physical disturbance of macrophytes or epilithic algae adds DOM and nutrients to the pelagic system.

The main movement of detritus through aquatic systems is determined by gravity, including transport from upstream to downstream, or from surface to deep layers. Of critical importance, however, are processes that work in the opposite direction, returning the nutrients derived from decomposition back to zones in which primary production can occur. Upwelling zones in oceans, mixing processes across thermoclines and movement upstream or into the terrestrial environment by animals, all contribute to this recirculation of nutrients.

9.3 Parallel processes

Only three of the six habitat chapters in this book cover either only marine systems (coastal seas, open ocean) or only freshwater systems (rivers). The other three chapters cover transitional environments (estuaries), freshwater and saline—albeit rarely marine—environments (lakes and ponds) and both marine and freshwater environments (wetlands). These three chapters display common patterns which occur irrespective of salinity, and the links among some of these habitat types have already been considered in Section 9.2.

Parallels among habitats

Across the separate habitats, we can identify parallel processes. The analogies between large, deep lakes and coastal seas are clear, both being strongly influenced by stratification, the importance of phytoplankton as the main primary producer, but also the role of the adjacent land as a source of energy. There are of course important differences, particularly the normal state for permanently stratified deep lakes to be deoxygenated below the thermocline. This is a result of the lack of deepwater circulation which, in deep oceans, keeps water oxygenated throughout its depth, but is not an inevitable feature of freshwater lakes, as the example of Lake Baikal shows. Neither is the marine environment any less susceptible to deoxygenation if isolated: the Black Sea extends over nearly

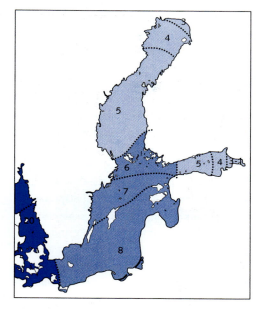

Figure 9.4 Salinity gradient in the Baltic Sea. Numerical values are PSUs.

440,000 km² and reaches a depth of 2200 m, but it is connected to the rest of the ocean only via the Bosphorus, a narrow, shallow channel not exceeding 125 m deep. Separated from deep ocean circulation, the Black Sea is anoxic below 200 m depth.

Estuaries are zones of transition between sea water and fresh water. Normally considered in terms of rivers entering coastal seas, estuarine conditions also occur throughout the Baltic Sea, which ranges from fully freshwater to around half of full oceanic salinity, due to high freshwater inputs and little opportunity for mixing with water from the North Sea (Figure 9.4). Conversely, the mouth of the Amazon discharges directly into the Atlantic Ocean, but inputs so much fresh water that freshwater conditions extend many kilometres from its mouth, and significant reductions in salinity during peak discharge season can extend up to 3000 km and encompass 2 million km². These two examples differ from true estuaries in that they lack the daily fluctuations associated with tidal movement; otherwise, they are effectively estuarine conditions but in a different structural environment to more typical estuaries.

Perhaps the clearest examples of parallel processes are in wetlands. Salt marshes and seasonally inundated freshwater wetlands show strong structural and functional similarities, the response of different species adapting to physically analogous conditions. High gradient rivers show some physical similarities with rocky sea shores, in that organisms inhabiting them need to be adapted to the strong physical perturbations that occur; the similarities are not exact of course, because the flow in a river is unidirectional and—from the perspective of its inhabitants—unpredictable, whereas that in the tidal sea shore is reversible and generally less variable. Tidal fluctuations are not replicated in fresh waters, except in tidally influenced areas close to river mouths, but neither are they of direct relevance to marine organisms beyond the littoral zone, and similar patterns of exposure and inundation have been created in fresh waters by pumped hydroelectric schemes.

Parallels among species traits

Parallel processes are not only identifiable at the habitat level, but also occur among species traits. Among functional feeding groups, the prevalence of generalist detritivores and scavengers in the benthos is not surprising, as detritus from the pelagic or the terrestrial catchment is such an important energy resource. The importance of grazers adapted to scraping solid surfaces and consuming the biofilm attests both to the relative paucity of multicellular plants in many photic environments and the good conditions for the development and maintenance of biofilm in aquatic habitats, in contrast to terrestrial environments.

Detritivory in soft substratum aquatic habitats often involves relatively immobile organisms that take advantage of water mobility to replenish food particles. Even on hard substrata, organisms can adopt this strategy, net-spinning caddis larvae, for example, building structures analogous to spiders' webs that capture particles from the water column. A natural

extension to this is the filter-feeding guild. Filter feeding is unique to aquatic environments, in which organisms are bathed in a medium both full of POM and, in most cases, continually replenishing itself. Furthermore, it is almost universal, wherever benthos can find suitable surfaces for attachment. A variation of filter feeding has developed in the pelagic zone among nekton, culminating in baleen whales and large planktivorous sharks.

9.4 Aquatic ecosystems and ecological functioning

9.4.1 Ecosystems and ecological functioning

In Chapter 2 we introduced the concept of ecological functioning, the role of the biota and biological processes in structuring ecosystems and in keeping them working. We identified eight types of ecological function delivered by aquatic systems (energy and elemental cycling, modification of physical processes, habitat/refugia provision, productivity, energy redistribution, community resistance and resilience, propagule export, adult immigration and emigration) and noted that most habitats would to some extent deliver all of them but that the contribution would vary.

In the habitat chapters we considered the main ecosystem functions each delivered. Primary production occurs in all aquatic environments, although in the aphotic deep ocean it is restricted to chemoautotrophy at hydrothermal vents. This production then supports productive food webs including export of production via living organisms (propagules, migration) and as detritus. In all environments decomposition of detritus occurs, so nutrients and carbon are recycled. Natural ecological communities will always include species that create habitat for other species, even if it is only for parasites, and such communities will have an ability to resist impacts and recover from perturbations. The more dynamic systems, such as those involving moving water, lentic systems, the coastal seas and the surface zones of lakes and the open ocean, are probably more robust than other less physically dynamic environments.

Such processes occur in terrestrial ecosystems. The greatest distinction between terrestrial and aquatic systems, therefore, and one which unites all aquatic systems, is not so much the presence of species or processes that deliver a particular type of ecological function but more the role of water as a medium for transporting the results of these processes around the biosphere. Moving water, whether in the currents of the oceans, the tides of the coastal seas, the circulation in lakes, the linear flows of rivers or the flood cycles of wetlands, moves production, nutrients, biogeochemically active processes, and organisms, thereby providing a link that ultimately connects the whole of the Earth ecosystem.

At this point it is worth noting how the terminology used with considerations of ecological functioning is often value laden. This is natural when considering ecological functions whose value to humankind is obvious, whether by production of goods such as food and raw materials or provision of valuable economic services such as waste assimilation and breakdown. However, the terminology is also used to imply value to the 'ecosystem' of processes that keep it 'working'. The ecosystem does not function as a 'super-organism', but is the product of individual organisms using resources, interacting with other species and the physical environment while seeking to maximize their 'fitness', and this results in the appearance of a 'designed' system. The appearance of design is the product of co-evolution and adaptation to create a dynamic stability, but change is inevitable, from short-term perturbation requiring an adjustment of species dominance through immigration and local extinction, to extremely long term processes measurable at the geolo-gical scale. For example, a group of organisms evolved to harness inorganic chemicals and light to synthesize organic matter. This produced oxygen as a waste product which over time built up in the environment. A new group of organisms evolved that could utilize this waste product to make their metabolism more efficient, a trait whose improved efficiency allowed it to persist and led to expansion and diversification of this group of organisms over time. The appearance is of a system designed to contain complementary groups, but it is simply the evolution of organisms into niches that allow them to exploit available resources, and competition, along with natural selection to ensure that the most efficient persist.

An ecosystem is an entirely human-defined construct. The definition of a species has a biological basis and hence can be seen as a 'natural' ecological unit. However, all other units, such as metapopulations, communities, habitats and ecosystems are subject to definition and different people may use vastly different definitions. For example, the North Atlantic Ocean is an ecosystem to some who work primarily at the ocean scale, but it is equally valid to refer to a 'rock pool ecosystem', of which there will be many along the ocean's boundaries. Equally, a pond can be seen as an ecosystem, or as simply a component of the larger terrestrial or wetland environment in which it sits, depending again upon the perspective of the viewer. To try to introduce some degree of consistency to this, ecosystems are often thought of as being delimited by units where the ecological linkages within the system are much larger than the linkages into and out of the system. For example the cycling of carbon within a large lake is much greater than the input and export of carbon from the lake, so the lake can be regarded as an ecosystem. However, as all the components of the biosphere are ultimately linked it still remains a matter of choice as to the cut-off point used.

Aquatic habitats that are clearly definable from a human perspective are, by this definition, not necessarily functioning as separate

9

ecosystems. For example, rivers are so closely linked with their catchments and sea shores with the adjacent marine pelagic, that to consider them as separate ecosystems would overlook the key linkages that maintain them.

9.4.2 Global-scale ecological functioning

It is an interesting thought experiment to consider a planet Earth without any ecological functions. Planetary scientists suggest that if there were no life on Earth, the temperature on the planet surface would be greater than 60°C, possibly as high as 100°C, due to the greenhouse effect of additional atmospheric carbon dioxide and methane emitted from volcanoes. There would therefore be no liquid water and the surface would be a giant arid desert. The residual heat in the planet's core would still drive a series of geochemical processes which would also gain energy from photochemical reactions. Without plants and photosynthesis, the atmosphere would be almost completely devoid of oxygen because unless molecular oxygen in the atmosphere is constantly replenished by photosynthesis, it is quickly consumed in chemical reactions, in the atmosphere, on land and in water. Nitrogen- and carbon-rich compounds such as ammonia and methane would be far more common in the environment. Combined with excess geothermal heat and lightning this would promote a chemically very active environment.

Thus biological processes, and their associated ecological functions, are the key to the production and maintenance of the biosphere. Extra-planetary scientists are now using spectra of planet chemistry to search for life on other worlds (Figure 9.5), as the presence of a large amount of oxygen in an extra-solar planet's atmosphere would provide strong evidence that it might host an ecosystem like present-day Earth's.

The only incontestable 'ecosystem' is planet Earth. Seen from space the Earth appears blue and white—the blue of the oceans and seas and the white of the clouds—and is therefore predominantly an aquatic ecosystem. While the mesopelagic zone of the oceans is the habitat with greatest volume on Earth and the deep sea floor the habitat with the greatest area, these environments are aphotic and so with the exception of the minor role of chemoautotrophy do not contribute to primary production. A comparison of the productivity of the marine, freshwater and terrestrial habitats shows that aquatic environments have, per unit area, comparable rates of carbon fixation to terrestrial systems (Table 9.1). The global carbon cycle has two elements: a short turnover cycle involving carbon cycling between atmospheric carbon dioxide and the biosphere (living organisms and organic matter) and a long term cycle including the geosphere. Despite their low rates of primary production, the oceans overwhelmingly dominate the long term carbon cycle. Atmospheric carbon dioxide is fixed into skeletal structures in marine organisms including coral reefs, shells and coccolith plates; these are deposited as carbonate sediments, becoming carbonate sedimentary rocks. Thus the carbon is sequestered away from the short term atmosphere–biosphere cycle. The carbonate

Table 9.1 Net primary productivity, as an index of carbon cycling, by a variety of aquatic and terrestrial habitats

	g C m^{-2} yr^{-1}
Desert scrub	10–250
Open ocean	2–400
Upwelling ocean	400–1000
Lake and stream	100–1500
Temperate grassland	200–1500
Tropical grassland	200–2000
Boreal forest	400–2000
Tropical rainforests	1000–3500
Cultivated land	100–4000
Estuaries	200–4000
Marine algal beds and reefs	500–4000
Freshwater wetlands*	100–5000

*See Table 8.1 for details.
Values given are normal ranges.

9

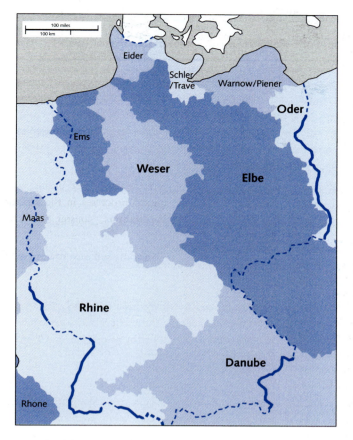

Figure 9.6 An example of the mismatch between natural and political boundaries: river basins in Germany with the international border superimposed. The border is shown as a solid blue line where it follows the course of a river; otherwise it is shown as a dashed line. The international border rarely follows the natural boundary between river catchments and in many cases follows rivers themselves, cutting the river basin into several national jurisdictions.

for environmental management. Amongst the key provisions was the recognition that people are part of the ecosystem and that, while use of ecological goods and services inevitably would lead to impacts on the system, the scale of these must be judged on the basis of their impact against their benefit and should include presumption not to change systems to such an extent that it constrained the options of future generations (the sustainability criteria). Furthermore there was a requirement for environmental management to follow an 'ecosystem approach', which requires management to give consideration to the full diversity of ecosystem components and to recognize the cumulative effects of the range of human pressures on the system. One of the main challenges to such integration in the past has been the lack of a common currency to understand ecosystem impacts from various sources. Recent advances in our understanding of the spatial structure of aquatic ecosystems, the development of novel techniques to link biological entities with the delivery of ecological functions and the emergence of spatial maps of human impacts mean that a common, spatial, framework now exists. We have plenty to learn about the ecology of aquatic systems, but we probably know enough about most systems in order to manage them sustainably; we simply need to apply the understanding that we have at our disposal.

9

Bulleted summary

- All aquatic habitats are interconnected. Even apparently physically isolated water bodies are connected through dispersive abilities of their inhabitants. Dispersal is an important component of aquatic biota, most species have a relatively (or absolutely) sessile or non-mobile life phase and a dispersive life phase. Migration provides another mechanism whereby aquatic organisms connect geographically separate habitats. Transfer of energy by movement of detritus is a key connection among different aquatic systems.

- It is difficult to identify any fundamental ecological distinctions between aquatic habitats. Any distinctions mask so much overlap in function and interconnection that aquatic habitats should be considered to be a single aquatic system, analogous to the single hydrological cycle.

- The Earth's surface is essentially an aquatic environment, with global scale ecological functions. Different aquatic ecosystems are identifiable, but boundaries between them will always be determined by opinion.

- Management of aquatic systems for human benefit must take into consideration the natural ecological processes and functions that generate the goods and services important for humanity. Management must take place within units defined by natural processes, such as catchments rather than political boundaries

9

Glossary

■■

A

AABW – Antarctic Bottom Water. Deep oceanic water which downwells off Antarctica.

Abiotic – referring to the non-living component of the environment; e.g. climate, temperature, pH (cf. **Biotic**).

Abyssal – referring to the deep ocean, 4000–6000 m in depth.

Acidification – increase in the acidity of water. Normally refers to anthropogenic activity.

Acidity – the concentration of alkalinity-buffering ions in water. Normally refers to the concentration of hydrogen ions, expressed as pH *(see Section 1.3)*.

Advect – to move laterally within a water column.

Alga (plural **algae**) – photosynthetic organism in the phylum Protoctista. Includes unicellular forms (e.g. diatoms) and multicellular forms (e.g. seaweeds).

Alkalinity – the concentration of acidity-buffering ions in water *(see Section 1.3)*.

Allochthonous – originating from outside a given site; e.g. allochthonous detritus (cf. **autochthonous**).

Alluvium – sediment deposited by a river.

Amphibyte – an organism whose life cycle involves stages in both groundwater and surface environments.

Amplitude – the difference in height between the crest and the trough of a wave.

Anadromy – see **Diadromy**.

Anoxic – deprived of oxygen.

Anthropogenic – caused by human activity.

Athalassic – referring to saline lakes whose saline water derives from non-marine sources (cf. **Thalassic**).

Attenuation – reduction in intensity of light by absorption or reflection as it passes through water *(see Box 1.2)*.

Aufwuchs – see **Biofilm**.

Autochthonous – originating from within a given site; e.g. autochthonous primary production (cf. **Allochthonous**).

Autotroph – an organism that obtains energy from inorganic substances through primary production (cf. **Heterotroph**).

B

Bedload – sediment, transported by a current, which moves by rolling or sliding and generally, therefore, remains in contact with the bed.

Benthic – referring to organisms living on or in the bed of a water body. **Benthos** – collective term for such organisms.

Biofilm – the layer of bacteria, algae, FPOM and extracellular bacterial secretions which develops on solid surfaces in water. Occasionally referred to as 'aufwuchs'.

Biogenic – created by biological activity.

Biome – a climatically determined vegetation zone defined by structural vegetation type rather than the actual plant species present; e.g. hot desert, tropical rainforest, tundra.

Biota – the living organisms within an area.

Biotic – referring to living components of the environment, or to products derived from living components; e.g. detritus (cf. **Abiotic**).

Biotic index – a water quality scoring system based upon the presence of living organisms.

Bioturbation – disturbance caused by biological activity.

Boundary layer – a layer of high-shear stress created by friction between a moving fluid and a stationary surface. Its upper limit occurs where the speed of the current is no longer influenced by the presence of the solid surface and may, in slow-flowing water bodies, extend to the surface. See **Current**.

BP – years before present. Used to denote dates which have been inferred from radiocarbon dating, for which the 'present day' is 1950.

Broadcast breeding – production of a dispersive reproductive stage e.g. spores, planktonic larvae.

C

Catadromy – see **Diadromy**.

Channelization – engineering a river channel to make it straighter and more homogeneous in structure than the natural channel.

Chemosynthesis – see **Primary production**.

Conditioning – the process by which detritus becomes more palatable to detritivores through the activity of decomposers.

Conductivity – a measure of salinity of water, through its ability to conduct electricity. Can be used to determine concentration of solutes in water *(see Section 1.3)*.

Coriolis force – the apparent force which deflects fluids moving over the surface of the rotating Earth from a linear to a circular trajectory *(see Box 1.1)*.

CPOM – Coarse particulate organic matter. Detrital particles retained by a 1 mm sieve.

Current – continuous movement of water within a water body. Current velocity—the speed of movement of water, normally expressed as m s^{-1}. The presence of a viscous (or laminar) sublayer, a layer of zero flow between a solid surface and a moving fluid, means that there must also be a gradient in current velocity, creating the boundary layer, close to the bed, where the change is most rapid. In a river, current velocity is measured at a point 0.6 of the depth of the channel (see figure below).

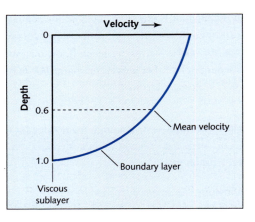

CWD – Coarse woody debris. Large pieces of wood of terrestrial origin.

Cyanobacteria – 'Blue-green algae': photosynthetic bacteria formerly known as Cyanophyta and classed as algae.

D

Dead zone – a portion of the river channel where current is reduced by the presence of a barrier to free water movement.

Deep ocean – see **hypolimnion**.

Demersal – living adjacent to, and in association with, the sea floor. e.g. demersal fish.

Detritus – particulate organic matter derived from formerly living organisms.

DHM – diel horizontal migration. The daily pattern of active horizontal movement in the water column by pelagic organisms.

Diadromy – referring to a life cycle which involves stages in both freshwater and marine habitats. **Anadromy** – migration from the sea to fresh water to breed; **catadromy** – migration from fresh water to the sea to breed.

Dimictic – referring to a lake whose mixing cycle has two periods of stratification per year.

Discharge – the total volume of water moving past a point, usually expressed in $m^3 \, s^{-1}$ (or cumecs—cubic metres per second), or, for low discharges, $l \, s^{-1}$. For a river, discharge (Q) may be approximated from:

$$Q = WDU$$

where W = stream width, D = mean depth and U = current velocity.

Disturbance – a discrete event which removes, damages or impairs the normal function of organisms.

DOC – dissolved organic carbon.

DOM – dissolved organic matter.

Downwelling – the movement of surface water into the body of the water column.

Drift – passive movement of living organisms with the current of a water body.

DVM – diel vertical migration. The daily pattern of active vertical movement in the water column by pelagic organisms.

E

Epibenthic – attached to or living on the bed of a water body.

Epifauna – organisms living or active on the surface of a water body.

Epilimnion – the surface layer of a stratified water body, above the thermocline. A limnological term equivalent to the mixed layer in marine systems.

Epilithic – attached to stones.

Epiphytic – attached to plants.

Epiphyte – an organism growing on a plant.

Euhydrophyte – a fully submerged macrophyte, including those with floating leaves.

Euphotic zone – see Photic zone.

Euryhaline – tolerant of a wide range of salinities (cf. **Stenohaline**).

Eutrophic – referring to a water body containing high concentrations of nutrients, relative to oligotrophic and mesotrophic water bodies. **Eutrophication** – the increase in nutrient concentrations in a water body, normally used to refer to the effects of anthropogenic pollution *(see Box 7.3)*.

Evaporation – the transition of a liquid or solid phase into a gaseous state. Can be used to refer to loss of surface water to the atmosphere.

Evapotranspiration – the sum of water lost from the land through evaporation and transpiration.

F

Flocculation – aggregation of small particles into larger particles.

Floodplain – the area flooded by a river with a predictable, normally seasonal frequency.

Fluvial – referring to flowing water.

FPOM – fine particulate organic matter. Detritus passing through a 1 mm sieve, but retained by a 0.45 μm sieve.

G

Glide (Run) – a stretch of river where flow is rapid but not turbulent.

Groundwater – water in sediments beneath the Earth's surface.

H

Hadal – referring to the deep ocean beneath 6000 m depth.

Halocline – a zone of abrupt change in salinity.

Heterotroph – an organism that obtains its energy from eating other living organisms or their by-products (cf. **Autotroph**).

Holoplankton – see **Plankton**.

Hydrology – the study of water and its properties on the Earth's surface, including atmospheric and groundwater.

Hydrosphere – the waters on the Earth's surface, including all water bodies which act as habitat for biota.

Hypertrophic – extremely eutrophic *(see Box 7.3)*.

Hypolimnion – the deep layer of a stratified lake, below the thermocline. A limnological term equivalent to the deep ocean in marine systems.

Hyporheic – referring to the sediments and interstitial water beneath a water body. **Hyporheos** – organisms inhabiting this zone.

Hypotonic – a solution of lower osmotic pressure (i.e. a relative term).

I

Infauna – benthos living within the bed sediments, rather than on its surface.

Interstitial – between the particles of sediments.

Isohaline – at the same salinity.

K

Karst – limestone-dominated landscape, typically containing caves and associated subterranean aquatic habitats.

L

Labile – easily assimilated or modified (cf. **Refractory**).

Laminar sublayer – see **Current**.

Levée – a raised embankment along the edge of a river.

Limnology – the study of inland waters.

Littoral zone (1) – in lakes, that part of the water body in which the photic zone extends to the bed. Also referred to as the **lake littoral**.

Littoral zone (2) – in the sea, the intertidal zone.

M

Ma – millions of years ago.

Macrofauna (macroinvertebrates) – animals retained on a 0.5 mm (500 µm) sieve.

Macrophyte – a large multicellular photo-synthesizing organism; generally applied to vascular plants but also refers to multicellular algae.

Marine snow – particulate detritus descending through the water column to the ocean bed.

Maximum sustainable yield (MSY) – the greatest harvest of a wild living resource (e.g. fish) which can be taken each year while still leaving a viable population to harvest the following year.

Meiofauna – animals retained on a 10 µm sieve but which pass through a 500 µm sieve.

Meroplankton – see **Plankton**.

Mesotrophic – referring to a water body containing a concentration of nutrients intermediate between that of eutrophic and oligotrophic water bodies *(see Box 7.3)*.

Metapopulation – a group of spatially distinct populations of a species which are linked by dispersal of individuals between them.

MHW – mean high water. The average height on the sea shore to which the water rises at high tide. **MHWS** – the average height of high tides during spring tides; **MHWN** – the average height of high tides during neap tides *(see Figure 5.1)*.

MLW – mean low water. The average height on the sea shore to which the water falls at low tide. **MLWS** – the average height of low tides during spring tides; **MHWN** – the average height of low tides during neap tides *(see Figure 5.1)*.

Microphyte – small, single-celled or filamentous photosynthesizer – algae and autotrophic bacteria. **Microphytobenthos** – microphyte inhabiting the bed of a water body.

Minerotrophic – see Mire.

Mire – a wetland whose water supply is independent of surface water bodies. **Minerotrophic** – referring to a mire fed by groundwater or overland flow. **Ombrotrophic** – referring to a mire fed only by rainwater and atmospheric inputs.

Mixed layer – see **Epilimnion**.

Monomictic – referring to a lake whose mixing cycle has one period of stratification per year.

N

NADW – North Atlantic Deep Water. Deep oceanic water which downwells in the North Atlantic.

Nauplius (plural **nauplii**) – the free-swimming first stage of certain crustaceans.

Nekton – organisms living within the water column which are able to swim or otherwise move independently of the current.

Neuston – organisms inhabiting the surface film of water.

Nutricline – a region of rapid change in concentration of one or more nutrients.

Nutrient cycle – the pathway followed by a nutrient as it changes from an inorganic to an organic form and back. **Nutrient spiral** – a condition in which the cycle is completed only when the nutrient has moved to a different geographical location.

O

Oceanography – the study of marine systems.

Oligohaline – tolerant only of low salinities (cf. **Stenohaline** and **Euryhaline**).

Oligomictic – referring to a lake with infrequent, aseasonal mixing.

Oligotrophic – referring to a water body containing low concentrations of nutrients, relative to mesotrophic or eutrophic water bodies *(see Box 7.3)*.

Ombrotrophic – see **Mire**.

Osmoconformer – a species whose internal salt concentrations fluctuate in response to changes in the external medium.

Osmoregulation – maintaining constant internal salt concentrations despite fluctuations in the external medium.

Oxycline – a region of rapid change in oxyygen concentration.

P

Paludification – development of a peat mire on dry land.

Pelagic – referring to the water column, as opposed to its bed or edges.

Periphyton – unicellular algae attached to surface layers.

pH – see Acidity.

Photic zone (euphotic zone) – the depth of water to which sunlight penetrates. The bottom of the euphotic zone is defined as the depth at which only 1% of light intensity at the surface remains.

Photosynthesis – see **Primary production**.

Phytoplankton – see **Plankton**.

Plankter – an individual member of the plankton.

Plankton (**planktonic**) – Organisms inhabiting the pelagic zone which have no independent means of propulsion, or are too small and weak to swim in the horizontal plane. **Holoplankton** – organisms which spend their entire lives as members of the plankton. **Meroplankton** – organisms which spend only part of their life cycle as members of the plankton. **Phytoplankton** – the photosynthesizing component of the plankton, mainly single-celled algae but including autotrophic bacteria. **Potamoplankton** – phytoplankton species which are endemic to river systems. **Zooplankton** – animal components of the plankton.

Pleistocene – the geological period incorporating the most recent Ice Ages. It began around 1.6 Ma and ended with the onset of the Holocene, or Recent, period about 10,000 BP.

POC – particulate organic carbon.

Polymictic – referring to a lake with frequent but aseasonal mixing.

POM – particulate organic matter.

Pool – a stretch of river where flow is relatively slow.

Potamoplankton – see **Plankton**.

Precipitation (1) – water that falls from the atmosphere to the surface of land or sea. including rain, snow etc.

Precipitation (2) – the process by which a substance dissolved in water separates out of solution and becomes a solid.

Primary production – production of organic compounds from inorganic components, using energy fixed from an external source. **Photosynthesis** – primary production in which energy is obtained from sunlight. **Chemosynthesis** – primary production in which energy is derived from chemical oxidation of simple inorganic compounds.

Pycnocline – a region of rapid change in density.

R

Redox – an abbreviation for oxidation-reduction, a chemical reaction in which an atom or ion loses electrons to another atom or ion. **Redox discontinuity layer (RDL)** – the layer at which a sediment changes from an oxygenated to an anoxic state.

Refractory – not readily assimilated and therefore slow to decompose (cf. **Labile**).

Rheotaxis – movement of an organism in response to current.

Rheotrophic – relying upon current to provide food.

Riffle – a stretch of river where flow is relatively rapid and turbulent.

Riparian – referring to the bank of a river.

Run – see **Glide**.

Run-off – rainfall which is not absorbed by soil, but passes into surface water bodies.

S

Salinity – the concentration of dissolved ions in water *(see Section 1.3)*.

Seston – suspended particles in water, including both living (plankton) and non-living (tripton) components.

Shear stress – the effect of a series of forces of different magnitude acting simultaneously at different parts of an object.

Sill – a shallow ridge which separates relatively deep parts of a water body.

Solute – a substance dissolved in another substance.

Stenohaline – adapted to a narrow range of salinities (cf. **Euryhaline**).

Stratification – development of discrete vertical layers within a water body *(see Box 1.3)*.

Stygobyte – an organism which is a permanent inhabitant of groundwater, including caves.

Succession – directional change through time at a site by means of colonization and local extinction (NB succession is a controversial process with several, often contradictory, definitions in ecology).

T

Taxon (plural **taxa**) – organisms classed together within the same taxonomic group (differs from species in that the taxonomic group may be genus, family, order etc.).

Terrestrialization – conversion of open water to dry land through successional processes.

Terrigenous – derived from land.

Thalassic – referring to saline lakes whose water is marine in origin (e.g. coastal lagoons) (cf. **Athalassic**).

Thermocline – zone between the epilimnion (or mixed layer) and hypolimnion (or deep ocean) within which the temperature changes abruptly *(see Box 1.3)*.

Tidal limit – the point furthest up an estuary (i.e. inland) where tides cause periodic vertical displacement of the water surface.

Tidal range – The height difference in sea level between high and low water at a given point.

Transpiration – transfer of water from soil to the atmosphere by living plants.

Tripton – non-living particles suspended in water.

Turbidity – the concentration of suspended particles in water.

U

Upwelling – the appearance at the surface of a water mass previously within the depths of the water body.

W

Water mass – a body of water characterized by certain values of its conservative properties (i.e. those not altered except by mixing), e.g. salinity, temperature.

Wave – a periodic displacement in a medium, normally used to refer to the succession of peaks and troughs on the surface of a water body. Temporally, there are two components to a wave. **Period** (T) – the time interval between successive peaks (or troughs) passing a fixed point. **Frequency** (f) – the number of peaks (or troughs) passing a fixed point per second (see figure below).

Z

Zooplankton – see Plankton.

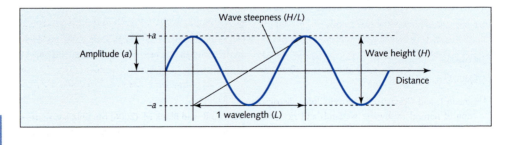

References

A

Adis J. and Junk W.J. (2002) Terrestrial invertebrates inhabiting lowland river floodplains of Central Amazonia and Central Europe: A review. *Freshwater Biology*, **47**, 711–731.

Aladin N., Crétaux J.-F., Plotnikov I.S., Kouraev A.V., Smurov A.O., Cazenave A., Egorov A.N. and Papa F. (2005) Modern hydro-biological state of the Small Aral Sea. *Environmetrics*, **16**, 375–392.

Allan J.D. (1995) *Stream ecology. Structure and function of running waters.* Chapman and Hall, London.

Allison G. (2004) The influence of species diversity and stress intensity on community resistance and resilience. *Ecological Monographs*, **74**, 117–134.

Ambrose W.G. (1984a) Role of predatory infauna in structuring marine soft-bottom communities. *Marine Ecology Progress Series*, **17**, 109–115.

Arnold S.L. and Ormerod S.J. (1997) Aquatic macroinvertebrates and environmental gradients in *Phragmites* reedswamps: Implications for conservation. *Aquatic Conservation: Marine and Freshwater Ecosystems*, **7**, 153–163.

Arrington D.A. and Winemuller K.A. (2006) Habitat affinity, the seasonal flood pulse, and community assembly in the littoral zone of a Neotropical floodplain river. *Journal of the North American Benthological Society*, **25**, 126–141.

Aselmann I. and Crutzen P.J. (1989) Global distribution of natural freshwater wetlands and rice paddies, their net primary productivity, seasonality, and possible methane emissions. *Journal of Atmospheric Chemistry*, **8**, 307–358.

Attrill M.J. and Thomas R.M. (1996) Long term distribution patterns of mobile estuarine invertebrates (Ctenophora, Cnidaria, Crustacea: Decapoda) in relation to hydrological parameters. *Marine Ecology Progress Series*, **143**, 25–36.

Azam F., Fenchel T., Field J.G., Gray J.S., Meyerreil L.A. and Thingstad F. (1983) The ecological role of water-column microbes in the sea. *Marine Ecology Progress Series*, **10**, 257–263.

B

Babcock R.C., Wills B.L. and Simpson C.J. (1994) Mass spawning of corals on a high-latitude coral-reef. *Coral Reefs*, **13**, 161–169.

Backéus I. (1990) The cyclic regeneration on bogs—a hypothesis that became an established truth. *Striae*, **31**, 33–35.

Badri A., Giudicelli J. and Prévot G. (1987) Effets d'une crue sur la communauté d'invertebrés benthiques d'une rivière méditerranéene, Le Rdat (Maroc). *Acta Oecologia, oecologia generalis*, **8**, 481–500.

Balirwa J.S., Chapman C.A., Chapman L.J., Cowx I.G., Geheb K., Kaufman L., Lowe-McConnell R.H., Seehausen O., Wanink J.H., Welcomme R.L. and Witte F. (2003) Biodiversity and fishery sustainability in the Lake Victoria basin: an unexpected marriage? *Bioscience*, **53**, 703–715.

Balls H., Moss B. and Irvine K. (1989) The loss of submerged plants with eutrophication I. Experimental design, water chemistry, aquatic plant and phytoplankton biomass in experiments carried out in Norfolk Broadland. *Freshwater Biology*, **22**, 71–87.

Barnes H. and Marshall S.M. (1951) On the variability of replicate plankton samples and some application of 'contagious' series to the statistical distribution of catches over restricted periods. *Journal of the Marine Biological Association of the UK*, **30**, 233–263.

Bastrop R. and Blank M. (2006) Multiple invasions—a polychaete genus enters the Baltic Sea. *Biological Invasions*, **8**, 1195–1200.

Beaugrand G., Reid P.C., Ibanez F., Lindley J.A. and Edwards M. (2002) Reorganization of North Atlantic marine copepod biodiversity and climate. *Science*, **296**, 1692–1694.

Bedford A.P. and Powell I. (2005) Long-term changes in the invertebrates associated with the litter of *Phragmites australis* in a managed reedbed. *Hydrobiologia*, **549**, 267–285.

Benedetti-Cecchi L. (2006) Understanding the consequences of changing biodiversity on rocky shores:

How much have we learned from past experiments? *Journal of Experimental Marine Biology and Ecology*, **338**, 193–204.

Berner E.K. and Berner R.A. (1987) *The global water cycle*. Prentice-Hall, Englewood Cliffs, New Jersey.

Blom C.W.P.M. and Voesenek L.A.C.J. (1996) Flooding: the survival strategies of plants. *Trends in Ecology and Evolution*, **11**, 290–295.

Bolger T. (2001) The functional value of species bio-diversity—a review. *Biology and Environment: Proceedings of the Royal Irish Academy*, **101B(3)**, 199–234.

Bondarenko N.A., Guselnikova N.E., Logacheva N.F. and Pomazkina G.V. (1996) Spatial distribution of phytoplankton in Lake Baikal, Spring 1991. *Freshwater Biology*, **35**, 517–523.

Boulton A.J. (2003) Parallels and contrasts in the effects of drought on stream macroinvertebrate assemblages. *Freshwater Biology*, **48**, 1173–1185.

Bravard J.P., Roux A.L., Amoros C. and Reygrobellet J.L. (1992) The Rhône River: a large alluvial temperate river. In P. Calow and G.E. Petts (eds) *The rivers handbook, volume 1*, pp. 426–447. Blackwell Science, Oxford.

Brönmark C. and Hanssen L.-A. (2005) *The biology of lakes and ponds*, 2nd Edn. Oxford: Oxford University Press.

Bruno J.F., Boyer K.E., Duffy J.E., Lee S.C. and Kertesz J.S. (2005) Effects of macroalgal species identity and richness on primary production in benthic marine communities. *Ecology Letters*, **8**, 1165–1174.

Burgis M.J. and Morris P. (2007) *The world of lakes*. Freshwater Biological Association, Ambleside.

Burton T.M., Stricker C.A. and Uzarski D.G. (2002) Effects of plant community composition and exposure to wave action on invertebrate habitat use of Lake Huron coastal wetlands. *Lakes and Reservoirs: Research and Management*, **7**, 255–269.

C

Cahoon L.B. (1999) The role of benthic microalgae in neritic ecosystems. *Oceanography and Marine Biology Annual Review*, **37**, 47–86.

Cameron W.M. and Pritchard D.W. (1963) Estuaries. In M.N. Hill (ed.) *The sea*, vol. **2**, pp. 306–324. Wiley and Sons, New York.

Carpenter S.R. and Lodge D.M. (1986) Effects of submerged macrophytes on ecosystem processes. *Aquatic Botany*, **26**, 341–370.

Carpenter S.R., Kitchell J.F. and Hodgson J.R. (1985) Cascading trophic interactions and lake productivity. *Bioscience*, **35**, 634–639.

Carrick H.J., Aldridge F.J. and Schelske C.L. (1993) Wind influences phytoplankton biomass and composition in a shallow, productive lake. *Limnology and Oceanography*, **38**, 1179–1192.

Cassie R.M. (1963) Micro-distribution of plankton. *Oceanography and Marine Biology, an Annual Review*, **1**, 223–252.

Cavanaugh C.M. (1994) Microbial symbiosis: Patterns of diversity in the marine environment. *American Zoologist*, **34**, 79–89.

Chapman J.L. and Reiss M.J. (1992) *Ecology, principles and applications*. Cambridge University Press, Cambridge.

Chapman L.J., Chapman C.A. and Chandler M. (1996) Wetland ecotones as refugia for endangered fishes. *Biological Conservation*, **78**, 263–270.

Chapman L.J., Chapman C.A., Schofield P.J., Olowo J.P., Kaufman L., Seehausen O. and Oguto-Ohwayo R. (2003) Fish faunal resurgence in Lake Nabugabo, East Africa. *Conservation Biology*, **17**, 500–511.

Chapman V.J. (ed.) (1977) *Ecosystems of the world 1. Wet coastal ecosystems*. Elsevier, Amsterdam.

Chavez F.P., Buck K.R., Coale K., Martin J.H., DiTullio G.R., Welschmeyer N.A., Jacobsen A.C. and Barber R.T. (1991) Growth rates, grazing, sinking, and iron limitation of equatorial Pacific phytoplankton. *Limnology and Oceanography*, **36**, 1816–1833.

Cherrill A.J. and James R. (1987) Character displacement in *Hydrobia*. *Oecologia*, **71**, 618–623.

Clymo R.S. and Hayward P.M. (1982) The ecology of *Sphagnum*. In A.J.E. Smith (ed.) *Bryophyte ecology*, pp. 229–289. Chapman and Hall, London.

Collie J.S., Richardson K. and Steele J.H. (2004) Regime shifts: can ecological theory illuminate the mechanisms? *Progress in Oceanography*, **60**, 281–302.

Connell J.H. (1961) The influence of interspecific competiiton and other factors on the distribution of the barnacle *Chthamalus stellatus*. *Ecology*, **42**, 710–723.

Connell J.H. (1978) Diversity in tropical rainforests and coral reefs. *Science*, **199**, 1302–1309.

Connor V.M. and Quinn J.F. (1984) Stimulation of food species growth by limpet mucus. *Science*, **225**, 843–844.

Corbet P.S. (1964) Temporal patterns of emergence in aquatic insects. *Canadian Entomologist*, **96**, 264–279.

Cryer M.G., Pierson G. and Townsend C.R. (1986) Reciprocal interactions between roach (*Rutilus rutilus* L.) and zooplankton in a small lake: prey dynamics and fish growth and recruitment. *Limnology and Oceanography*, **31**, 1022–1038.

Cummins K.W. (1974) Structure and function of stream ecosystems. *Bioscience*, **24**, 631–641.

D

Davy-Bowker J., Murphy J.F., Rutt G.P., Steel J.E.C. and Furse M.T. (2005) The development and testing of a macroinvertebrate biotic index for detecting the impact of acidity on streams. *Archiv für Hydrobiologie*, **163**, 383–403.

Dayton P.K. (1984) Processes structuring some marine communities: are they general? In D.R. Strong Jnr, D. Simberloff, L.G. Abele and A.B. Thistle (eds) *Ecological communities: Conceptual issues and the evidence*, pp. 181–197. Princeton University Press, Princeton, NJ.

Dayton P.K. and Hessler R.R. (1972) Role of biological disturbance in maintaining diversity in the deep sea. *Deep-Sea Research*, **19**, 199–208.

De Forges B.R., Koslow J.A. and Poore G.C.B. (2000) Diversity and endemism of the benthic seamount fauna in the southwest Pacific. *Nature*, **405**, 944–947.

DEFRA (2005) *Charting progress: An integrated assessment of the state of UK seas*. London: Department of Environment, Food and Rural Affairs, 130.

Denman K.L. (1994) Scale determining biological-physical interactions in oceanic food webs. In P.S. Giller, A.G. Hildrew and D.G. Raffaelli (eds) *Aquatic ecology. Scale, pattern and process. 34th Symposium of the British Ecological Society*, pp. 377–403. Blackwell Science, Oxford.

Denny M.W. (1987) Life in the maelstrom: The biomechanics of wave swept rocky shores. *Trends in Ecology and Evolution*, **2**, 61–66.

Denny M.W. (2006) Ocean waves, nearshore ecology, and natural selection. *Aquatic Ecology*, **40**, 439–461.

Denny P. (1995) Wetlands of Africa. Eastern Africa. In D.F. Whigham, D. Dyjová and S. Hejný (eds) *Wetlands of the world I: Inventory, ecology and management*, pp. 1–31. Kluwer Academic Publishers, Dordrecht.

Descy J.P., Gosselain V. and Evrard F. (1994) Respiration and photosynthesis of river phytoplankton. *Verhandlungen der Internationale Vereinigung für Theoretische und Angewandte Limnologie*, **25**, 1555–1560.

Devine J.A. and Vanni M.J. (2002) Spatial and seasonal variation in nutrient excretion by benthic invertebrates in a eutrophic reservoir. *Freshwater Biology*, **47**, 1107–1121.

Dobson M., Mathooko J.M., Ndegwa F.K. and M'Erimba C. (2004) Leaf litter processing rates in a Kenyan highland stream, the Njoro River. *Hydrobiologia*, **519**, 207–210.

Dodson S. (1989) Predator-induced reaction norms. *Bioscience*, **39**, 447–452.

Domis L.N.D., Mooij W.M. and Huisman J. (2007) Climate-induced shifts in an experimental phytoplankton community: a mechanistic approach. *Hydrobiologia*, **584**, 403–413.

Doney S.C. (2006) The dangers of ocean acidification. *Scientific American*, **294**, 58–65.

Doody J.P. (1992) Sea defence and nature conservation: Threat or opportunity? *Aquatic Conservation: Marine and Freshwater Ecosystems*, **2**, 275–283.

Duarte C.M. and Chiscano C.L. (1999) Seagrass biomass and production: a reassessment. *Aquatic Botany*, **65**, 159–174.

Duffy J.E. and Stachowicz J.J. (2006) Why biodiversity is important to oceanography: potential roles of genetic, species, and trophic diversity in pelagic ecosystem processes. *Marine Ecology Progress Series*, **311**, 179–189.

Dugdale R.C. and Goering J.J. (1967) Uptake of new and regenerated forms of nitrogen in primary productivity. *Limnology and Oceanography*, **12**, 196–206.

E

Ehrenfeld J.G. and Schneider J.P. (1991) *Chamaecyperis thyoides* wetlands and suburbanization: Effects on hydrology, water quality and plant community composition. *Journal of Applied Ecology*, **28**, 467–490.

Elser J.J. and Hassett R.P. (1994) A stoichiometric analysis of the zooplankton-phytoplankton interaction in marine and freshwater ecosystems. *Nature*, **370**, 211–214.

Engel A., Zondervan I., Aerts K., Beaufort L., Benthien A., Chou L., Delille B., Gattuso J.P., Harlay J., Heemann C., Hoffmann L., Jacquet S., Nejstgaard J., Pizay M.D., Rochelle-Newall E., Schneider U., Terbrueggen A. and Riebesell U. (2005) Testing the direct effect of CO_2 concentration on a bloom of the coccolithophorid *Emiliania huxleyi* in mesocosm experiments. *Limnology and Oceanography*, **50**, 493–507.

F

Falkowski P., Scholes R.J., Boyle E., Canadell J., Canfield D., Elser J., Gruber N., Hibbard K., Högberg P., Linder S., Mackenzie F.T., Moore B. III, Pedersen T., Rosenthal Y., Seitzinger S.,

Smetacek V. and Steffen W. (2000) The global carbon cycle: a test of our knowledge of Earth as a system. *Science*, **290**, 291–296.

Fauchald K. and Jumars P.A. (1979) The diet of worms: a study of Polychaete feeding guilds. *Oceanography and Marine Biology Annual Review*, **17**, 193–284.

Fenchel T. (1975) Character displacement and co-existence in mud snails (Hydrobiidae) *Oecologia*, **20**, 19–32.

Finn R.N. and Kristoffersen B.A. (2007) Vertebrate vitellogenin gene duplication in relation to the '3R hypothesis': correlation to the pelagic egg and the oceanic radiation of teleosts. *Public Library of Science ONE*, **2**, e169. doi:10.1371/journal.pone.0000169.

Flecker A.S. and Allan J.D. (1988) Flight direction in some Rocky Mountain mayflies (Ephemeroptera), with observations of parasitism. *Aquatic Insects*, **10**, 33–42.

Fojt W. (1994) The conservation of British fens. *British Wildlife*, **5**, 355–366.

Forster R.M., Créach V., Sabbe K., Vyverman W. and Stal L.J. (2006) Biodiversity-ecosystem function relationship in microphytobenthic diatoms of the Westerschelde estuary. *Marine Ecology Progress Series*, **311**, 191–201.

Frid C. and Dobson M. (2002) *Ecology of aquatic management*. Harlow, Pearson Education Limited.

Frid C.L.J. and Huliselan N.V. (1996) Far field control of long term changes in Northumberland (NW North Sea) coastal zooplankton. *ICES Journal of Marine Science*, **53**, 972–977.

Frid C.L.J. and James R. (1988) The role of epibenthic predators in structuring the marine invertebrate community of a British coastal salt marsh. *Netherlands Journal of Sea Research*, **22**, 307–314.

Frid C.L.J., Paramor O.A.L., Brockington S. and Bremner, J. (2008) Incorporating ecological functioning into the designation and management of marine protected areas. *Hydrobiologia*, **606**, 69–79.

Fromentin J.M. and Powers J.E. (2005) Atlantic bluefin tuna: population dynamics, ecology, fisheries and management. *Fish and Fisheries*, **6**, 281–306.

G

Gallais J. (1984) *Hommes du Sahel*. Flammarion, Paris.

Garthe S., Camphuysen C.J. and Furness R.W. (1996) Amounts discarded by commercial fisheries and their significance as food for seabirds in the North Sea. *Marine Ecology Progress Series*, **136**, 1–11.

Gibbard P.L. (1988) The history of the great northwest European rivers during the past three million years. *Philosophical Transactions of the Royal Society of London B*, **318**, 559–602.

Giller P.S. and Malmqvist B. (1998) *The biology of streams and rivers*. Oxford, Oxford University Press.

Giller P.S., Sangpradub N. and Twomey H. (1991) Catastrophic flooding and macroinvertebrate community structure. *Verhandlungen der Internationale Vereinigung für Theoretische und Angewandte Limnologie*, **24**, 1724–1729.

Gophen M., Ochumba P.B.O. and Kaufman L.S. (1995) Some aspects of perturbation in the structure and biodiversity of the ecosystem of Lake Victoria (East Africa) *Aquatic Living Resources*, **8**, 27–41.

Guralnick R. (2004) Life-history patterns in the brooding freshwater bivalve *Pisidium* (Sphaeridae). *Journal of Molluscan Studies*, **70**, 341–351.

H

Hairston N.G., Smith F.E. and Slobodkin L.B. (1960) Community structure, population control, and competition. *American Naturalist*, **94**, 421–425.

Hall R.J., Likens G.E., Fiance S.B. and Hendrey G.R. (1980) Experimental acidification of a stream in the Hubbard Brook Experimental Forest, New Hampshire. *Ecology*, **61**, 976–989.

Harris L.D. (1988) The nature of cumulative impacts on biotic diversity of wetland invertebrates. *Environmental Management*, **12**, 675–693.

Healey M.C. and Richardson J. (1996) Changes in the productivity base and fish populations of the lower Fraser River associated with historical changes in human occupation. *Archiv für Hydrobiologie Supplement 113* (Large Rivers 10), 279–290.

Hemminga M.A., Slim F.J., Kazungu J., Ganssen G.M., Nieuwenhuize J. and Kruyt N.M. (1994) Carbon outwelling from a mangrove forest with adjacent seagrass beds and coral-reefs (Gazi Bay, Kenya). *Marine Ecology-Progress Series*, **106**, 291–301.

Hiddink J.G., Jennings S. and Kaiser M.J. (2006) Indicators of the ecological impact of bottom-trawl disturbance on seabed communities. *Ecosystems*, **9**, 1190–1199.

Hildrew A.G. and Edington J.M. (1979) Factors facilitating the coexistence of hydropsychid caddis larvae (Trichoptera) in the same river system. *Journal of Animal Ecology*, **48**, 557–576.

Hildrew A.G. and P.S. Giller (1994) Patchiness, species interactions and disturbance in the stream benthos. In Giller, P.S., A.G. Hildrew and D.G. Raffaelli (eds) *Aquatic ecology. Scale, pattern and process. 34th Symposium of the British Ecological Society*, pp. 21–62. Blackwell Science, Oxford.

Hoarau G., Piquet A.M.T., van der Veer H.W., Rijnsdorp A.D., Stam W. and Olsen J.L. (2004) Population structure of plaice (*Pleuronectes platessa* L.) in northern Europe: A comparison of resolving power between microsatellites and mitochondrial DNA data. *Journal of Sea Research*, **51**, 183–190.

Hohensinner S., Habersack H., Jungwirth M. and Zauner G. (2004) Reconstruction of the characteristics of a natural alluvial river–floodplain system and hydromorphological changes following human modifications: the Danube River (1812–1991) *River Research and Applications*, **20**, 25–41.

Hudson I.R., Wigham B.D., Billett D.S.M. and Tyler P.A. (2003) Seasonality and selectivity in the feeding ecology and reproductive biology of deep-sea bathyal holothurians. *Progress in Oceanography*, **59**, 381–407.

Hughes R.G. and Paramor O.A.L. (2004) On the loss of saltmarshes in south-east England and methods for their restoration. *Journal of Applied Ecology*, **41**, 440–448.

Hughes T.P. (1994) Catastrophes, phase shifts and large-scale degradation of a Caribbean coral reef. *Science*, **265**, 1547–1551.

Humpesch U.H. (1992) Ecosystem study Altenwörth: impacts of a hydroelectric power-station on the River Danube in Austria. *Freshwater Forum*, **2**, 33–58.

Huryn A.D., Butz Huryn V.M., Arbuckle C.J. and Tsomides L. (2002) Catchment land-use, macroinvertebrates and detritus processing in headwater streams: taxonomic richness versus function. *Freshwater Biology*, **47**, 401–415.

Hutchinson G.E. (1961) The paradox of the plankton. *American Naturalist*, **95**, 137–145.

Hylleberg J. (1975) Selective feeding by *Abarenicola pacifica* with notes on *Abarenicola vagabunda* and a concept of gardening in lugworms. *Ophelia*, **14**, 113–137.

I

Ieno E.N., Solan M., Batty P. and Pierce G.J. (2006) How biodiversity affects ecosystem functioning: Roles of infaunal species richness, identity and density in the marine benthos. *Marine Ecology– Progress Series*, **311**, 263–271.

Irvine K., Moss B. and Balls H. (1989) The loss of submerged plants with eutrophication II. Relationships between fish and zooplankton in a set of experi-

mental ponds, and conclusions. *Freshwater Biology*, **22**, 89–107.

Iversen T.M., Thorup J. and Scriver J. (1982) Inputs and transformations of allochthonous particulate organic matter in a headwater stream. *Holarctic Ecology*, **5**, 10–19.

J

Jamieson C.D. (2005) Coexistence of two similar copepod species, *Eudiaptomus gracilis* and *E. graciloides*: the role of differential predator avoidance. *Hydrobiologia*, **542**, 191–202.

Jenkins S.R. (2005) Larval habitat selection, not larval supply, determines settlement patterns and adult distribution in two chthamalid barnacles. *Journal of Animal Ecology*, **74**, 893–904.

Jennings, S., Alvsvag, J., Cotter, A.J.R., Ehrich, S., Greenstreet, S.P.R., Jarre-Teichmann, A., Mergardt, N., Rijnsdorp, A.D. & Smedstad, O., 1999. Fishing effects in northeast Atlantic shelf seas: patterns in fishing effort, diversity and community structure. III. International trawling effort in the North Sea: an analysis of spatial and temporal trends. *Fisheries Research*, **40**, 125–134.

Jeppesen E., Jensen J.P., Søndergaard M., Lauridsen T. and Landkildehus F. (2000) Trophic structure, species richness and biodiversity in Danish lakes: Changes along a phosphorus gradient. *Freshwater Biology*, **45**, 201–218.

Jónasson P.M. (1996) Limits for life in the lake ecosystem. *Verhandlungen der Internationale Vereinigung für Theoretische und Angewandte Limnologie*, **26**, 1–33.

Jonsson M. and Malmqvist B. (2000) Ecosystem process rate increases with animal species richness: Evidence from leaf-eating, aquatic insects. *Oikos*, **89**, 519–523.

Junk W.J. (2005) Flood pulsing and the linkages between terrestrial, aquatic, and wetland systems. *Verhandlungen der Internationale Vereinigung für Theoretische und Angewandte Limnologie*, **29**, 11–38.

Junk W.J., Bayley P.B. and Sparks R.E. (1989) The Flood Pulse Concept in river–floodplain systems. In D.P. Dodge (ed.) *Proceedings of the International Large River Symposium*, pp. 110–127. National Research Council, Ottawa, Canadian Special Publications in Fisheries and Aquatic Sciences, 106.

K

Kaiser M.J. and Spencer B.E. (1996) The behavioural response of scavengers to beam trawl disturbance. In S.P.R. Greenstreet and M.L. Tasker (eds) *Aquatic*

predators and their prey, pp. 117–123. Blackwell Science, Oxford.

Kamykowski D. and Zentara S.J. (2005) Changes in world ocean nitrate availability through the 20th century. *Deep-Sea Research Part I – Oceanographic Research Papers*, **52**, 1719–1744.

Keith P. (2003) Biology and ecology of amphidromous Gobiidae of the Indo-Pacific and the Caribbean regions. *Journal of Fish Biology*, **63**, 831–847.

Kiørboe T. (1993) Turbulence, phytoplankton cell size and the structure of pelagic food webs. *Advances in Marine Biology*, **29**, 1–72.

Kosarev A.N. and Yablonskaya E.A. (1994) *The Caspian Sea*. SPB Academic Publishing, The Hague.

Kratz T.K. and DeWitt C.B. (1986) Internal factors control peatland-lake ecosystem development. *Ecology*, **67**, 100–107.

Kristensen R.M. (2002) An introduction to Loricifera, Cycliophora, and Micrognathozoa. *Integrative and Comparative Biology*, **42**, 641–651.

Künitzer A., Basford D., Craeymeersch J.A., Dewaramez J.M., Dörjes J., Duineveld G.C.A., Eletheriou A., Heip C., Herman P., Kingston P., Niermann U., Rachor E., Rumohr H. and deWilde P.A.J. (1992) The benthic fauna of the North Sea: Species distribution and benthic assemblages. *ICES Journal of Marine Science*, **49**, 127–143.

L

Lake P.S. (2003) Ecological effects of perturbation by drought in flowing waters. *Freshwater Biology*, **48**, 1161–1172.

Lalli C.M. and Parsons T.R. (1993) *Biological oceanography: An introduction*. Butterworth-Heinemann, Oxford.

Lampert W. (1989) The adaptive significance of diel vertical migration in zooplankton. *Functional Ecology*, **3**, 21–27.

Lancaster J. and Belyea L.R. (1997) Nested heirarchies and scale-dependence of mechanisms of flow refugium use. *Journal of the North American Benthological Society*, **16**, 221–238.

Laws R.M. (1985) The ecology of the Southern Ocean. *American Scientist*, **73**, 26–40.

Lecerf A., Dobson M., Dang C. and Chauvet E. (2005) Riparian plant diversity loss affects stream ecosystem function through direct and indirect effects in detrital food webs. *Oecologia*, **146**, 432–442.

Lees K., Pitois S., Scott C., Frid C. and Mackinson S. (2006) Characterizing regime shifts in the marine environment. *Fish and Fisheries*, **7**, 104–127.

Lenssen J.P.M., van de Steeg H.M. and de Kroon H. (2004) Does disturbance favour weak competitors? Mechanisms of changing plant abundance after flooding. *Journal of Vegetation Science*, **15**, 305–314.

Lévêque C. (1995) Role and consequences of fish diversity in the functioning of African freshwater ecosystems: a review. *Aquatic Living Resources*, **8**, 59–78.

Levinton J.S. (1995) *Marine biology: Function, biodiversity and ecology*. Oxford University Press, Oxford.

Longhurst A.R. (1976) Vertical migration. In D.H. Cushing and J.J. Walsh (eds) *The ecology of the sea*, pp. 116–137. Oxford: Blackwell Scientific.

Lowe-McConnell R. (1994) The changing ecosystem of Lake Victoria, East Africa. *Freshwater Forum*, **4**, 76–89.

Lynch M. (1980) The evolution of cladoceran life histories. *The Quarterly Review of Biology*, **55**, 23–42.

M

March J.G., Pringle C.M., Townsend M.J. and Wilson A.I. (2002) Effects of freshwater shrimp assemblages on benthic communities along an altitudinal gradient of a tropical island stream. *Freshwater Biology*, **47**, 377–390.

McAuliffe J.R. (1984) Competition for space, disturbance, and the structure of a benthic community. *Ecology*, **65**, 894–908.

McCallister S.L., Bauer J.E., Ducklow H.W. and Canuel E.A. (2006) Sources of estuarine dissolved and particulate organic matter: a multi-tracer approach. *Organic Geochemistry*, **37**, 454–468.

McCann L.D. and Levin L.A. (1989) Oligochaete influence on settlement, growth and reproduction in a surface deposit feeding polychaete. *Journal of Experimental Marine Biology and Ecology*, **131**, 233–253.

McGowan J.A. and Walker P.W. (1985) Dominance and diversity maintenance in an oceanic ecosystem. *Ecological Monographs*, **55**, 103–118.

McLachlan A. and Jaramillo E. (1995) Zonation on sandy beaches. *Oceanography and Marine Biology, an Annual Review*, **33**, 305–335.

McLusky D.S. (1989) *The estuarine ecosystem*, 2nd Edn. Blackie, Glasgow.

McQuatters-Gollop A., Raitsos D.E., Edwards M., Pradhan Y., Mee L.D., Lavender S.J. and Attrill M.J. (2007) A long-term chlorophyll dataset reveals regime shift in North Sea phytoplankton biomass unconnected to nutrient levels. *Limnology Oceanography*, **52**, 635–648.

Meerhoff M., Fosalba C., Buuzzone C., Mazzeo N., Noordoven W. and Jeppesen E. (2006) An experimental study of habitat choice by *Daphnia*: plants signal danger more than refuge in subtropical lakes. *Freshwater Biology*, **51**, 1320–1330.

Meijer M.-L., de Haan M.W., Breukelaar A.W. and Buiteveld H. (1990) Is reduction of the benthivorous fish an important cause of high transparency following biomanipulation in shallow lakes? *Hydrobiologia*, **200/201**, 303–315.

Melack J.M. (2002) Ecological dynamics in saline lakes. *Verhandlungen der Internationale Vereinigung für Theoretische und Angewandte Limnologie*, **28**, 29–40.

Menge B.A. (1976) Organisation of the New England rocky intertidal community: role of predation, competition and environmental heterogeneity. *Ecological Monographs*, **46**, 355–393.

Menge B.A. and Sutherland J.P. (1976) Species diversity gradients: synthesis of the roles of predation, competition and temporal heterogeneity. *American Naturalist*, **110**, 351–369.

Menge B.A. and Sutherland J.P. (1987) Community regulation: variation in disturbance, competition, and predation in relation to environmental stress and recruitment. *American Naturalist*, **130**, 730–757.

Michinton T.E., Simpson J.C. and Bertness M.D. (2006) Mechanisms of exclusion of native coastal marsh plants by an invasive grass. *Journal of Ecology*, **94**, 342–354.

Mikolajewski D.J. and Johansson F. (2004) Morphological and behavioral defenses in dragonfly larvae: trait compensation and cospecialization. *Behavioral Ecology*, **15**, 614–620.

Milner A.M. (1987) Colonization and ecological development of new streams in Glacier Bay National Park, Alaska. *Freshwater Biology*, **18**, 53–70.

Minshall G.W., Cummins K.W., Petersen R.C., Cushing C.E., Bruns D.A., Sedell J.R. and Vannote R.L. (1985) Developments in stream ecosystem theory. *Canadian Journal of Fisheries and Aquatic Sciences*, **42**, 1045–1055.

Mirabdullayev I.M., Joldasova I.M., Mustafaeva Z.A., Kazakhbaev S., Lyubimova S.A. and Tashmukhamedov B.A. (2004) Succession of the ecosystems of the Aral Sea during its transition from oligohaline to polyhaline water body. *Journal of Marine Systems*, **47**, 101–107.

Moreira F. (1997) The importance of shore birds to energy fluxes in a food web of a south European estuary. *Estuarine, Coastal and Shelf Science*, **44**, 67–78.

Müller K. (1954) Investigations on the organic drift in north Swedish streams. *Annual Report of the Institute of Freshwater Research, Drottningholm*, **35**, 133–148.

N

Naeem S., Loreau M. and Inchausti P. (2004) Biodiversity and ecosystem functioning: the emergence of a synthetic ecological framework. In M. Loreau, S. Naeem and P. Inchausti (eds) *Biodiversity and ecosystem functioning*, pp. 3–11. Oxford University Press, Oxford.

Nehring S. (2006) Four arguments why so many alien species settle into estuaries, with special reference to the German river Elbe. *Helgoland Marine Research*, **60(2)**, 127–134.

Neill W.E. (1990) Induced vertical migration in copepods as a defence against invertebrate predation. *Nature*, **345**, 524–526.

Neill W.E. (1994) Spatial and temporal scaling and the organization of limnetic communities. In P.S. Giller, A.G. Hildrew and D.G. Raffaelli (eds) *Aquatic ecology. Scale, pattern and process. 34th Symposium of the British Ecological Society*, pp. 189–231. Blackwell Science, Oxford.

Nelson-Smith A. (1977) Estuaries. In R.S.K. Barnes (ed.) *The coastline*, pp. 123–146. John Wiley and Sons, London.

Njiru M., Waithaka E., Muchiri M., van Knaap M. and Cowx I.G. (2005) Exotic introductions to the fishery of Lake Victoria: what are the management options? *Lakes and Reservoirs: Research and Management*, **10**, 147–155.

Nogata Y. and Matsumura K. (2006) Larval development and settlement of a whale barnacle. *Biological Conservation*, **2**, 92–93.

O

Olafsson E.B., Peterson C.H. and Ambrose W.G. (1994) Does recruitment limitation structure populations and communities of macroinvertebrates in marine soft sediments—the relative significance of presettlement and postsettlement processes. *Oceanography and Marine Biology*, **32**, 65–109.

Olowo J.P. and Chapman L.J. (1999) Trophic shifts in predatory catfishes following the introduction of Nile perch into Lake Victoria. *African Journal of Ecology*, **37**, 457–470.

Olsen J.L., Stam W.T., Coyer J.A., Reusch T.B.H., Billingham M., Bostrom C., Calvert E., Christie H.,

Granger S., La Lumiere R., Milchakova N., Oudot-Le Secq M.P., Procaccini G., Sanjabi B., Serrao E., Veldsink J., Widdicombe S. and Wyllie-Echeverria S. (2004) North Atlantic phylogeography and large-scale population differentiation of the seagrass *Zostera marina* L. *Molecular Ecology*, **13**, 1923–1941.

Orr J.C., Fabry V.J., Aumont O., Bopp L., Doney S.C., Feely R.A., Gnanadesikan A., Gruber N., Ishida A., Joos F., Key R.M., Lindsay K., Maier-Reimer E., Matear R., Monfray P., Mouchet A., Najjar R.G., Plattner G.K., Rodgers K.B., Sabine C.L., Sarmiento J.L., Schlitzer R., Slater R.D., Totterdell I.J., Weirig M.F., Yamanaka Y. and Yool A. (2005) Anthropogenic ocean acidification over the twenty-first century and its impact on calcifying organisms. *Nature*, **437**, 681–686.

Ott J.A. and Novak R. (1989) Living at the interface: Meiofauna at the oxygen/sulphide boundary of marine sediments. In J.S. Ryland and P.A. Tyler (eds) *Reproduction, genetics and distribution of marine organisms*, pp. 415–422. Olsen and Olsen, Denmark.

P

Paine R.T. (1966) Food web complexity and species diversity. *American Naturalist*, **100**, 65–75.

Palmer M.A., Bely A.E. and Berg K.E. (1992) Response of macroinvertebrates to lotic disturbance: test of the hyporheic refuge hypothesis. *Oecologia*, **89**, 182–194.

Parsons T.R., Takahashi M. and Hargrave B. (1984) *Biological oceanographic processes*. Pergamon Press, Oxford.

Pearson T.H. and Rosenberg R. (1978) Macrobenthic succession in relation to organic enrichment and pollution of the marine environment. *Oceanography and Marine Biology Annual Review*, **16**, 229–311.

Pearson T.H. and Rosenberg R. (1986) Feast and famine: Structuring factors in marine benthic communities. In J.H.R. Gee and P.S. Giller (eds) *Organization of communities: Past and present*, pp. 373–398. Blackwell Scientific, Oxford.

Petersen R.C. and Cummins K.W. (1974) Leaf processing in a woodland stream. *Freshwater Biology*, **4**, 343–363.

Philippart C.J.M., Beukema J.J., Cadeé G.C., Dekker R., Goedhart P.W., Iperen J.M.V., Leopold M.F. and Herman P.M.J. (2007) Impacts of nutrient reduction on coastal communities. *Ecosystems*, **10**, 95–118.

Pimm S.L. and Kitching R.L. (1987) The determinants of food chain lengths. *Oikos*, **50**, 302–307.

Pinay G. and Décamps H. (1988) The role of riparian woods in regulating nitrogen fluxes between the alluvial aquifer and surface water: a conceptual model. *Regulated Rivers: Research and Management*, **2**, 507–516.

Pingree R.D. (1978) Mixing and stabilisation of phytoplankton distributions on the Northwest European Continental shelf. In J.H. Steele (ed.) *Spatial patterns in plankton communities*, pp. 181–220. Plenum Press, New York.

Ponton F., Lebarbenchon C., Lefèvre Th., Biron D.G., Duneau D., Hughes D.P. and Thomas F. (2006) Parasitology: parasite survives predation on its host. *Nature*, **440**, 756.

Por F.D. (1978) *Lessepsian migration. The influx of Red Sea biota into the Mediterranean by way of the Suez Canal*. Ecological Studies, 23. Springer-Verlag, Berlin.

Pozo J., González E., Díez J.R., Molinero J. and Elósegui A. (1997) Inputs of particulate organic matter to streams with different riparian vegetation. *Journal of the North American Benthological Society*, **16**, 602–611.

Pretty J.L. and Dobson M. (2004) Leaf transport and retention in a high gradient stream. *Hydrology and Earth System Sciences*, **8**, 560–566.

Pretty J.L., Giberson D.J. and Dobson M. (2005) Resource dynamics and detritivore production in an acid stream. *Freshwater Biology*, **50**, 578–591.

Probert P.K. (1984) Disturbance, sediment stability and trophic structure of soft-bottom communities. *Journal of Marine Research*, **42**, 893–921.

R

Raffaelli D., Balls P., Way S., Patterson I.J., Hohmann S. and Corp N. (1999) Major long-term changes in the ecology of the Ythan estuary, Aberdeenshire, Scotland; how important are physical factors? *Aquatic Conservation – Marine and Freshwater Ecosystems*, **9**(2), 219–236.

Raffaelli D.G. and Hawkins S. (1996) *Intertidal ecology*. Chapman and Hall, London.

Raffaelli D.G. and Moller H. (2000) Manipulative experiments in animal ecology—do they promise more than they can deliver? *Advances in Ecological Research*, **30**, 299–330.

Raymont J.E.G. (1983) *Plankton and productivity in the oceans. Volume 2 zooplankton*. Pergamon Press, Oxford.

Reid P.C., Borges M.F. and Svendsen E. (2001) A regime shift in the North Sea circa 1988 linked to changes in the North Sea horse mackerel fishery. *Fisheries Research*, **50**, 163–171.

Reid P.C., Edwards M., Hunt H.G. and Warner A.J. (1998) Phytoplankton change in the North Atlantic. *Nature*, **391**(6667), 546.

Reid P.C., Lancelot C., Gieskes W.W.C., Hagmeier E. and Weichert G. (1991) The phytoplankton of the North Sea and its dynamics: a review. *Netherlands Journal of Sea Research*, **26**, 295–331.

Reissen H.P. and Young J.D. (2005) *Daphnia* defense strategies in fishless lakes and ponds: one size does not fit all. *Journal of Plankton Research*, **27**, 531–544.

Reynolds C.S. (1987) Community organisation in the freshwater plankton. In J.H.R. Gee and P.S. Giller (eds) *Organization of communities, past and present, 27th Symposium of the British Ecological Society*, pp. 297–325. Blackwell Science, Oxford.

Rhoads D.C. and Young D.K. (1970) The influence of deposit-feeding organisms on sediment stability and community trophic structure. *Journal of Marine Research*, **28**, 150–178.

Rice A.L. and Lambshead P.J.D. (1994) Patch dynamics in the deep-sea benthos: the role of a heterogeneous supply of organic matter. In P.S. Giller, A.G. Hildrew and D.G. Raffaelli (eds) *Aquatic ecology. Scale, pattern and process. 34th Symposium of the British Ecological Society*, pp. 469–497. Blackwell Science, Oxford.

Roughgarden J., Gaines S.D. and Pacala S.W. (1987) Supply-side ecology: the role of physical transport processes. In J.H.R. Gee and P.S. Giller (eds) *Organisation of communities past and present*, pp. 491–518. Blackwell Science, Oxford.

Roy P.S., Williams R.J., Jones A.R., Yassini I., Gibbs P.J., Coates B., West R.J., Scanes P.R., Hudson J.P. and Nichol S. (2001) Structure and function of south-east Australian estuaries. *Estuarine Coastal and Shelf Science*, **53**(3), 351–384.

Rudnick D.L. and Davis R.E. (2003) Red noise and regime shifts. *Deep-Sea Research Part I – Oceanographic Research Papers*, **50**, 691–699.

Ruesink J.L., Feist B.E., Harvey C.J., Hong J.S., Trimble A.C. and Wisehart L.M. (2006) Changes in productivity associated with four introduced species: ecosystem transformation of a 'pristine' estuary. *Marine Ecology Progress Series*, **311**, 203–215.

Rydin H. (1993) Mechanisms of interactions among *Sphagnum* species along water-level gradients. *Advances in Bryology*, **5**, 153–185.

Rzóska J. (1974) The Upper Nile swamps, a tropical wetland study. *Freshwater Biology*, **4**, 1–30.

S

Sale P.F. (1984) The structure of communities of fish on coral reefs and the merit of a hypothesis testing manipulative approach to ecology. In D.R. Strong, Jnr, D. Simberloff, L.G. Abele and A.B. Thistle (eds) *Ecological communities: Conceptual issues and the evidence*, pp. 478–490. Princeton University Press, Princeton, NJ.

Saltonstall K. (2003) Genetic variation among North American populations of *Phragmites australis*: implications for management. *Estuaries*, **26**, 444–451.

Sanders H.L. (1968) Marine benthic diversity: a comparative study. *American Naturalist*, **102**, 243–282.

Schindler D.W. and Fee E.J. (1974) Experimental Lakes area: whole-lake experiments in eutrophication. *Journal of the Fisheries Research Board of Canada*, **31**, 937–953.

Sedell J.R., Ridley J.E. and Swanson F.J. (1989) The River Continuum concept: a basis for the expected ecosystem behaviour of very large rivers? In D.P. Dodge (ed.) *Proceedings of the International Large River Symposium*, pp. 49–55. Canadian Special Publications in Fisheries and Aquatic Sciences, 106.

Seehausen O., van Alphen J.J.M. and Witte F. (1997) Cichlid diversity threatened by eutrophication that curbs sexual selection. *Science*, **277**, 1808–1811.

Self M. (2005) A review of management for fish and bitterns, *Botaurus stellatus*, in wetland reserves. *Fisheries Management and Ecology*, **12**, 387–394.

Shank T.M., Fornari D.J., Von Damm K.L., Lilley M.D., Haymon R.M. and Lutz R.A. (1998) Temporal and spatial patterns of biological community development at nascent deep-sea hydrothermal vents (9°N, East Pacific Rise). *Deep-sea Research II*, **45**, 465–516.

Sherr E.B. and Sherr B.F. (1991) Planktonic microbes: tiny cells at the base of the ocean's food webs. *Trends in Ecology and Evolution*, **6**, 50–54.

Sigee D.C. (2005) *Freshwater microbiology. Biodiversity and dynamic interactions of microorganisms in the aquatic environment*. Chichester, John Wiley and Sons Ltd.

Smith C.R. (1992) Whale falls: chemosynthesis on the deep seafloor. *Oceanus*, **35**, 74–78.

Smock L.A., Smith L.C., Jones J.B., Jr and Hooper S.M. (1994) Effects of drought and a hurricane on a coastal headwater stream. *Archiv für Hydrobiologie*, **131**, 25–38.

Sobczak W.V., Cloern J.E., Jassby A.D., Cole B.E., Schraga T.S. and Arnsberg A. (2005) Detritus fuels ecosystem metabolism but not metazoan food webs in San Francisco estuary's freshwater delta. *Estuaries*, **28**, 124–137.

Sousa W.P. (1979) Disturbance in marine intertidal boulder fields: the non-equilibrium maintenance of species diversity. *Ecology*, **60**, 1225–1239.

Sousa W.P. (1984) The role of disturbance in natural communities. *Annual Review of Ecology and Systematics*, **15**, 353–391.

Southward A.J. (1974) Changes in the plankton community of the Western English Channel. *Nature*, **249**, 180–181.

Stavn R.H. (1971) The horizontal–vertical distribution hypothesis: Langmuir circulation and *Daphnia* distributions. *Limnology and Oceanography*, **16**, 453–466.

Stevens P.W., Montague C.L. and Sulak K.J. (2006) Patterns of fish use and piscivore abundance within a reconnected saltmarsh impoundment in the Northern Indian River lagoon, Florida. *Wetlands Ecology and Management*, **14**, 147–166.

Stokstad E. (2005) After Katrina: Louisiana's wetlands struggle for survival. *Science*, **310**, 1264–1266.

Stone J.H., Bahr L.M., Jr, Day J.W. and Darnell R.M. (1982) Ecological effects of urbanization on Lake Pontchartrain, Louisiana, between 1953 and 1978, with implications for management. In J. A. Lee and M.R.D. Seaward (eds) *Urban ecology. Second European Ecological Symposium*, pp. 243–252. Blackwell Scientific Publications, Oxford.

Svensson B.M. (1995) Competition between *Sphagnum fuscum* and *Drosera rotundifolia*: a case of ecosystem engineering. *Oikos*, **74**, 205–212.

T

Taylor A.H. (1995) North–south shifts in the Gulf Stream and their climatic connection with the abundance of zooplankton in the UK and its surrounding seas. *ICES Journal of Marine Science*, **200**, 711–722.

Thomas F., Schmidt-Rhaesa A., Martin G., Manu C., Durand P. and Renaud F. (2002) Do hairworms (Nematomorpha) manipulate the water seeking behaviour of their terrestrial hosts? *Journal of Evolutionary Biology*, **15**, 356–361.

Thorp J.H. and Delong A.D. (2002) Dominance of autochthonous autotrophic carbon in food webs of heterotrophic rivers. *Oikos*, **96**, 543–550.

Thorson G. (1957) Bottom communities (sublittoral or shallow shelf) *Memoirs of the Geological Society of America*, **67**, 461–534.

Thurston M.H., Rice A.L. and Bett B.J. (1998) Latitudinal variation in invertebrate megafaunal abundance and biomass in the North Atlantic Ocean Abyss. *Deep-Sea Research Part II*, **45**, 203–224.

Townsend C.R. and Hildrew A.G. (1994) Species traits in relation to a habitat templet for river systems. *Freshwater Biology*, **31**, 265–275.

Tsuda A. and Miller C.B. (1998) Mate-finding behaviour in *Calanus marshallae* Frost. *Philosophical Transactions of the Royal Society London B*, **353B**, 713–725.

U

Usinger R.L. (1957) Marine insects. *Memoirs of the Geological Society of America*, **67**, 1177–1182.

Usio N. and Townsend C.R. (2002) Functional significance of crayfish in stream food webs: roles of omnivory, substrate heterogeneity and sex. *Oikos*, **98**, 512–522.

V

Van der Hage J.C.H. (1996) Why are there no insects and so few higher plants, in the sea? New thoughts on an old problem. *Functional Ecology*, **10**, 546–547.

Van Es F.B. (1977) A preliminary carbon budget for part of the Ems Estuary: The Dollard. *Helgolander wissenschaftliche Meeresuntersuchungen*, **30**, 283–291.

Vannote R.L., Minshall G.W., Cummins K.W., Sedell J.R. and Cushing C.E. (1980) The River Continuum concept. *Canadian Journal of Fisheries and Aquatic Science*, **37**, 130–137.

Vervier P., Dobson M. and Pinay G. (1993) Role of interaction zones between surface and ground waters in DOC transport and processing: considerations for river restoration. *Freshwater Biology*, **29**, 275–284.

Vinogradov M.E., Gitelzon I.I. and Sorokin Y.I. (1970) The vertical structure of a pelagic community in the tropical ocean. *Marine Biology*, **6**, 187–194.

Vollenweider R.A. (1975) Input output models with special reference to the phosphorus loading concept in limnology. *Schweizerische Zeitshrift für Hydrologie*, **37**, 53–84.

W

Wafar S., Untawale A.G. and Wafar M. (1997) Litter fall and energy flux in a mangrove ecosystem. *Estuarine, Coastal and Shelf Science*, **44**, 111–124.

Walker D. (1970) Direction and rate in some British post-glacial hydroseres. In D. Walker and R.G. West (eds) *Studies in the vegetational history of the British Isles*, pp. 117–139. Cambridge University Press, Cambridge.

Wallace J.B., Eggert S.L., Meyer J.L. and Webster J.R. (1999) Effects of resource limitation on a detrital-based system. *Ecological Monographs*, **69**, 409–443.

Wetzel P.R., van der Valk A., Newman S., Gawlik D.E., Gann T.T., Coronado-Molina C.A., Childers D.L. and Sklar F.H. (2005) Maintaining tree islands in the Florida Everglades: nutrient redistribution is the key. *Frontiers in Ecology and the Environment*, **3**, 370–376.

Wiebe P.H. (1970) Small-scale spatial distribution in oceanic zooplankton. *Limnology and Oceanography*, **15**, 205–217.

Wilcock H.R., Nichols R.A. and Hildrew A.G. (2003) Genetic population structure and neighbourhood population size estimates of the caddisfly *Plectrocnemia conspersa*. *Freshwater Biology*, **48**, 1813–1824.

Wildish D.J. (1977) Factors controlling marine and estuarine sublittoral macrofauna. *Helgolander wissenschaftliche Meeresuntersuchungen*, **30**, 445–454.

Williams A., Koslow J.A., Terauds A. and Haskard K. (2001) Feeding ecology of five fishes from the mid-slope micronekton community off southern Tasmania, Australia. *Marine Biology*, **139**(6), 1177–1192.

Williams W.D. (2002) Environmental threats to salt lakes and the likely status of inland saline ecosystems in 2025. *Environmental Conservation*, **29**, 154–167.

Williams W.D. and Sherwood J.E. (1994) Definition and measurement of salinity in lakes. *International Journal of Salt Lake Research*, **3**, 53–63.

Winder M., Boersma M. and Spaak P. (2003) On the cost of vertical migration: are feeding conditions really worse at greater depths? *Freshwater Biology*, **48**, 383–393.

Wirtz K.W. and Wiltshire K. (2005) Long term shifts in marine ecosystem functioning detected by inverse modelling of the Helgoland Roads time-series. *Journal of Marine Systems*, **56**, 262–282.

Witte F., Goldschmidt T., Goudswaard P.C., Ligtvoet W., van Oijen M.J.P. and Wanink J.H. (1992) Species extinction and concomitant ecological changes in Lake Victoria. *Netherlands Journal of Zoology*, **42**, 214–232.

Wolff W.J. (1977) A benthic food budget for the Grevelingen Estuary, The Netherlands, and a consideration of the mechanisms causing high benthic secondary production in estuaries. In B.C. Coull (ed.) *Ecology of marine benthos*, pp. 267–280. University of South Carolina Press, Charleston.

Woodin S.A. (1981) Disturbance and community structure in a shallow water sand flat. *Ecology*, **62**, 1052–1066.

Worm B., Sandow M., Oschlies A., Lotze H.K. and Myers R.A. (2005) Global patterns of predator diversity in the open oceans. *Science*, **309**(5739), 1365–1369.

Wurtsbaugh W.A. and Gliwicz Z.M. (2001) Limnological control of brine shrimp population dynamics and cyst production in the Great Salt Lake, Utah. *Hydrobiologia*, **466**, 119–132.

X

Xavier J.C., Croxall J.P., Trathan P.N. and Wood A.G. (2003) Feeding strategies and diets of breeding grey-headed and wandering albatrosses at South Georgia. *Marine Biology*, **143**(2), 221–232.

Z

Zaret T.M. and Paine R.T. (1973) Species introduction in a tropical lake. *Science*, **182**, 449–455.

Organism index

■■■■■■■■■■■■■■■■■■■■■■■■■■■■■■■■■■

Subject index

■■

DEDICATION

To all those who have helped me learn and with whom I've shared the journey towards understanding.

Problem-Based Learning

A Self-Directed Journey

Problem-Based Learning

A Self-Directed Journey

Sue E. Baptiste, MHSc, OT Reg (Ont)
McMaster University
Hamilton, Ontario

An innovative information, education, and management company
6900 Grove Road • Thorofare, NJ 08086

Cover illustration © Steve Dinnino/Images.com

As the author is based in Canada, this book uses Canadian spellings of some terms.

The procedures and practices described in this book should be implemented in a manner consistent with the professional standards set for the circumstances that apply in each specific situation. Every effort has been made to confirm the accuracy of the information presented and to correctly relate generally accepted practices. The author, editor, and publisher cannot accept responsibility for errors or exclusions or for the outcome of the application of the material presented herein. There is no expressed or implied warranty of this book or information imparted by it.

Any review or mention of specific companies or products is not intended as an endorsement by the author or publisher.

The work SLACK Incorporated publishes is peer reviewed. Prior to publication, recognized leaders in the field, educators, and clinicians provide important feedback on the concepts and content that we publish. We welcome feedback on this work.

Baptiste, Sue
Problem-based learning: a self-directed journey/Sue E. Baptiste.
p. cm.
Includes bibliographical references and index.
ISBN 1-55642-563-5 (alk. paper)
1. Medical education. 2. Problem-based learning. [DNLM: 1. Health Occupations--education--Outlines.
2. Problem-based Learning--Outlines. W 18.2 B222p 2003] I. Title
R834.B355 2003   610'.71'1--dc21   2003005236

Printed in the United States of America.

Published by: SLACK Incorporated
                     6900 Grove Road
                     Thorofare, NJ 08086 USA
                     Telephone: 856-848-1000
                     Fax: 856-853-5991
                     www.slackbooks.com

Contact SLACK Incorporated for more information about other books in this field or about the availability of our books from distributors outside the United States.

For permission to reprint material in another publication, contact SLACK Incorporated. Authorization to photocopy items for internal, personal, or academic use is granted by SLACK Incorporated provided that the appropriate fee is paid directly to Copyright Clearance Center. Prior to photocopying items, please contact the Copyright Clearance Center at 222 Rosewood Drive, Danvers, MA 01923 USA; phone: 978-750-8400; website: www.copyright.com; email: info@copyright.com.

For further information on CCC, check CCC Online at the following address: http://www.copyright.com.

Last digit is print number: 10   9   8   7   6   5   4   3   2   1

CONTENTS

ACKNOWLEDGMENTS

It is not often that we have the chance in our lives to take time to pull together something that is of great importance to us. I have been privileged to do this in the writing of this little book.

Douglas: thank you for not only putting up with many hours of me on the computer, but also with the many years of talking about and declaring my excitement with problem-based learning.

McMaster colleagues: thank you for giving me the chance to be a part of such a wonderful group of people who appreciate and celebrate innovation and the importance of functioning from principle.

Amy at SLACK Incorporated: thank you for your energy, support, and acceptance of eccentricity!

ABOUT THE AUTHOR

Over the past three decades, since emigrating to Canada from Britain, Sue Baptiste has significantly influenced Canadian occupational therapy through her unique and innovative approaches to practice, leadership, education, and research. Her work as leader of a large occupational therapy program through the 1970s and 80s led to the development of organizational models integrating practice, student education, and clinical research. The growth of many occupational therapists in Canada has been fostered through Sue's mentorship and enthusiastic support of excellence. An internationally renowned leader in problem-based education, Sue has facilitated the development of problem-based, self-directed learning across many disciplines. Her abiding interest in issues of culture and organizational change characterize her research.

Sue is an author of the *Canadian Occupational Performance Measure*, an outcome measure used throughout the world. Sue has an unequalled ability to challenge long held assumptions (with humour) while positively facilitating change to ensure excellence in occupational therapy.

Mary Law, PhD, OT Reg (Ont.)
Professor and Associate Dean, School of Rehabilitation Science
Co-director, CanChild Centre for Childhood Disability Research
McMaster University
Hamilton, Ontario, Canada

INTRODUCTION

Over the past 30 years, the notion of learning in a different way from the traditional teacher-centred methods has been growing rapidly in popularity. Educational programs that are designed in a problem-based, learner-centred manner are becoming more common; the models that emerge from curriculum reform number as many as the institutions adopting this approach to curriculum renewal. However, few high schools have followed the same path. Consequently, there are growing numbers of students who enter their university careers at either the undergraduate or graduate level to be faced with a totally new way of gaining knowledge; one for which their previous education has not prepared them very well. It is for these learners, and the faculty who are involved with helping them learn, that this book has been written.

I have been engaged with helping learners learn for several decades, most of which has been within a university setting known for its early innovations in the area of problem-based learning (PBL). I came from Britain as part of the exodus of the late 1960s. Having been faced with difficult career decisions, since I had been told many times that I would not be able to make it through university, I came to Canada. After a few years, I found McMaster University. I also found PBL, and, to my great pleasure, discovered that I was much more able to learn than I had ever thought possible. In fact, I actually enjoyed it, and relished the idea that I could find out about things of interest to me and about which I wished to learn. This was a more than pleasant change from the old ways of trying to cram into my head all the "stuff" that other people told me I should care about and needed to know. I became almost evangelical about this newfound lease on learning. I wanted to make sure that others appreciated the delight of it all; that others would become "converted" was one of my major goals.

Most recently, I have been involved in a curriculum renewal process within our School of Rehabilitation Science, as we changed to entry-level master's programs in occupational therapy and physiotherapy. It was with renewed energy and commitment that the decision was made to make no changes in the core pedagogy and educational philosophy. That is, we would remain fully committed to the inherent values, beliefs, and methods of PBL. These would remain the cornerstones and the foundation for the emerging curricula.

Despite the faculty's commitment to this approach to gaining knowledge, and the clear messages available to all in our marketing materials, some students still arrive at the door not really believing that learning this way will be all that different after all. Many have an erroneous sense of what it will mean, and expect that they will be able to manage somehow, rather than engage in a serious effort to internalize these new ways. Not so. This is indeed serious business—it is not Mickey Mouse—and yet, it is also fun. And this leads us directly to why I decided to write this book. I want to be able to help learners experience the process of PBL while learning about it.

True to its philosophy, learning about PBL cannot be done through didactic means; this is simply flying in the face of the point. Consequently, I have structured this book in a manner that follows the natural process of a tutorial group, from its first meeting to the end of the course of study. Obviously, it would become quite tedious if I maintained a truly conversational tone throughout this text, and therefore I have taken some license with this notion of an experiential "read"! From the beginning, I have assumed that this is an excellent group, thereby avoiding the need to engage in resolving poor group dynamics or disagreements between members. Similarly, I have clumped the reporting back and learning synthesis in each tutorial into one piece of narrative rather than emulate a true dialogue. Other than these adaptations, the overall process undertaken is closely related to what happens in a real tutorial, and the information gleaned from a wide range of resources is similar to the search and retrieve process undertaken by students in their tutorial groups.

In order to make a difference between the voice of the tutor and the voices of the group members in the text, we have used italics and smaller margins. I hope this helps in guiding you. And so, I would like to share a journey of discovery with you, the learner and reader.

"…it is remarkable how much well-motivated, bright students can learn in two days of self-study and how far along the route they can be two days later with a minimum of faculty guidance." (Federman, 1999, p. 94)

REFERENCE

Federman, D. D. (1999). Little-heralded advantages of problem-based learning. *Acad Med, 74,* 93-94.

Tutorial 1

Getting Started

"Implicit in PBL and the tutorial process is an awesome respect for the beginning student."
(Federman, 1999, p. 93)

Note to the Reader: In order to make a difference between the voice of the tutor and the voices of the group members in the text, we have used italics and smaller margins for the tutor, and smaller margins for the group members.

Welcome to our tutorial group. You have decided to participate in this small group experience because you want to learn about problem-based learning (PBL); and I am here because I am very committed to the ideas that are inherent within problem-based, learner-centred learning and I love to facilitate learning in small groups. So, I am sure we will share a very positive experience.

I know that I always work best when I am aware of any rules and guidelines that I should use or follow. In the case of small group learning, it is very useful to have some ground rules for how the group will run. It provides us with a sense of a framework for functioning together. So, if I may, I will suggest a few things that I know from experience work well. First, we should all be punctual, so if for any reason you find you will be late or cannot attend due to exceptional circumstances, then please call or leave a message. My contact information is in the handbook in front of you. Also, for a small group learning experience to work it is important to subscribe to a set of basic values and expectations of behaviour: honesty, open communication, and mutual respect are some of these. Subscribing to this way of working together will, by nature, result in the development of trust, which is so important when embarking on a shared journey of learning. Speaking of trust, we will not always get along famously; there will be times when we will become irritated with each other, and may well disagree quite strongly. That is fine. However, when these situations arise, I always find it best to keep such disagreements within the confines of the group, and not to share them broadly among friends and class colleagues. After all, we are hoping to develop a climate within which we can feel comfortable exploring information, being clear about what we don't know, and helping each other by sharing our different areas of skill and knowledge. I hope that this sounds all right to you, and that it helps to set the scene.

Yes, I think we will become quite comfortable with that, although it seems rather strange right now. Perhaps the best thing is to get started on actually doing whatever it is we are going to do, and then we will be able to see for ourselves how it works and how it feels.

Another way to help us move along smoothly is for me to share with you a suggested model of how the PBL process works. This is not intended as a recipe in any way, but I find it can be helpful as a process skeleton so that you feel less at sea about what is going to happen next.

A model for a problem-based small group process (after introductions are made and ground rules are set or reinforced):

- Choose the learning scenario.
- Define any unfamiliar language or concepts within the written scenario (and determine whether any of these are learning issues in themselves).
- Brainstorm around any issues that come to mind when reading the scenario.
- Distill from this open exploration key areas for potential learning.
- Organize these key areas within a logical, conceptual framework.
- Select priority issues for focused exploration.
- Develop a learning plan with specific questions.
- Define potential learning resources.
- Clarify that all group members understand and subscribe to the learning plan.
- Evaluate the learning experience of that particular session.

So the scenario that we have decided to explore in this group is as follows:

All the students entering the new Bachelor of Health Sciences program at Everytown College chose this program because the curriculum was designed according to principles of PBL. While they knew a little about what this would entail from reading the brochures and college calendar, they were unsure what was expected of them and what they could expect in return. The majority of incoming students come from traditional high schools where the teacher outlined the learning and there were always mid-term and final exams.

Are there any phrases or words that are unfamiliar?

Other than PBL, no—and that is the point of this whole thing! So perhaps we should move on. What happens next?

Now we brainstorm anything that comes to mind from reading the scenario. Remember that with brainstorming there are no wrong ideas; anything can be raised as long as it stems from the words in the scenario.

So, to begin the brainstorm, let's simply throw our ideas out in any order, in bullet form if we wish, as well as phrases and potential questions to address:

- Roles of students
- Roles of faculty
- What does it look like to be in a PBL environment?
- How different is it from a more traditional and familiar system?
- Do the rooms look the same? After all, if there are a lot of small groups then surely space cannot be configured exactly the same.
- Are there courses, the same as in the familiar system?
- Do we go to lectures and are we given lessons by professors? I seem to have heard that this is not the same at all.
- Is it mostly small groups, and what if I don't do well in that sort of a structure?
- How do we learn? I have always gone to lectures and been given notes, textbooks, and multiple-choice exams.
- How are we evaluated? Do we take exams, and how do we evaluate the small group learning?
- What do exams look like?
- How do we know when we have learned enough? This concerns me, since I know when I go onto the Internet I can easily get lost in a lot of very interesting things that are not directly relevant to what I was looking for! The same can go for when I am using the library.
- I have learned OK the old way, why change it? I wanted to come to this college, and I know the programs are very good here, but I really don't need to learn how to

learn all over again! It sounds as if learning this way could be rather easy, and I wonder if it is really a Mickey Mouse way to just work in groups. I want to learn, and I have my goals laid out—I am aiming for graduate school and don't have time to waste in "touchy-feely" group stuff.

- Do we have to do it this way? I wonder if we can change things when we are in the program. After all, from what I have heard, this kind of curriculum is learner- and student-centred, so surely that could mean that we can make changes as we see fit.

- How do we behave in small groups? I have always learned on my own very well. I have tended not to like group projects the way they are structured in high school, so I would hope that we can still learn on our own if we want to.

I wonder if these are plenty of ideas to work with for the moment. Perhaps it is time to try to put some order into the questions and concepts that have been raised. The best way to do that is to decide on a framework within which to place all the ideas. For example, when dealing with a health care scenario, using a model that reflects the biological, psychological, and social elements of health and illness can be helpful. In rehabilitation cases particularly, the use of the person-environment-occupation framework can provide some clarity. In this case, when we are exploring a broader based concept such as PBL, then it makes sense to create a framework that relates to the content most sensibly. Perhaps following a curriculum development or system framework would work here?

Yes, it would seem to make sense if we structure all our content into a model that reflects the developmental nature of creating a problem-based environment. To start:

What is PBL; what are the underlying principles and from where did they stem? That would make a good beginning and provide us with the necessary understanding of the theories and ideas that lead to this educational approach. Also, is it an educational approach, a methodology, or a philosophy? That is a very good question.

The next logical piece is to **describe a PBL learning environment**. It is interesting to ponder how a PBL curriculum differs from a more traditional one, and to try to appreciate what types of learning resources support a PBL curriculum best. Naturally then we need to **understand the differences between PBL and traditional learning environments, particularly when considering the different roles**—for learners, faculty, resource people, and experts.

Since we are about to enter into a PBL program, it would benefit us to have a clear idea of **the intricacies of being a learner** in such a culture. For example, how do we know when we know enough? Also, if the resources are so different and how we learn is so different, then it only seems logical to assume that how we are evaluated will also be very different. At the same time, it would be very useful to understand how best to survive in a PBL environment, and to gather some helpful strategies for coping.

From what we have heard and/or read, there seem to be varying views of whether a PBL method is better than the old way, worse, or just simply different. We need to find out **what kind of research findings there are to support or refute the utility and success of PBL as an approach to learning.** It would also be very helpful to try to gain an appreciation of what students and faculty think about their experiences with a reform curriculum model such as PBL.

It would appear that you have created a natural course of study for yourselves. From the way that you have grouped your issues, a learning plan can be created that would provide your agendas for each of the tutorials ahead. There seem to be about four or five sessions that fall out of the discussion you have just held. So, it would make sense now to create your specific learning issues and learning plan for the next time we meet. The best way to do that is to take another look at what you identified as the first issues you thought you would like to address.

I think we said:

What is PBL; what are the underlying principles and from where did they stem? That would make a good beginning and provide us with the necessary understanding of the theories and ideas that lead to this educational approach. Also, is it an educational approach, a methodology, or a philosophy?

So, can we define the learning issues more clearly from that statement? For example, an issue to address the first part of that paragraph would be simply:

What is PBL?

Yes, and I would add to it:

What are the underlying principles?

Where did it begin?

Good, now what resources would you look for, and where, to help you learn about these issues?

I think we should look at the literature—you know, journals, textbooks. Also, we could go to the Web to see what is there.

Which journals are you aware of that could help here? Also, what kinds of key words would you think would be most useful?

I think we should look at the educational literature, and perhaps the psychology abstracts. Someone on faculty must have some books and articles about this. Do you have any suggestions about some people we could contact?

That is very good. Yes, people resources are just as valuable as the written word sometimes, and occasionally more so. People who have a recognized expertise are to be considered an invaluable part of a PBL culture. I can give you some names to contact. The Programme for Faculty Development, for example, is a great resource, because they mount work-

shops each year and have a mass of literature, including archival and historical information as well as a current bibliography.

There is another step to this process that is very important: namely, evaluating how we have functioned during this first time together. Group evaluation is seen as an essential component of any PBL experience. It is only through consistently giving and receiving feedback that we can increase our awareness of how well we work as members of a small group or team, and also how well the group environment is working overall.

For the sake of this text, we will assume that this group is exceptional; that, from the beginning, each member internalizes the values that are expected, that feedback is given constructively, and is received in the spirit in which it is meant. Some useful resources about evaluating small group learning are cited in Appendix A.

So, I am sensing that everyone is feeling quite comfortable with these learning issues for getting started on this process. Good luck with your learning.

KEY POINTS

- All members introduce themselves.
- Group ground rules are set.
- The learning scenario for exploration is chosen.
- Brainstorming is undertaken.
- Potential learning issues are distilled from the brainstorming and organized within a logical framework.
- A learning plan is developed from these learning issues, taking into consideration the timeframe available, i.e., reasonable learning objectives that can be accomplished in the time between group sessions.
- Potentially useful resources are identified.
- Evaluation of the session is undertaken, focusing on how well the group functioned as a whole and how each individual functioned, based on self, peer, and tutor feedback.

REFERENCE

Federman, D. D. (1999). Little-heralded advantages of problem-based learning. *Acad Med, 74,* 93-94.

Tutorial 2

What Is Problem-Based Learning?

> "PBL is not the solution for all problems in education; however, it is a powerful tool for allowing students to actively build collaborative problem-solving skills that will be required in the work environment."
> (Kanter, 1998, p. 391)

OBJECTIVES

- What is problem-based learning?
- What are the underlying principles of problem-based learning?
- Where did it begin?

Good morning. You all seem quite cheerful, and in fact, quite animated, so I can only assume that you found some interesting and useful information to assist you in addressing your learning issues.

Yes, actually, it was exhilarating to explore the literature with such a high level goal in mind. This approach leaves us with such autonomy, but can be dangerous as you suggested, in that it leaves us open to go down all kinds of paths of inquiry that are not necessarily directly related to the goals at hand. Nevertheless, we found out a great deal.

I would like to suggest that we approach the process of you sharing your learning in a collaborative manner. Rather than each group member presenting what they found in a "show and tell" fashion, I find it supports the notion of synthesis and integration much more effectively if we combine our mutual learning under each objective and discuss openly what we have discovered and what we understand, defining areas of confusion as we go along. This way helps us raise our level of critical thinking and ensures that we push ourselves to understand rather than passively receive information, take notes, and move on. That approach only serves to minimize the potentially rich opportunity of group debate; this seems rather a waste to me. This does not take away from the importance of individual learning objectives for each student. In fact, the ideal balance is to have one or two group learning goals and then add to the richness of the learning agenda with other objectives that are important to individual students in the group. We should all be encouraged and able to share our independent learning as it fits into the discussion.

Yes, this makes sense, although I am not sure that we are really aware of what you mean. Again, it would probably be best if we moved on and then we can experience it for ourselves.

Yes, let's reprise the learning issues from last time:
- *What is problem-based learning?*
- *Where did problem-based learning begin?*
- *What are the underlying principles of problem-based learning?*

WHERE DID PROBLEM-BASED LEARNING BEGIN?

PBL began at McMaster University in Hamilton, Ontario, Canada, where it was created as the guiding philosophy for the development of a new medical school. PBL was introduced as a new method of learning that promoted student-centred education, building on adult learning principles of self-directed learning, and supporting the development of life-long learning skills (Barrows & Tamblyn, 1980). These core elements were identified as critical to the preparation of physicians for the emerging new era of evolving consumer awareness. Information was growing at such an incredible pace that it was becoming impossible, and also undesirable, for health care practitioners to attempt to "learn" in the traditional sense everything there was to learn (Neufeld, 1983; Neufeld, Woodward, & MacLeod, 1989).

McMaster's model became one from which others developed. For example, Case Western Reserve University incorporated similar methods of instruction into a learning laboratory setting for many disciplines. From the late 1960s onward, many other universities (for example, in Maastricht, the Netherlands and Newcastle, Australia) began to adopt this approach to enhance their medical education. Back in North America, even more centres known for their educational excellence were developing their own "spin" on the notion; from Harvard to Boston, from Missouri to Hawaii, from Pittsburgh to Texas, and internationally from Sweden to South Africa (Johnson, Finucane, & Prideaux, 1999; Schmidt & Bouhuijs, 1980).

Schmidt (1983) described a seven-step process that illustrates the problem-based process to be undertaken in small and large groups, and in individualized evaluation/assessment experiences. Research has also shown that the seven steps are not accomplished in a strictly linear fashion, but rather groups tend to go back and forth as needed to clarify and redefine learning as they move through the problem scenario (Goodall, 1990; Jensen & Chilberg, 1991). Similarly, Walton and Matthews (1989) provide another schematic for structuring the PBL process.

This progression illustrates particularly the movement of this educational philosophy through the medical school network.

However, there were parallel initiatives underway within other health care and service-based disciplines. The rehabilitation disciplines of occupational therapy and physiotherapy in Hamilton, Ontario were following closely the evolution of their colleagues in medicine. The programs in these disciplines, initially housed within Mohawk College from 1978 to 1989, were relocated and upgraded at McMaster University by 1990. From the onset, these curricula were designed using a problem-based philosophy and framework (Saarinen & Salvatori, 1994). Since that time, many other programs in these rehabilitation disciplines across North America and Europe have embraced their own versions of this approach. More recently, other countries are embracing the idea because it fits well with the emerging health care system delivery principles of client-centred practice and attention to consumer satisfaction. Other parallel initiatives in the earlier years were ongoing, involving many other disciplines, such as veterinary medicine (Texas A & M, Cornell), dentistry (University of Minneapolis, Minnesota), and chiropractic (Pacific University, Canadian Memorial College). The whole set of ideas exemplified by a problem-based approach have relevance across many areas of educational endeavour (Sciarra, personal communication, 2002). More recent initiatives have included an interdisciplinary view of applying PBL principles, including faculties and departments of Art, Mathematics, English, and Education. Most recent developments have shown the adoption of PBL approaches within undergraduate curricula in programs of combined Arts and Sciences, and in Health Sciences (Harnish, personal communication, 2002).

WHAT ARE THE UNDERLYING PRINCIPLES OF PROBLEM-BASED LEARNING?

The whole discussion around principles seems to be divided into two main components. The first addresses the values upon which this kind of learning is based, and the second outlines the characteristics that are now seen to be inherent within any iteration of a PBL curriculum.

As we mentioned in the first tutorial, there are some key underlying values and assumptions about how people are going to work together within a PBL culture.

Partnership

As with concepts of client-centred service, PBL and student-centred learning environments are deeply invested in the notion of partnership. PBL encourages those involved in the learning process to see each other in a collegial, collaborative way and to minimize ideas of competition. It is felt that faculty and teachers are part of the same system as the students, simply with different roles (see Tutorial 4). Both parties are invested in the same desired outcomes: namely, that the learners learn and become whatever they desire to become. The faculty simply facilitate that happening. This is best achieved from a place of sharing that common vision, and from an expectation of a collaborative relationship.

Honesty and Openness

From the beginning of a PBL experience, students are aware of everything that they need to know to learn what they need to learn. The structure of each learning experience, from individual courses to overall curriculum objectives, is transparently available to all learners. The path to be taken through the educational "quicksand" is well-marked so that individual learners may constantly refer to the next steps, refresh their view of objectives, and check on deadlines for evaluation events and submission dates. Similarly, interpersonal interactions should also reflect the intentions of openness and honesty. The climate within small groups generates opportunities for sound and open relationships to develop between faculty and learners, and among the learners themselves. Learning from a basis of honest disclosure around strengths and knowledge gaps can only serve the collective well. In this way, group members can shine in their areas of experience and excellence, and gain from those of their peers. Faculty tutors also can benefit from the knowledge of the students in their group; this is a wonderful way to stay current and become reenergized. In large group learning experiences, the same applies.

Respect

One of the most central of these values is respect. If a learning system is going to maintain integrity, it is critical that the players and partners respect one another and behave congruently. Respect manifests itself in punctuality, paying attention, asking pertinent and thoughtful questions, and, in essence, being engaged in the learning process. The task of providing constructive feedback is facilitated only when respect is present between group members, within a large or small group environment, but with particular relevance to the small group. When respect is apparent, it is much more likely that difficult feedback will be given in an acceptable manner, and will be received with serious attention rather than dismissed with discomfort and potential anger.

Trust

Once all the preceding values are integrated into the environment, trust can begin to grow and, in turn, become established. Trust is a critical element of a successful PBL context, because it is only when individuals trust each other that risk-taking becomes perceived as more possible by those wishing to take those risks, and acceptable, even desirable, to others within the group. When risk-taking becomes a natural dynamic within a group, this is an indicator that the group is functioning well, with naturally supportive relationships in place.

> Over the course of my work in consulting with people in other educational centres about PBL, I have come across some very powerful examples of the shifting of power from a teacher-centred system to one of collaboration between teachers and learners. In one particular university, the whole culture was formal and precise, all the way down to the beautiful boardroom where the consultation workshop was held. When I introduced myself and asked the participants to do the same, despite me using my first name, everyone returned the gesture using their academic credential. I didn't mention it initially, but once the whole table of people had completed their introductions, I mentioned that they could perhaps let

me know the names by which they would like to be called for the duration of our time together. The responses were the same. This gave me an excellent entry point to the discussion of underlying principles of PBL. While using an honorific is not necessarily a problem, it does give a certain tone to a climate and culture. For example, calling faculty members Dr. Smith or Professor Jones may appear outwardly to indicate respect, but I question whether this is necessarily so. Similarly, calling the same faculty members Fred and Sheila does not necessarily indicate the reverse. However, how the informal rules of relationships are created within an environment can give cues and clues to how things are done around there. It is important, therefore, to think about the impact of titles or the lack of them. This is just an example of how a simple matter of addressing each other can set the scene for how an educational experience unfolds.

PBL is called everything from a strategy to a philosophy, and is used to influence the development of cultures, curricular models, programs, and course and learning modules. However it is termed, and at whatever level of conceptual reasoning it is applied, the basic characteristics that support it are similar.

- Learning is student/learner-centred.
- Faculty assumes the role of facilitator or guide.
- Problems or learning scenarios form the basis, focus, and stimulus for learning.
- New information and understanding is acquired through self-directed learning.

There are many other elements of a PBL environment that could be discussed here, but perhaps such an in-depth exploration fits best in the tutorials where we will be looking closely at what a PBL environment looks like and the roles assumed by faculty and learners. In fact, the whole discussion would best come in the next tutorial, when we are looking more closely at the details of a PBL environment.

*Shall we just reflect for a moment on what you have dis-
covered about the history of PBL and what some of the most
important lessons for you have been so far?*

KEY POINTS

- It began back in the late 1960s in Canada.
- Since that time, it has become an important alternative to more traditional learning and educational models.
- It began as a reform movement to educate medical students.
- Now it is used across many other professional groups.
- Underlying values and assumptions include partnership, honesty, openness, respect, and trust.
- Common characteristics include a learner-centred focus, faculty acting more as guides, scenarios being a springboard to learning, and new knowledge being acquired through self-directed learning.

---

*That is great. Just to check, how do you feel about your
first attempt to find out information and then synthesize it
within the context of a small group?*

We seem to have managed very well, although it was
extremely difficult to sift through everything that we
found. There is a massive amount of information in the
literature with PBL as one of the key words. Also, much
of the literature tends to describe particular experiences
of developing a PBL curriculum and that is not what we
needed this time. We have kept track of some of those
references for further along in our process of discovery!

We managed to work well as individuals when searching for the information and also as a group when we put
all of our findings together. It is particularly interesting
to see how most of us found similar things, and yet we all
also found some items that were unique in some way. We
can only hope that this quality of experience continues.

Since we are getting close to the end of this tutorial, perhaps we should revisit what the learning objectives are for next time.

- *What does a PBL learning environment look like?*
- *How does a PBL curriculum differ from a more traditional one?*
- *How do the roles differ for faculty and learners?*

These were the objectives we identified right at the beginning. I wonder though if this is too much to undertake at once, and perhaps we should consider looking just at student and faculty roles for next time. When we were discussing exploring student roles, we also included wanting to talk to students about how best to survive in such an unknown environment. This adds a lot onto your agenda for preparing for next time, so perhaps it is best to keep all the faculty and student information for the tutorial after that?

Yes, perhaps that does make the best sense. So we have two main objectives for the next tutorial.

Are you comfortable with the resources that you will access to help with this particular learning plan?

Perhaps a more important focus for this time will be to talk with faculty about how they see their roles, also obtain examples of what some PBL curriculum outlines look like, as well as going back to the literature. See you next time.

References

Barrows, H. S., & Tamblyn, R. M. (1980). *Problem-based learning: An approach to medical education.* [Series: Medical Education, Vol.1.] New York: Springer Publishing Company.

Goodall, H. L., Jr. (1990). *Small group communication in organizations* (2nd ed.). Dubuque, IA: Wm. C. Brown Publishers.

Jensen, A. D., & Chilberg, J. C. (1991). *Small group communication: Theory and application.* Belmont, CA: Wadsworth Publishing Company.

Johnson, S. M., Finucane, P. M., & Prideaux, D. J. (1999). Problem-based learning: Process and practice. *Aust N Z J Med, 29,* 350-354.

Kanter, S. L. (1998). Fundamental concepts of problem-based learning for the new facilitator. *Bulletin of the American Library Association, 86*(3), 391-395.

Neufeld, V. R. (1983). Adventures of an adolescent: Curriculum changes at McMaster University. In C. Friedman & E. S. Purcell (Eds.), *New Biology and Medical Education* (pp. 256-270). New York: Josiah Macy Jr. Foundation.

Neufeld, V. R., Woodward, C. A., & MacLeod, S. M. (1989). The McMaster M.D. program: A case study of renewal in medical education. *Acad Med, 64,* 423-447.

Saarinen, H. & Salvatori, P. (1994). Dialogue: Educating occupational and physiotherapists for the year 2000: What, no anatomy courses? *Physiotherapy Canada, 46*(2), 81-86.

Schmidt, H. (1983). Problem-based learning: Rationale and description. *Med Educ, 17,* 11-16.

Schmidt, H. G., & Bouhuijs, P. A. J. (1980). Onderwijs in taakgerichte groepen (Task-oriented small group learning). Het Spectrum, Utrecht.

Walton, H. J. & Matthews, M. B. (1989). Essentials of problem-based learning. *Med Educ, 23,* 542-558.

LEARNING RESOURCES

Hak, T., & MaGuire, P. (2000). Group process: The black box of studies on problem-based learning. *Acad Med, 75*(7), 769-772.

Lloyd-Jones, G., Margetson, D., & Bligh, J. G. (1998). Problem-based learning: A coat of many colours. *Med Educ, 32,* 492-494.

Marchese, T. (1998). The new conversations about learning. Retrieved April 23, 2003 from http://www.aahe.org/pubs/TM-essay.htm

Margetson, D. (1996). Beginning with essentials: Why problem-based learning begins with problems. *Education for Health, 1,* 61-69.

Maudsley, G. (1994). Do we all mean the same thing by "problem-based learning"? A review of the concepts and a formulation of the ground rules. *Acad Med, 74*(2), 178-185.

Norman, G. (1993). Where is the learning in problem-based learning? *Pedagogue* (4), 2.

Norman, G. (1988). Problem-solving skills, solving problems and problem-based learning. *Med Educ, 22,* 279-286.

Programme for Faculty Development. (1997). *Problem-based learning.* Internal document, Faculty of Health Sciences. Hamilton, Ontario: McMaster University.

Thomas, R. E. (1997). Problem-based learning: Measurable outcomes. *Med Educ, 31,* 320-329.

Walton, H. J., & Matthews, M. B. (Eds.). (1987). Medical education booklet No. 23: *Essentials of problem-based learning.* World Federation for Medical Education. Dundee: UK: Ninewells Hospital and Medical School.

The Learning Environment in Problem-Based Learning

"The content of the curriculum becomes that of joint study and discovery, rather than static delivery and receipt." (Federman, 1999, p. 94)

OBJECTIVES

- What does a problem-based learning environment look like?
- How does a problem-based curriculum differ from a more traditional one?

RECOGNIZING PROBLEM-BASED LEARNING WHEN YOU SEE IT

When we speak of "what an environment looks like", there are several core components of that discussion. One focuses on the actual physical environment, incorporating space, furniture, equipment, and so on. Another centers around how the curriculum is designed, thus reflecting the educational environment. Yet a third component introduces the matter of the interactional or relationship environment; that is, how people relate to each other and what roles they assume. Finally, there is the importance of the institutional context: the environment of rules, regulations, guidelines, and structures that enable successful management and operations.

THE PHYSICAL ENVIRONMENT

Traditional educational environments, and university environments in particular, are often imagined as classic buildings covered in ivy with stained glass or leaded glass windows, paths edged by ancient shrubs and trees, and wood paneled lecture halls with blackboards and carved lecterns from which the professor intones his wisdom during lectures. Such is the dream of academia. However, even in the reality of modern educational settings, at least in those that support traditional curriculum models, there are the large lecture halls, the well-equipped computer labs, and the more formal smaller classrooms with seats fixed to the ground together with fold-down writing tablets on the arms. Not so with PBL environments. In a setting ideally configured for reform curricular models such as PBL, the focus is on many smaller rooms utilized for small group tutorial sessions. Within these rooms, the furniture is easily movable, composed of small tables that can be configured into bigger ones. The chairs are lightweight and stackable, yet comfortable. There are flip charts with pens and black or white boards with swing-back sides for extra space when brainstorming and restructuring ideas into a logical framework. When larger, lecture-style spaces are designed, they too, ideally, allow for furniture to be moved into different constellations of seating, because the application of PBL in large groups is a growing art. Some approaches to incorporating PBL into larger groups involve the convening of many students into triads (groups of three) and quads (groups of four) to facilitate self-directed learning, despite the apparent overwhelming number of learners in the class. There are still flip charts regardless of the size of the room.

And so, the very first clues that we are working in a PBL environment reside within the chairs in which we sit, and the rooms in which we gather. Such flexible spaces serve to set the scene for approaching our learning in a manner that suits the people and the task.

THE EDUCATIONAL/CURRICULAR ENVIRONMENT

As we mentioned in an earlier discussion, the true PBL curriculum is transparent—nothing is hidden, everything is open and declared. This includes all objectives, the resources available for learning, and the evaluation/assessment methods. In order to organize a PBL curriculum, it is useful, and perhaps even necessary, to develop a model for recognizing the layout of content, to embark upon the development of learning scenarios to reflect that content, and to develop evaluation tools and methods that are congruent with the way things are learned.

As with any curriculum, there is a taxonomy of objectives beginning with the program and moving downward in detail through years of study, terms/semesters, courses, and individual learning modules. Each level bears relation to the one above and the one below. In order to fulfill the true essence of PBL, it is critical that the learners are aware of all this information, that it is clearly articulated and readily available to them. Often, one of the best ways to ensure that is to have curriculum and term handbooks, within which is outlined all the information necessary to ensuring that the objectives are met, the learning resources needed are readily accessed, and the evaluation of progress and learning is successfully accomplished. A useful image to help to see how a PBL curriculum is created is to imagine objectives at one end, and evaluation methods at the other; all around are resources from which learners can draw. In the centre is a large space, waiting to be filled with different constellations of knowledge, skills, and professional behaviours that are defined by the specific learner(s) as essential to meeting the objectives and successfully completing the assessment expectations.

A term handbook becomes the day-to-day "bible" for students, since it contains everything they need to know to be successful in their given program. A handbook contains:

- Information on faculty members associated with the particular learning unit, including their office address, phone and fax numbers, and email. Often there is a small paragraph outlining the faculty members' areas of interest and expertise, which helps when students are seeking specific resources to address a learning issue.
- Term and course objectives.
- The learning grid.
- Brief statements encapsulating each potential learning scenario available to be considered in small group tutorial.
- Timetables for each week, outlining the regular expectations, special events, and timing of evaluations.
- Details of evaluations, giving grading schemes together with clear outlines of each evaluation experience delineating expectations.
- Listing of required and additional/suggested texts and readings.

One key strategy for graphically depicting the content of a course is the development of a learning grid. A learning grid plots the areas of content against the overarching themes or threads in that particular section of the curriculum. Learning scenarios are developed to fill the cells of the created matrix, with close attention being paid to ensuring that there is overlap throughout the grid, thus not creating discrete learning problems that are obviously meant to illustrate one specific cell. Through this strategy, there can be a clear profile of what the scope of learning is intended to be for that course, semester, or program.

The box on page 27 shows how the learning grid is described within the handbook. An example of a learning grid is provided in Table 3-1.

The handbook is not only a resource for students, but also for faculty, including visiting faculty such as tutors. The handbook

> The objective coverage grid should be used to guide the process of problem selection throughout the term and will enable your tutorial group to track the objectives that you have met for each problem. It should also allow you to plan to a certain extent what aspects of any given problem you might choose to explore. Although the tutorial group will establish their own learning objectives for each problem, the guide may assist you in identifying learning issues which require investigation at some point during the term. By the end of the term, your tutorial group should have addressed each of the content-oriented objectives listed in the grid to some degree.
>

becomes a central information tool for everyone involved with the curriculum. Similarly, for tutors there are tutor manuals, which provide an overview of the educational philosophy of the program, as well as specific details concerning the particular term of study in which the current tutorial resides.

THE RELATIONSHIP ENVIRONMENT

If we reflect back to the very first tutorial, we can recall that we set the scene for our new group by developing guidelines and ground rules for how we expect our group to function. Similar strategies are in place throughout a PBL curriculum. Much of how people relate to one another is mediated by the underlying principles and values of respect, trust, openness, and honesty. This does not always create a comfortable environment for the immediacy but does provide a rich grounding from which to commence a professional practice life in human service. Many of the skills modeled within a problem-based, small group are those that are needed when working within practice teams, establishing rapport with clients and families, and creating goodwill and support for launching a private practice or consultation business. From the outset, there is an implicit set of expectations regarding how one should

Table 3-1

Objective Coverage Grid

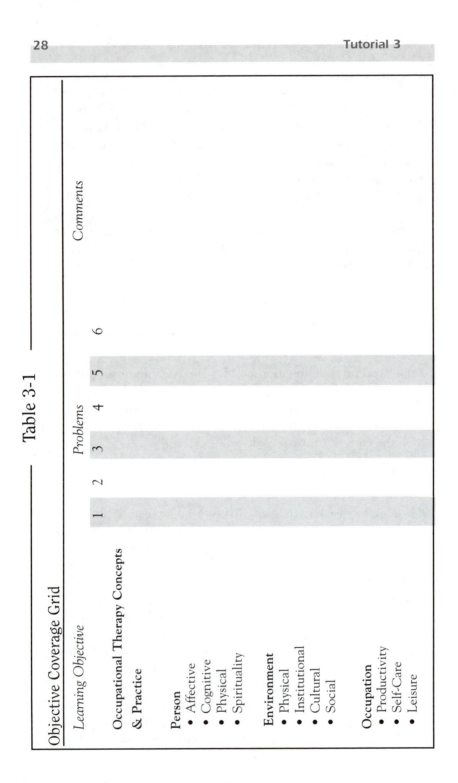

Learning Objective	Problems						Comments
Occupational Therapy Concepts & Practice	1	2	3	4	5	6	
Person							
• Affective							
• Cognitive							
• Physical							
• Spirituality							
Environment							
• Physical							
• Institutional							
• Cultural							
• Social							
Occupation							
• Productivity							
• Self-Care							
• Leisure							

Table 3-1, continued

Objective Coverage Grid

Learning Objective	Problems						Comments
	1	2	3	4	5	6	
Occupational Performance Process							
• Name, validate, prioritize OP issue							
• Select theoretical approach							
• Identify OP components & environmental conditions							
• Identify strengths & resources							
• Negotiate targeted outcomes & develop action plans							
• Implement plans through occupation							
• Evaluate OP outcomes							
Practice Environments (e.g., community/home care, hospitals, rehabilitation centres, private practice, schools, industry, rural/under-serviced)							

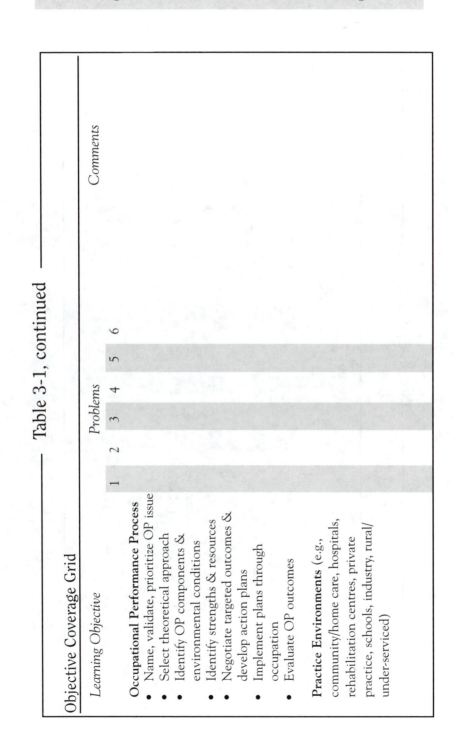

—— Table 3-1, continued ——

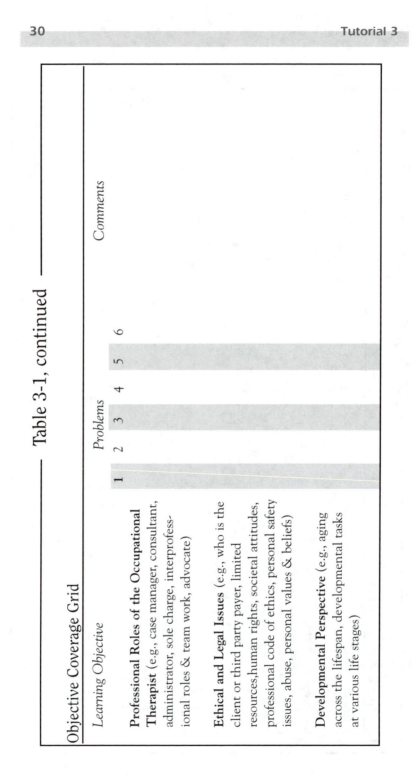

Objective Coverage Grid

Learning Objective	Problems						Comments
	1	2	3	4	5	6	
Professional Roles of the Occupational Therapist (e.g., case manager, consultant, administrator, sole charge, interprofessional roles & team work, advocate)							
Ethical and Legal Issues (e.g., who is the client or third party payer, limited resources, human rights, societal attitudes, professional code of ethics, personal safety issues, abuse, personal values & beliefs)							
Developmental Perspective (e.g., aging across the lifespan, developmental tasks at various life stages)							

Table 3-1, continued

Objective Coverage Grid

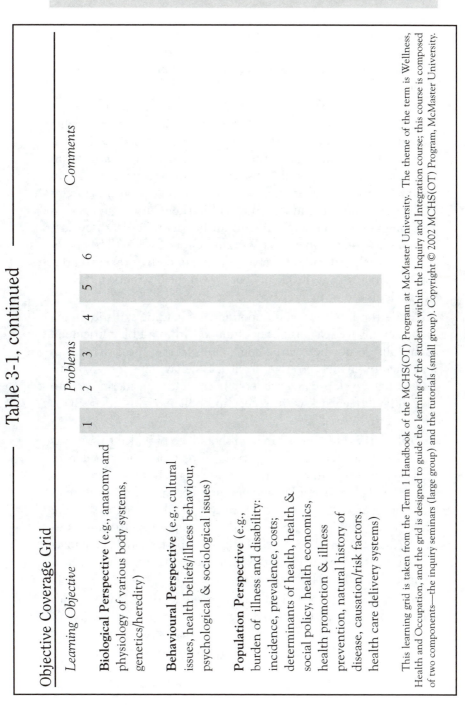

Learning Objective	Problems						Comments
	1	2	3	4	5	6	
Biological Perspective (e.g., anatomy and physiology of various body systems, genetics/heredity)							
Behavioural Perspective (e.g., cultural issues, health beliefs/illness behaviour, psychological & sociological issues)							
Population Perspective (e.g., burden of illness and disability: incidence, prevalence, costs; determinants of health, health & social policy, health economics, health promotion & illness prevention, natural history of disease, causation/risk factors, health care delivery systems)							

This learning grid is taken from the Term 1 Handbook of the MCHS(OT) Program at McMaster University. The theme of the term is Wellness, Health and Occupation, and the grid is designed to guide the learning of the students within the Inquiry and Integration course; this course is composed of two components—the inquiry seminars (large group) and the tutorials (small group). Copyright © 2002 MCHS(OT) Program, McMaster University.

behave when functioning within a professional context. In many problem-based curricula, there are explicit criteria against which students are measured when their small group skills and progress are evaluated.

The guide in Table 3-2 was developed within the Faculty of Health Sciences at McMaster University, using an expert panel Delphi approach for selection of top-ranking items. This guide is provided to students within the occupational therapy program in each term handbook, and is used as a self- and peer-evaluation measure. Group members are encouraged to reflect on these items in the context of their functioning within their current group, thus providing a sound underpinning to further, more in-depth evaluation (Professional Behaviours Working Group, 1994). The use of this guide helps to remind everyone of the underlying expectations for behaviour within the faculty but also in the discipline for which the students are being prepared.

While the intention and importance of gaining skills in giving and receiving feedback is not questioned within a PBL curriculum, such skills do not necessarily come easily. It is critical that tutors and faculty model positive feedback behaviours for students. Lehner has developed an excellent checklist for considering how to give difficult feedback with respect and honesty. This discussion is included in Tutorial 5.

Both of these resources can, and should, be provided to students at frequent and consistent intervals throughout the program of study, perhaps within all handbooks, as well as to tutors for ease of reference. These sets of behaviours then become second nature as the learner develops a natural facility with them, skills that are often coveted by others not exposed to this approach to professional education. A third document focused on professional behaviours in practice settings is also provided to students from the onset of their studies (MScOT Program, 2002).

Yet another focus of a PBL curriculum model that seems very different from a more traditional view involves the integration of learning. This integrated approach to course structure celebrates horizontal synthesis of content rather than vertical preservation of precise subject matter. In traditional curricula, students take specific courses to introduce them to the study of individual subjects; this

―――――――― Table 3-2 ――――――――

Guide to Professional Behaviours in Tutorial Meetings

Respect
1. Listens, and indicates so with appropriate verbal or nonverbal behaviour.
2. Verbal and nonverbal behaviour is not rude, arrogant, or patronizing.
3. Does not humiliate or denigrate group members for their opinions or information.
4. Differentiates value of information from value of person.
5. Acknowledges group members' contributions.
6. Does not interrupt inappropriately.
7. Participates in discussion of differences in moral values.
8. Apologizes when late or gives reason for being so.

Communication Skills
1. Speaks directly to group members.
2. Uses words that group members understand.
3. Presents clearly.
4. Uses open-ended questions appropriately.
5. Uses nonjudgmental questions.
6. Identifies misunderstanding between self and others or among other group members.
7. Attempts to resolve misunderstanding.
8. Tests own assumptions about group members.
9. Accepts and discusses emotional issues.
10. Able to express own emotional state in appropriate situations.
11. Nonverbal behaviour is consistent with tone and content of verbal communication.
12. Verbal or nonverbal behaviour indicates that statements have been understood.
13. Recognizes and responds to group members' nonverbal communication.

Responsibility
1. Is punctual.
2. Completes assigned tasks.
3. Presents relevant information.

Table 3-2, continued

Guide to Professional Behaviours in Tutorial Meetings

4. Identifies irrelevant or excessive information.
5. Takes initiative or otherwise helps to maintain group dynamics.
6. Takes initiative or otherwise helps to define group goals.
7. Advances discussion by responding to or expanding on relevant issues.
8. Identifies own emotional or physical state when relevant to own functioning or group dynamics.
9. Accepts priority of tutorial time over other activities.
10. Identifies lack of honesty in self or others that interferes with group dynamics or attainment of group goals.
11. Describes strengths and weaknesses of group members in a supportive manner.
12. Gives prior notice of intended absence.
13. Negotiates alternatives if unable to complete assigned tasks.

Self-Awareness/Self-Evaluation
1. Acknowledges own difficulty in understanding.
2. Acknowledges own lack of appropriate knowledge.
3. Acknowledges own discomfort in discussing or dealing with a particular issue.
4. Identifies own strengths.
5. Identifies own weaknesses.
6. Identifies means of correcting deficiencies or weaknesses.
7. Responds to fair negative evaluative comment without becoming defensive or blaming others.
8. Responds to fair negative evaluative comment with reasonable proposals for behavioural changes.

---

is not so in PBL curricula. Courses are exemplified by titles that reflect high-level concepts rather than discrete collections of facts. For example, there are no isolated anatomy or basic sciences courses; rather, the basic science content is the second level of detail and the "meat" for the "problems" or learning scenarios that are developed. Pallie and Brain (1978) and Pallie and Miller (1982) describe the rich learning resources that were developed within the Department of Anatomy at McMaster University in order to support the problem-based approach to learning for health sciences students. In essence, the anatomy laboratory has been configured into self-directed learning modules, triggered from a learning scenario, and can include specimens, histological materials, videos, X-ray films, CD-ROMs, and so on. The idea is that learners go to the lab and access the modules that will assist them in addressing the learning issues they have identified from their tutorial groups, to address personal learning needs as well as specific group objectives. Often, students attend the lab in their tutorial group formations and work together in common goals for knowledge and understanding. While such resources are not available everywhere, most PBL environments have developed special resources that address the needs of students to learn in different ways from simply going to the library and reading texts and journals. Such resources range from innovative videotapes to interactive computer tools, and from rich compendia of resources available outside the immediate system to details for accessing individuals with specific expertise and skill.

Given that the curriculum content is not delivered in discrete discipline-based courses, a PBL curriculum can appear quite strange in the university calendar. Courses are frequently known by the functions that they play rather than the content they evoke. For example, the central component of most PBL curricula is the small group tutorials; however, often those external to the actual learning environment believe that is all there is. Not so at all; there are other "courses" that provide different milieux for learning, e.g., large group seminars, skills laboratories, and practica (fieldwork experiences). While the tutorial groups are the hub from which other learning evolves, the alternative contexts provide different opportunities to learn in various ways. The large group seminar sessions promote conceptual understanding through guest speakers who work together with students, presenting their area of particular expertise, then facilitating discussion and clarification of content and learning objectives. In the skills labs, students are given the chance to learn hands-on skills as needed, and to build their individual professional tool kits that will act as the underpinning for real world practice. Here, students learn skills such as interviewing, assessment, intervention planning, and evaluation. In the practica, or fieldwork experiences, students engage in real world practice under the guidance of a clinical preceptor, thus being presented with a chance to integrate all areas of learning from theory to practice. All of these learning experiences are grounded in problem-based, self-directed learning principles.

> *About a decade ago, we worked with colleagues in a particular medical school in Germany to develop a problem-based stream to their undergraduate medical curriculum. One of the most fascinating elements of this experience was watching the faculty struggling with the concept of horizontal integration of course content. At the outset, basic science faculty rarely, if ever, related to clinical faculty. However, by the time that the curriculum was becoming more than an idea, there were many faculty members who formed themselves into working groups to create learning scenarios and other learning resources. These groups were representative of the basic sciences, the applied and clinical sciences, as well as senior students. Through this strategy, the faculty as a whole heralded their readiness to embrace the notion of PBL and to accept an integrated approach to the development of learning resource materials.*

THE INSTITUTIONAL ENVIRONMENT

Given that PBL is an intimate and personally invested approach to education, each curriculum will be developed in congruence with the particular institutional environment. Therefore, curriculum details will differ, but intention and principle should remain constant and held in common across any program using the PBL designation. We are describing a pure PBL model in this tutorial. We recognize that there are as many hybrid versions of PBL models as there are facilities embracing the idea. However, it is also recognized that the core concepts of PBL should be present for a curriculum to be termed a PBL curriculum. There are many other educational methodologies, such as case method, that are learner-centred and consider the real life context, but they are not the same as PBL.

Perhaps one of the major differences in institutional PBL environments is the presence of learners at many committees and task forces that may not be the case within more traditional contexts. For example, in a well-established PBL culture, students will be active within all key functions of the organizational structure: from admissions to education committees, from curriculum evaluation to search committees for key leadership roles. This is one way of living the philosophy of partnership and openness. Transparent management reinforces a sense of involvement and perceived worth. Involving students in the ongoing evaluation of the curriculum, of faculty, and of staff pays tribute to the staunch belief that students, faculty, and staff play different but fiercely interdependent roles.

All policies, guidelines, and procedures should reflect the basic values upon which the curriculum is built, and the philosophy that overarches the entire enterprise. For example, having a sense of secrecy invades the manner in which a particular committee does its business and flies in the face of open communication. Perhaps this, of all elements, can illustrate most clearly the underlying principles of PBL if they are integrated well. However, by subscribing to these principles, faculties tend to become vulnerable. In many ways, it is much easier to do things in a traditional manner. That way, teachers teach and are rarely challenged; exams are developed, students write them and they are graded; students pass or fail with some appeals, but mostly a tacit acceptance of what is. Once a system is opened up to the challenges and chances of a problem-based system, then rules change and relationships shift. Power becomes shared; faculty and students together engage in the process of learning and the journey of enrichment. Within a system that is open and honest, there has to be a willingness to take risks and to accept feedback, whether it is readily palatable or not. This is not easy for anyone. In essence then, it is critical that both parties enter into this "contract" with appreciation of the elements of shared risk and the central importance of mutual regard and respect. It is only then that the trust that we spoke of earlier can become a cornerstone.

Now, how are you feeling about the learning you have accomplished today? It seems that you accessed a great deal of information, but did you manage to glean some important pieces of knowledge and understanding to help you appreciate how PBL learning can look and feel different?

Yes, indeed! There is a lot more out there to read and to use for resources. However, it is important to recognize quality resources from articles that are simply interesting or compelling to read. Talking with faculty proved to be difficult, so we wondered if it would be better if we planned to send out a brief questionnaire to some faculty members and to students who are currently enrolled in the program? This way, it would be considerate of everyone's time and ensure that people did not get irritated at being asked questions by a lot of students individually. We will send that out now so that we can get the answers back in time for accumulating the responses before the tutorial, when we had planned to address faculty and student opinions on their PBL experiences.

That is an excellent idea, and truly exemplifies the underlying principles of partnership within PBL.

So now, perhaps we should set the scene for the next tutorial and determine what the focus for the learning plan will be.

Yes, when we looked over the original tutorial plan, the following were the issues that we agreed we would explore for the next session.
• What are the key differences in the roles of faculty in a PBL environment?
• What are the main differences in the roles of students in a PBL environment?
• What are the intricacies of being a learner in a PBL environment?

That sounds fine to me; a good balance to look at this time.

REFERENCES

Federman, D. D. (1999). Little-heralded advantages of problem-based learning. Acad Med, 74, 93-94.

> ## KEY POINTS
>
> - The physical environment is flexible, easily changeable.
> - The curriculum looks different, with integrated courses incorporating content from many distinct bodies of knowledge, using problems/learning scenarios as a springboard for learning, and large group sessions as a means to refine questions and acquire knowledge in response to the questions.
> - Faculty and students relate as partners, exemplifying behaviors that reflect honesty, openness, respect, and trust.
> - Institutional systems, including policies, procedures, and guidelines, illustrate the application of PBL principles.

MScOT Programme. (2002). *Term 1 Handbook*. Hamilton, Ontario: McMaster University.

Pallie, W., & Brain, E. (1978). "Modules" in morphology for self-study: A system for learning in an undergraduate medical programme. *Med Educ, 12,* 107-113.

Pallie, W., & Miller, D. (1982). Communicating morphological concepts in health sciences. *J Biocommun, 9*(3), 26-32.

Professional Behaviours Working Group. (1994). *Guide to Professional Behaviours in Tutorial Meetings*. Hamilton, Ontario: McMaster University.

LEARNING RESOURCES

Abdul-Ghaffar, T. A., Lukowiak, K., & Nayar, N. (1999). Challenges of teaching physiology in a PBL school. *Adv Physiol Educ, 22*(1), S140-S147.

Branda, L. (1990). Implementing problem-based learning. *J Dent Educ, 54*(9), 548-549.

Ferrier, B. (1990). Problem-based learning: Does it make a difference? *J Dent Educ, 54*(9), 550-551.

Greening, T. (1998). Scaffolding for success in problem-based learning. Retrieved April 23, 2003 from *Medical Education Online*: www. med-ed-online.org/f0000012.htm

Jacobs, T. (1997). Developing integrated education programmes for occupational therapy: The problem of subject streams in a problem-based course. *British Journal of Occupational Therapy, 60*(3), 134-138.

Salvatori, P. (2000). Implementing a problem-based learning curriculum in occupational therapy: a conceptual model. *Australian Occupational Therapy Journal, 47,* 119-133.

Tutorial 4

Roles in a Problem-Based Learning Environment

"The benefits of contact between student and student, and between students and tutor, need no statistical confirmation."

(Federman, 1999, p. 93)

OBJECTIVES

- What are the key differences in the roles of faculty in a problem-based learning environment?
- What are the main differences in the roles of learners in a problem-based learning environment?
- What are the intricacies of being a learner in a problem-based learning environment?

Faculty Roles

Professionals tend to choose an academic career because they wish to help others learn or they wish to pursue a lifetime of inquiry, research, and writing. Perhaps some educators combine part of both of these ambitions; however, there is without a doubt a sense of great achievement when an academic succeeds in obtaining a faculty appointment, is granted tenure, and is promoted up the ranks to the status of full professor. Such success is something of which to be very proud. In more traditional systems, teaching expectations tend to centre around courses taught, lectures given, graduate students supervised, and workshops led. While many of these expectations are very similar within a PBL environment, the manner in which they are fulfilled tends to be very different. The role of expert becomes a secondary one, with the role of guide or facilitator taking precedence. This facilitating role does not come naturally to everyone, nor does everyone wish to assume this responsibility. While the role of tutor or facilitator can be seen as central to a successful PBL initiative, there are other roles of equal importance that are necessary to maintain the overall structure. For example, the role of "expert" is still needed, but in response to questions raised by the learners rather than in providing a lecture to begin a process of knowledge acquisition. In a PBL context, the lecture is best placed later along the continuum of discovery. It is here that lecturers who see the large gathering as a large group session can use their skills of theatre and creativity to reinforce problem-based principles within a large group context. Many faculty have worked their entire lifetime to gain the regard and respect of their colleagues and the field in which they specialize. By no means should a PBL curriculum minimize or denigrate that excellence and renown. The expectation then is on the shoulders of the particular expert faculty to revisit the way in which he or she delivers knowledge. Lectures become large group sessions, during which students can gain information based on questions they have identified through independent or small group study. The expert "teacher" then should embrace the challenge of responding to the learning needs as expressed by the learners and not simply deliver a lecture that may have been delivered in a similar manner many times before. To see the problem as a springboard for learning, and to see the large group session as providing the more finite and precise knowledge to answer the pivotal questions raised through the learning scenario, is the ideal manner in which PBL principles can be supported and realized.

I can share a very good example of someone I know who managed to do exactly that. For many years, a senior faculty member in the biochemistry department of a faculty of health sciences taught a course for nursing students. It was not unusual to have at least 120 students in this course, and the challenge for this faculty member was to translate his commitment and belief in PBL principles into a workable model for such a large learner group. The course took place twice a week for periods of 2 hours each. On the first session of the whole course, after setting the scene and providing some basic information, just prior to the students leaving for the day, the professor would give them a three page document to take home. This document contained details of a case scenario, followed by two key questions: what learning issues do you see emerging for you from this scenario, and what are the priority questions that you feel are important for the class to address. The students would return for the next session with the brief questionnaire completed. They would be asked to provide their opinions. The professor would fill one blackboard with the learning issues that class members would call out, and the other board would be used to develop the learning agenda for the next two to three class sessions. This way, the students become the creators of their own learning plan, and the professor pulls from his repertoire of knowledge to provide "lectures" in response to the questions. This approach has become used more broadly since then, and not only at the original university. I think this is a very sound and powerful example of the application of problem-based principles within a large group context.

Yes, indeed, that is very helpful and certainly moves the whole notion of PBL out of simply the small group context. It also points out clearly how faculty expertise can become a true asset for student learning, rather than just a repository of knowledge to be acquired.

Another critical role is that of developer of learning resources. As is now becoming very clear, a PBL curriculum is strongly dependent upon a rich offering of learning resources tailored to meet the needs of active, interactive learners rather than of passive recipients of information. Also, the learning resources should be tied closely to the real world of practice; consequently, the involvement of front-line practitioners as resource developers is an excellent opportunity to engage the practice community more closely with the educational context. Similarly, there are faculty members who in more formal teaching models would be lecturing, and can now utilize their specific skills and knowledge in the development of specialized resources. Often, these individuals can form interdisciplinary planning teams to create rich resources that reflect an integrated approach to, for example, basic sciences, clinical sciences, or societal trends of policy, planning, and service delivery. For those who truly excel at the lectern, or in a situation where practice skills are being introduced and practiced, these individuals can best utilize their talents by being guest presenters or leaders for special skills laboratories. Perhaps a key strategy for assigning faculty to these innovative roles is to recognize each person's key strengths and ensure that these can be highlighted. Trying to force someone to become a facilitator assuming that this is the role of choice can only end up being a pointless battle.

Having said all of this about a broader look at roles within a PBL context, I think it is very helpful to look more closely at the detailed role of being a tutor in a small group. Without question, being a tutor is an important job and it seems equally important to understand how this pivotal role works at its best. "The tutor is an educator who aids a task-oriented group to achieve the objectives of the teaching programme" (Branda & Sciarra, 1995). This sounds relatively easy, but, upon examination, it is clear that the role of tutor involves many skill sets, including sound application of the principles and practice of PBL, knowledge of group dynamics, comfort with student assessment incorporating methods congruent with PBL principles, the use and design of learning resources, ease of leadership and facilitation within a small group, and the management of group tasks to meet objectives within a given time frame (Branda & Sciarra, 1995, p. 198). A good tutor will accept the

The tutor/facilitator should possess the skills to:

- Facilitate learning by asking nondirective questions, avoiding lecturing, guiding towards resources, and managing the interpersonal relationships in the group to minimize conflict or misunderstandings that can impede learning.
- Promote group problem solving and critical thinking.
- Promote efficient group function by assisting the group to set goals and create a plan, sensing group problems and helping deal with them, ensuring evaluation of group process, and serving as a role model for giving feedback.
- Promote individual student learning.
- Coordinate evaluation of student performance.

(Adapted from Barrows & Branda, 1975, rev. 1991, 1994)

value of a problem-based approach together with a firm commitment to self-directed learning for the students. The tutor will see the small group tutorial as a place for integrating knowledge and understanding; directing further, more in-depth learning; and providing constructive feedback to enhance future growth. The box above provides a quick reference guide to the detailed role components of a tutor within a PBL group.

Recent literature has indicated that tutors with content expertise are helpful only if they also possess skills in the management of group dynamics and group learning (De Grave, Dolmans, & van der Vleuten, 1999). Therefore, it is critical that faculty engage in a process of training and enhanced self-awareness to ensure that they possess the necessary skills to be facilitators of learning, not group therapists! One metaphor that describes well the importance of managing a small group environment designed to achieve specific learning objectives, is to imagine oneself as having an "iron fist in a velvet glove". Faculty are encouraged to understand group dynamics, to learn to appreciate the power of relationships in creating group climates, but to constantly refocus on the task at hand—namely, learning together.

When we look at all the information out there addressing the differences between more traditional and PBL learning environments, there is often a main focus on the differences that are manifested within the roles assumed by, and functions that are expected of, faculty members (Finucane, Allery, & Hayes, 1995; Mayo, Donnelly, & Schwartz, 1995; De Grave et al., 1999). There are also roles that address needs or special input and resources. Learners will require support in learning sessions designed to enable practical skill acquisition; also, there are needs for practice supervision, supervision in attaining research and evidence-based practice skills, and so on. While it would be grand to imagine that every faculty member is interested in engaging in any role that is needed, this is by no means a realistic approach in any educational environment, not simply one within which there are newly identified challenges. There are helpful graphics depicting the shift from teacher-centred to learner-centred educational models (Branda, 1990). Essential elements of assuming the new faculty roles and responsibilities are many and varied. In essence, a faculty member is indeed a facilitator and guide. He or she must gain comfort in relinquishing the safety of the podium in order to join the troops on the front line, rolling up sleeves to engage directly in the business of learning and understanding, of gaining essential knowledge and skills, of appreciating and adopting desired professional practice behaviours.

Overall, when addressing curricular change, it is essential that faculty are assisted in embracing the desired innovations, exhibiting a clear belief in a philosophy of problem-based, self-directed learning, as well as an appreciation of the importance of different venues for learning—from small to large groups, from labs, lecture halls, and tutorial rooms to practice settings.

Learner Roles

Much has been written concerning the reactions of students to their PBL experiences, but we will focus on the formal, peer-reviewed literature here. Opinions will be addressed in Tutorial 7, together with the opinions and stories from academic and clinical faculty members.

There have been many attempts made to identify the skill sets that are essential to the successful completion of a PBL program.

Essentially, PBL curricula are developed with the following areas for skill development as overall expectations and objectives. Students are prepared with the abilities to:

- Be self-directed in their learning, leading to adoption of life-long learning.
- Seek, select, and utilize the best and most appropriate resources.
- Think critically and reason clinically.
- Behave in an appropriately professional manner.
- Embrace ethical and legal principles in practice.
- Work with others in group and team environments.
- Lead when appropriate and follow when appropriate.
- Communicate clearly and professionally in oral and written forms.
- Think proactively.

Biley and Smith (1998) completed a study of nursing students' perceptions of their PBL following graduation. These graduates reported that they valued the skills they had gained in accepting autonomy and valuing their abilities as "reflexive, independent professionals rather than unthinking assistants" (p. 1024). Similarly, the graduates found it simple to transfer to the clinical setting their skills of finding out what they need to know and not relying on others to tell them. While these graduates often felt as if they were not seen by practicing nurses to be as well prepared as other graduates, this was thought to be related to the fact that they were graduates from a university working with others who had obtained their training in community (technical) colleges.

Peterson (1997) provides a rich discussion regarding the skills that are required in order to enhance the potential of a problem-based curriculum. He speaks of several categories of skill development that include consensual decision-making skills, dialogue and discussion skills, maintenance skills, conflict resolution skills, and team leadership skills.

Consensual decision-making requires that students have patience and the ability to listen and learn from others. They must be prepared to adjust their own needs in the interests of the group as a whole. Consensus demands that individuals are prepared to engage with each other in order to understand one another, thus leading to the need for skills in dialogue and discussion. "Dialogue is a process that builds shared meanings and definitions of the problem [or issue] between students within a group" (Peterson, 1997, p. 3). Processes that can help to support and reinforce constructive dialogue are brainstorming and clarification, both of which are inherent components of a problem-based tutorial process. Brainstorming must be uncensored and clarification must be open and last as long as necessary to enable all students to acquire the degree of clarity they seek. On the other hand, discussion is very different. Discussion should come after a period of dialogue, but should not be used as a means of debate or airing of disagreements. A well-directed discussion will focus on issues, not people, and will provide the chance for group members to challenge one another in an open and direct manner. Skills for group maintenance include being able to manage oneself, as well as the group as a whole; this includes determining that tasks are completed in the time frame allotted, ensuring feedback is given and well received, while paying attention to the needs of the group for remediation of any faulty or destructive dynamics. This is where well-honed conflict resolution skills can be most valuable. Conflict is not something to avoid, since it is helpful and healthy when handled well. However, in emerging groups, it is imperative that all members accept the responsibility of dealing openly with differences. It is also essential that the group as a whole engages in developing a group learning style that accommodates all members through compromise and understanding, while still accomplishing the task of learning in relation to knowledge acquisition, skill development, and the internalization of appropriate professional behaviours. Responsibility for team leadership can and should be adopted by every member according to the task at hand and the congruence with every member's skills and preferences. We all learn differently, as well as having varying preferences for being in the limelight or taking a stand. However, students engaged in PBL should embrace the opportunities to gain experience in all team roles, which will stand them in good stead for future professional responsibilities. In summary, Peterson (1997) stresses the importance of providing an environment within which students can sharpen previously owned skills while adopting and adapting others within a small group, team context. This viewpoint is reinforced by Tipping, Freeman, and Rachlis (1995) in their article highlighting the importance of a commitment by faculties to training students for their roles in self-directed, student-centred learning environments.

Table 4-1 helps to synopsize the differences between the learning environments, and the roles inherent within them, in a traditional context and in a PBL world; it has been adapted from the Samford University Web site (http://www.samford.edu/pbl/what3.html).

This is looking very good to me – how are you feeling about the learning that is happening here?

Well, it seems that we are uncovering a lot of information and we seem to be able to pull out a lot of what we need to know in order to answer our questions; however, there is always so much more that we could explore and much more detail. Anyway, it seems as if we are getting better at focusing in on the learning objective. Perhaps we should make sure we know what the focus of the next tutorial is going to be. From what we remember, we should be looking at the evaluation and student assessment methods that are congruent with a PBL curriculum.

I think that is right; how do you want to describe those objectives?

OK, how about: "What assessment methods have been developed that are congruent with PBL principles?" Also, "Have these tools been evaluated, and if so, what are the results of those studies?"

Excellent; good luck with your searching.

Table 4-1

Comparison of Traditional and Problem-Based Learning Environments

Traditional Classroom	PBL Environment
Instructor assumes role of expert or formal authority.	Faculty member is a facilitator, guide, co-learner, mentor, coach.
Faculty members tend to work independently.	Faculty members work in teams together and with others from outside the discipline. Faculty structure is supportive and flexible. Faculty members are involved in changing the instructional culture through the development of assessment tools that are congruent with PBL principles, including peer review.
Faculty members transmit information to learners.	Students take responsibility for learning, creating partnerships with teachers.
Faculty members organize course content into lectures based on discipline content.	Faculty members develop learning scenarios designed to empower students to seek information and to integrate what they find. Student motivation is enhanced through providing scenarios from real life, and thereby activating prior knowledge.
Students are viewed as passive recipients of information.	Faculty members support students, encouraging initiative, guiding learning to enable students to transfer knowledge.
Students work mostly independently and often in isolation.	Students interact with faculty and peers, thus facilitating the provision of immediate feedback and leading to remediation and improvement.

Table 4-1, continued

Comparison of Traditional and Problem-Based Learning Environments

Traditional Classroom	PBL Environment
Students absorb, transcribe, memorize, and repeat information to accomplish content-specific tasks such as tests, exams, and quizzes.	Faculty members design course materials based on case scenarios, thus creating flexible learning environments for students.
Learning is individualistic and competitive.	Students experience learning in a collaborative and supportive environment.
Students look for the "right answer" in order to succeed in an exam-driven context.	Faculty members discourage one "right answer", assisting students to frame questions, formulate learning issues, and explore alternatives.
Performance is measured on content-specific tasks.	Students identify, analyze, and resolve learning issues using knowledge from previous experiences and learning, not relying solely on recall.
Grading is summative and the instructor is the only evaluator.	Students evaluate their own contributions as well as those of other group members.
Lectures are based on one-way communication with information being conveyed to a student group.	Students work in groups of varying sizes to approach the required learning task. Students acquire and apply knowledge in a variety of contexts. Students discover resources with faculty guiding them to the best information. Students seek useful and relevant knowledge to be able to apply into the future.

Adapted from *Traditional Versus PBL Classroom.* Copyright 2003, Center for Problem-Based Research and Communications. Birmingham, AL.

KEY POINTS

- The process is student-led and not teacher-centred, therefore students are expected to be active participants in identifying their own learning needs and pursuing the needed knowledge and skills.
- Faculty are facilitators and guides, relinquishing control of teaching and joining in partnership with the learners in a mutual journey of exploration.

REFERENCES

Barrows, H., Branda, L. (1975; revised Branda, L. 1991; 1994). Role of the tutor/facilitator on problem-based, small group and self-directed learning. Internal document, copyright 1993, McMaster University: Hamilton.

Biley, F. C., & Smith, K. L. (1998). "The buck stops here": Accepting responsibility for learning and actions after graduation from a problem-based learning nursing education curriculum. *J Adv Nurs, 27*, 1021-1029.

Branda, L. (1990). Implementing problem-based learning. *J Dent Educ, 54*(9), 548-549.

Branda, L. A., & Sciarra, A. F. (1995). Faculty development for problem-based learning. *Annals of Community-Oriented Education, 8*, 195-208.

De Grave, W. S., Dolmans, D. H. J. M., & van der Vleuten, C. P. M. (1999). Profiles of effective tutors in problem-based learning: Scaffolding student learning. *Med Educ, 33*, 901-906.

Federman, D. D. (1999). Little-heralded advantages of problem-based learning. *Acad Med, 74*, 93-94.

Finucane, P., Allery, L. A., & Hayes, T. M. (1995). Comparison of teachers at a "traditional" and an "innovative" medical school. *Med Educ, 29*, 104-109.

Mayo, W. P., Donnelly, M. B., & Schwartz, R. W. (1995). Characteristics of the ideal problem-based learning tutor in clinical medicine. *Evaluation & the Health Professions, 18*(2), 124-136.

Peterson, M. (1997). Skills to enhance problem-based learning. Retrieved April 23, 2003 from *Medical Education Online*: www.med-ed-online.org/f0000009.htm

Tipping, J., Freeman, R. F., & Rachlis, A. R. (1995). Using faculty and student perceptions of group dynamics to develop recommendations for PBL training. *Acad Med, 70*(11), 1050-1052.

LEARNING RESOURCES

Hitchcock, M. A., & Mylona, Z.-H. E. (2000). Teaching faculty to conduct problem-based learning. *Teach Learn Med, 12*(1), 52-57.

Kaufman, D. M., Mensink, D., & Day, V. (1998). Stressors in medical school: Relation to curriculum format and year of study. *Teach Learn Med, 10*(3), 138-144.

Maudsley, G. (1999). Roles and responsibilities of the problem based learning tutor in the undergraduate medical curriculum. *British Medical Journal, 318*, 657-661.

Solomon, P., & Crowe, J. (2001). Perceptions of student peer tutors in a problem-based learning programme. *Medical Teacher, 23*(2), 181-186.

Solomon, P., & Finch, E. (1998). A qualitative study identifying stressors associated with adapting to problem-based learning. *Teach Learn Med, 10*(2), 58-64.

Assessment Methods for Problem-Based Learning

"PBL is a primary learning method in which new material is presented, and is not a euphemism for afternoon review sessions with voluntary attendance."
(Kanter, 1998, p. 394)

OBJECTIVES

- What assessment methods have been developed that are congruent with PBL principles?
- Have these tools been evaluated, and if so, what are the results of those studies?

After all the research we have done so far, it is becoming quite clear that traditional methods of evaluating student performance are not helpful in a PBL environment. Since PBL is a process-driven learning philosophy, it becomes critical that there are tools that can measure process across the continuum of an educational program. There are also needs for more summative evaluations because students within PBL curricula have to relate to the larger context by passing certification examinations and obtaining other practice credentials as required. Also, many students wanting to pursue graduate work need to be able to prove their abilities to other educational institutions, which frequently demand a transcript that reports grades not simply as pass/fail or satisfactory/unsatisfactory ratings. This interface in and of itself is a hard one to debate. However, we should start by illustrating what we know of the special kinds of evaluation tools developed to respond to the particular needs of PBL students.

Throughout the literature examining and reporting upon PBL, there are many examples of articles and resources that describe approaches to evaluation. Nendaz and Tekian (1999) provide an excellent overview of the literature in this area using Swanson's categorization of outcome- and process-oriented instruments (Nendaz & Tekian, 1999; Swanson, Case, & van der Vleuten, 1991).

Outcome-Oriented Tools

These tools are designed to evaluate the end-point of students' learning and do not look at how the results were reached. Traditional curricula rely heavily on multiple choice examinations (MCQs) to test knowledge acquisition. MCQs have been criticized for assessing isolated facts and testing only recall (Nendaz & Tekian, 1999, p. 234). However, there is a combined approach in which MCQs are grafted together with short answer questions (SAQs) that provides a neat, test-based model, which allows for a relatively quick marking process, and also is more responsive to illustrating students' reasoning abilities. This approach also reduces the risk of guessing! The writing of academic papers and essays is a well-tested method of student evaluation. This method is particularly sensitive to reflecting problem-solving skills and conceptual reasoning abilities in the writer. Incorporation of several shorter essays addressing multiple scenarios is better, enhancing the reliability of the measure and reducing the chances of subjective scoring (Swanson, Case, & van der Vleuten, 1991). Oral assessment is also a common method employed in many professional preparation programs. Standardization of these exams has been attempted by preparing set modules of learning scenarios complete with particular questions and well-articulated scoring criteria (e.g., Rendas, Pinto, & Gamboa, 1998). Modified essay questions (MEQs) provide learners with the opportunity to follow the same process as in the small group tutorials, but on paper and as an individual exercise. While this is appealing, it is essential that the bank of items used for MEQs has scope and depth to reduce the chance of recall rather than reasoning (Neville, 1995). Another instrument developed at McMaster University to confront the need to test students in a reliable fashion is the clinical reasoning exercise (CRE). This tool can be administered orally or in written form and contains between 10 and 20 questions (case scenarios) to be answered within 1 hour at the most. This tool has been shown to have acceptable reliability and sound correlation with objective tests (Neville, Cunnington, & Norman, 1996). Assessment of the students' abilities in the management of direct client service is accomplished in many different ways; these include paper case scenarios, computer simulations, as well as the use of standardized patients. Standardized patients are trained to assume the symptoms and signs of a given condition, together with learning the background of the given patient or client, thus presenting themselves in an authentic manner as the individual in question. This is a powerful method of becoming familiar and comfortable in working directly with clients (Barrows, 2000).

PROCESS-ORIENTED TOOLS

These tools focus on the acquisition of skill sets that are deemed to be core to the successful application of PBL principles, including problem solving, self-directed learning, and self-assessment. The triple-jump exercise was developed to provide students with an individual opportunity for assessing their abilities to engage in the problem-based reasoning process as illustrated in the small group tutorials. While this approach has been used quite widely, there are basic flaws related to the small number of practice scenarios addressed by this means and which preclude its use as part of a reliable summative evaluation (Smith, 1993; Foldevi, Sommanssen, & Trell, 1994). Efforts to increase the reliability have included the use of videotapes of real patients for students to view and critique, offering professional opinions regarding assessment and intervention planning. These are then evaluated by a family physician and a basic scientist (Foldevi et al., 1994) and the combination of written responses to the issue generation stage with subjective scores by faculty following student observation (Smith, 1993). Formative (process) assessment is a main focus of PBL systems. In order to be truly helpful to both learners and faculty, it must be continuous and a process that is "integrated with learning, ...favors self-assessment and assists students in defining their own learning needs" (Nendaz & Tekian, 1999, p. 233). Throughout a PBL learning experience, there should be expectations that students engage in a process of self-reflection that is guided by faculty, in any of the roles of tutor, advisor, seminar leader, or resource person. Input to this ongoing process is received from peers and others, such as clinical preceptors, as learning progresses. Students can be required to maintain a personal portfolio from the first day of their academic program that provides a tracking of the learning they accomplish throughout their educational experience. This portfolio should contain all the assignments, complete with grades/ratings, together with an overview of the scope of learning plotted against a grid representing the developmental lifespan, practice settings, and therapist roles. During each term, and within each course in each term, each student completes many evaluations, designed to provide input regarding personal learning accomplishments, as well as to give feedback to faculty on their level of accomplishment in the relevant role (e.g., tutor, course planner).

Examples of such forms are provided in Tables 5-1 through 5-3. There are six main areas for consideration in the student evaluation: learning skills and functioning, group skills and functioning, knowledge development and objective achievement, critical thinking and clinical reasoning skills, feedback and evaluation skills, and professional behaviours.

—————————— **Table 5-1** ——————————

Sample of In-Tutorial Evaluation Form

In-Tutorial Performance Evaluation

1. **Group Skills and Functioning**   Grade:_____
   - Actions that assist and facilitate group functioning
   - Sensitivity to needs of individuals in group
   - Demonstrates respect for others
   - Initiates development of group goals
   - Assists in the clarification of group goals
   - Willing to accept group goals even when not a personal learning need
   - Is aware of strengths and weaknesses of other group members regarding group process
   - Listens and is attentive to all members

   Comments:

2. **Learning Skills and Preparation**   Grade:_____
   - Identifies personal learning objectives
   - Identifies appropriate learning strategies and resources
   - Accesses a variety of resources (e.g., books, journals, audio-visual materials, clinical experts, community resources, anatomy lab, etc.)
   - Selects fundamental information sufficient for the full exploration of the learning issue(s)
   - Comes prepared to share information in an organized and useful manner
   - Uncovers additional, unexpected information through a diligent effort

   Comments:

3. **Knowledge Development and**
   Objective Achievement   Grade:_____
   - Demonstrates increased knowledge and understanding in relation to learning objectives
   - Provides good quality information to the group
   - Able to meet group demands regarding knowledge/information
   - Relates some previously acquired knowledge in a new context
   - Links new knowledge to the current problem
   - Explains some information to the group, which helps everyone understand
   - Shares complex information in an understandable manner
   - Uses appropriate language/professional terminology

   Comments:

------------------------ Table 5-1, continued ------------------------

Sample of In-Tutorial Evaluation Form

4. Critical Thinking and Clinical Reasoning Skills   Grade:_____
- Offers plausible hypotheses for group to consider
- Able to critically evaluate information
- Able to synthesize information and identify inconsistencies
- Able to develop and defend an argument or position based on theory, research evidence, and/or expert opinion
- Able to apply new information to the case under study (i.e., clinical reasoning) e.g., choice of assessment tools, explanation of functional deficits, identification of intervention objectives, program planning, discharge planning, evaluation of client outcomes
- Asks provocative or challenging questions
- Is willing to be convinced of some position but asks for sufficient evidence first

Comments:

5. Feedback and Evaluation Skills   Grade:_____
- Able to provide feedback to others in a tactful and sensitive manner
- Identifies actions which have both assisted and hindered learning within the group
- Provides constructive suggestions to others to help them improve their performance
- Demonstrates insight regarding personal performance (i.e., strengths and weaknesses)
- Able to identify personal learning objectives that build on strengths and also address areas of weakness
- Responds to feedback from others in attempting to change behaviour

Comments:

—————— **Table 5-2** ——————

PBL Tutor Performance Evaluation

In order to maintain high educational standards and to facilitate group functioning, the Programme requires feedback from students regarding the performance
of faculty. This information will be used in considering reappointments, to guide
faculty development, and to assist in considerations of promotion and tenure.
These evaluations are an integral part of the feedback that tutors receive and
will enable tutors to continue their pursuit of excellence by the identification of
both strengths and areas in need of improvement.

Use the following scale for your ratings:

1	2	3	4	5	6	7	8	9	10
very good	poor		acceptable			good		very good	exc.

Indicate your ratings on the optic scan sheet.

This tutor:
1. Assisted students in planning to meet course objectives.
2. Assisted the group in determining when an appropriate level of understanding had been reached.
3. Facilitated consistent provision of feedback (positive/constructive,
   oral/written).
4. Facilitated discussion of students' learning needs and expectations.
5. Was enthusiastic about educational role.
6. Provided a reasonable degree of student autonomy regarding learning
   objectives, strategies, and resources.
7. Communicated clearly with students.
8. Facilitated the development of students' ability to acquire, apply, and synthesize knowledge (e.g., promoted critical thinking, evidence-based practice).
9. Overall, what is your opinion of the effectiveness of the tutor as a facilitator of learning?

Comments:
Please add your comments in the space provided on the optic scan sheet in
terms of strengths (Section A), areas requiring attention (Section B), and additional comments (Section C).

Table 5-3

Oral Three-Phase Examination Evaluation Form

Student: Tutor:
Date:

Part One   *Comments*

Hypothesis Generation

Key aspects of problem and
possible explanations (comprehensive,
clear, relevant).

Immediate Data Search

Appropriate and efficient data gathering.

Problem Formulation

Final problem formulation should
adequately characterize client problems.

Learning Issues

Description of key learning issues raised
by problem.

Learning Plan

Outline of plan to pursue these issues.

Step One Grade: _____   (weighted 40% of overall grade)

Part Two   *Comments*

Independent Data Search

Description of process of implementing learning
plan (efficient use of time and resources).

Written reference list
(variety of useful resources).

Step One Grade: _____   (weighted 10% of overall grade)

─────── Table 5-3, continued ───────

Oral Three-Phase Examination Evaluation Form

Part Three   *Comments*

Synthesis

Presentation of cumulative knowledge
(relevance, quantity, quality, understanding).
Student should incorporate new information
into problem and demonstrate an understanding of the presenting issues and the OT role.

Responses to questions.

Self-assessment of strengths and weaknesses.

Step One Grade: _____   (weighted 50% of overall grade)

Overall Grade and Comments: _____
                              (Letter Grade)

Copyright © 2002, McMaster University MCHS(OT) Programme.

When rating performance in each of the six performance areas, the following statements are used as criteria to assign a grade of Satisfactory / Unsatisfactory. Both the current level of function that the student now demonstrates as well as any improvements that have occurred in that particular skill area over time (i.e., since the beginning of the term or since the mid-term evaluation) should be considered.

Satisfactory

Student consistently demonstrates most of the key skill behaviours within the performance area(s). Performance is of good-excellent quality, reflecting continued growth and improvement throughout the term. Student responds quickly to change performance following feedback and demonstrates responsibility and self-direction.

Unsatisfactory

Minimal expectations are not met. An Unsatisfactory grade in any performance area may be assigned if the student:

- Demonstrates some of the key skill behaviours on occasion but performance is not consistent.
- Does not demonstrate some key skill behaviours and further skill development is required.
- Demonstrates sporadic participation and there is limited evidence on which to base evaluation.
- Demonstrates minimal evidence of change in response to feedback.
- Does not demonstrate responsibility and self-direction.

An overall grade of the In-Tutorial Performance is assigned, taking into consideration the student's performance on each of the six performance areas.

The evaluation process within the tutorial is undertaken collaboratively between the student being evaluated, his or her peers, and the group tutor. Most often, the ideal approach is to take time at the end of each tutorial to gain an overall sense of how well the group is working together to achieve the learning objectives and what areas for improvement exist for individual students, the tutor, and the group as a whole. The more formalized evaluation process incorporating the completion of forms and written feedback should take place at mid-term and at term end. Similar forms are available for students to provide feedback to the tutor. However, it should always be remembered that truly constructive feedback should be given face-to-face and then followed up by written comments together with a plan to reinforce strengths and remediate weaknesses.

> *So, do you think you have an idea of what it feels like to be evaluated as a learner within a PBL environment?*
> Perhaps; it will probably become much clearer once we are started into our educational program. It is going

> ## KEY POINTS
>
> - Evaluation methods must be congruent with PBL principles; therefore methods used in traditional systems should not be utilized.
> - Examples of process-oriented methods include the triple-jump exercise, self-assessment, peer assessment, and self-reflection.
> - Examples of outcome-oriented methods include academic papers, modified essay questions, practica experiences, and the clinical reasoning exercise.

to be rather difficult to make the switch, I think, from relying on grades to give us a sense of what we know and how well we are doing. The idea appears very appealing, to be in control of one's own learning and to be able to steer how we find out what we need to know. However, it also seems very scary. In many ways, it was very reassuring to know that we were being taught what the faculty knew we needed to know. In fact, it will be difficult to get a sense of whether we are studying at the right level of detail, and whether we know enough or what people are expecting us to know.

Yes, and that is when it is critical to go back to the PBL principles—openness, honesty, and the trust that is forming around you. Look at your handbooks and reread the term and course objectives. This will provide you with a firm anchor again, and remind you of the scope of what you are being asked to explore for this term. Also, check the scope of the evaluation assignments. This will also provide you with a clear picture of what is expected. It is very important to rely on each other as well. This is a time when the whole notion of collaboration comes to the fore. Share resources you find, and support each other in filling the gaps of learning and understanding that are emerging.

I hope this has helped a little. Perhaps we can go back to these issues after you have talked with faculty and students, and seen the results of your survey.

Thanks, that is helpful. I guess we have to keep going back to basics! Now, what is the focus for the next tutorial?

If I recall, it is to explore the literature to see what evidence is out there to support the idea that PBL is a good way to learn (and teach).

Yes, that's right; perhaps we can frame the question this way: Is there evidence in the literature to support or refute the utility and success of PBL as an approach to learning? There is some literature that will help in the health professional education journals.

AIDS FOR GIVING AND RECEIVING FEEDBACK

George F.J. Lehner, Ph.D.
Professor of Psychology
University of California, Los Angeles

Some of the most important data we can receive from others (or give to others) consists of feedback related to our behaviour. Such feedback can provide learning opportunities for each of us if we can use the reactions of others as a mirror for observing the consequences of our behaviour. Such personal data feedback helps to make us more aware of what we do and how we do it, thus increasing our ability to modify and change our behaviour and to become more effective in our interactions with others.

To help us develop and use the techniques of feedback for personal growth, it is necessary to understand certain characteristics of the process. The following is a brief outline of some factors that may assist us in making better use of feedback, both as the giver and the receiver of feedback. This list is only a starting point. You may wish to add further items to it.

(continued)

1. Focus feedback on behaviour rather than the person.

It is important that we refer to what a person does rather than comment on what we imagine he is. This focus on behaviour implies that we use adverbs (which relate to qualities) when referring to a person. Thus we might say a person "talked considerably in this meeting", rather than that this person "is a loudmouth". When we talk in terms of "personality traits" it implies inherited, constant qualities that are difficult, if not impossible, to change. Focusing on behaviour implies that it is something related to a specific situation that might be changed. It is less threatening to a person to hear comments about his behaviour than his "traits".

2. Focus feedback on observations rather than inferences.

Observations refer to what we can see or hear in the behaviour of another while inferences refer to interpretations and conclusions that we make from what we see or hear. In a sense, inferences or conclusions about a person contaminate our observations, thus clouding the feedback for another person. When inferences or conclusions are shared and it may be valuable to have this data, it is important that they be so identified.

3. Focus feedback on description rather than judgement.

The effort to describe represents a process for reporting what occurred, while judgement refers to an evaluation in terms of good or bad, right or wrong, nice or not nice. The judgements arise out of a personal frame of reference or values, whereas description represents neutral (as far as possible) reporting.

4. Focus feedback on descriptions of behaviour, which are in terms of "more or less" rather than in terms of "either/or".

The "more or less" terminology implies a continuum on which any behaviour may fall, stressing quantity, which is objective and meaningful, rather than quality, which is subjective and judgemental. Thus, participation of a person may fall on a continuum from low participation to high participation, rather than "good" or "bad" participation. Not to think in terms of "more or less" and the use of continua is to trap ourselves into thinking in categories, which may then represent serious distortions of reality.

(continued)

5. *Focus feedback on behaviour related to a specific situation, preferably to the "here and now", rather than to behaviour in the abstract, placing it in the "there and then".*

What you and I do is always tied in some way to time and place, and we increase our understanding of behaviour by keeping it tied to time and place. Feedback is generally more meaningful if given as soon as appropriate after the observation or reactions occur, thus keeping it concrete and relatively free of distortions that come with the lapse of time.

6. *Focus feedback on the sharing of ideas and information rather than on giving advice.*

By sharing ideas and information we leave the person free to decide for himself, in the light of his own goals in a particular situation at a particular time, how to use the ideas and the information. When we give advice we tell him what to do with the information, and in that sense we take away his freedom to determine for himself what is for him the most appropriate course of action.

7. *Focus feedback on exploration of alternatives rather than answers or solutions.*

The more we can focus on a variety of procedures and means for the attainment of a particular goal, the less likely we are to accept our particular problem. Many of us go around with a collation of answers and solutions for which there are no problems.

8. *Focus feedback on the value it may have to the recipient, not on the value or "release" that it provides the person giving the feedback.*

The feedback provided should serve the needs of the recipient rather than the needs of the giver. Help and feedback need to be given and heard as an offer, not an imposition.

9. *Focus feedback on the amount of information that the person receiving it can use, rather than on the amount you have that you might like to give.*

To overload a person with feedback is to reduce the possibility that he may use what he receives effectively. When we give more than can be used we may be satisfying some need for ourselves rather than helping the other person.

(continued)

10. Focus feedback on time and place so that personal data can be shared at appropriate times.

Because the reception and use of personal feedback involves many possible emotional reactions, it is important to be sensitive to when it is appropriate to provide feedback. Excellent feedback presented at an inappropriate time may do more harm than good.

11. Focus feedback on what is said rather than why it is said.

The aspects of feedback that relate to the what, how, when, and where of what is said are observable characteristics. The why of what is said takes us from the observable to the inferred, and brings up questions of "motive" or "intent".

It is maybe helpful to think of why in terms of a specifiable goal or goals, which can then be considered in terms of time, place, procedures, probabilities of attainment, etc. To make assumptions about the motives of the person giving feedback may prevent us from hearing or cause us to distort what is said. In short, if I question "why" a person gives me feedback, I may not hear what he says.

In short, the giving (and receiving) of feedback requires courage, skill, understanding, and respect for self and others.

From Programme for Educational Development (1987). *Evaluation methods: A research manual.* Hamilton, Ontario: McMaster University. Used with permission.

REFERENCES

Barrows, H. S. (2000). *Practice-based learning applied to medical education.* Springfield, IL: Southern Illinois University School of Medicine.

Foldevi, M., Sommansson, G., & Trell, E. (1994). Problem-based medical education in general practice: Experience from Linkoping, Sweden. *Br J Gen Pract, 44,* 473-476.

Kanter, S. L. (1998). Fundamental concepts of problem-based learning for the new facilitator. *Bulletin of the American Library Association, 86*(3), 391-395.

Nendaz, M. R., & Tekian, A. (1999). Assessment in problem-based learning medical schools: A literature review. *Teach Learn Med, 11*(4), 232-243.

Neville, A. (1995). Student evaluation in problem-based learning. *Pedagogue, 5,* 2-7.

Neville, A., Cunnington, J., & Norman, G. R. (1996). Development of clinical reasoning exercises in a problem-based curriculum. *Acad Med, 71*, S105-S107.

Programme for Educational Development (1987). *Evaluation methods: A research manual.* Hamilton, Ontario: McMaster University.

Rendas, A. B., Pinto, P. R., & Gamboa, T. (1998). Problem-based learning in pathophysiology: Report of a project and its outcome. *Teach Learn Med, 10,* 34-39.

Smith, R. M. (1993). The triple-jump examination as an assessment tool in the problem-based medical curriculum at the University of Hawaii. *Acad Med, 68,* 366-372.

Swanson, D. B., Case, S. M., & van der Vleuten, C. P. M. (1991). Strategies for student assessment. In D. Boud, D. B. G. Feletti (Eds.). *The challenge of problem-based learning* (pp. 260-273). New York: St. Martin's Press.

Learning Resources

Alpert, I. L., & Perkins, R. M. (1999). Should students in problem-based learning sessions be graded by the facilitator? Research brief. *Acad Med, 74*(11), 1256.

De Grave, W. S., Dolmans, D. H. J. M., & van der Vleuten, C. P. M. (1998). Tutor intervention profile: Reliability and validity. *Med Educ, 32,* 262-268.

Faculty of Health Sciences (1994). *Guide to professional behaviours in tutorial meetings.* Hamilton, Ontario: McMaster University.

Jacobs, T. (1997). Integrating assessment in problem-focused curricula. *British Journal of Occupational Therapy, 60*(4), 174-178.

Lehner, G. F. J. (1975). *Aids for giving and receiving feedback.* Los Angeles: University of California.

Mennin, S. P., & Kalishman, S. (1998). Student assessment. *Acad Med, 73*(9), S46-S54.

Norman, G. R. (1991). What should be assessed? In D. Boud & G. Feletti (Eds.). *The challenge of problem-based learning* (pp. 254-259). New York: St. Martin's Press.

Programme for Faculty Development. (2001). *Problem based learning in small groups workshop evaluation resource package.* Hamilton, Ontario: Faculty of Health Sciences, McMaster University.

Sullivan, M. E., Hitchcock, M. A., & Dunnington, G. L. (1999). Peer and self-assessment during problem-based tutorials. *Am J Surg, 177,* 266-269.

Valle, R., Petra, I., Martinez-Gonzales, A., Rojas-Ranirez, J. A., Morales-Lopes, S., & Pina-Garza, B. (1999). Assessment of student performance in problem-based learning tutorial sessions. *Med Educ, 33,* 818-822.

Tutorial 6

Researching the Effectiveness of Problem-Based Learning

"Proving that PBL confers any educational advantage has
not been easy. Zealous defenders and hostile critics have lined
up around the still-meager literature…"
(Federman, 1999, p. 94)

OBJECTIVE

- Is there evidence in the literature to support or refute the utility and success of PBL as an approach to learning?

From what we have read and what we understand so far, it seems that a problem-based approach to learning is a very good one. At least, when we look at the kinds of values and beliefs that support it, it is almost like motherhood and apple pie! After all, how can anyone deny that it is right to respect one another, relate in an open and honest manner, collaborate one with the other, and learn to trust? There are some concerns that surface though, when we wonder how well we can learn on our own, and also if the system is a fair one; with so many tutors, how do we know that each group gets to cover the same information and learn the same things? Therefore, is the system reliable, or is it dependent upon the quality of the incumbents within it; that is, the quality and commitment of the faculty to the principles, and the readiness of the students to jump into a new initiative, putting their future careers in the hands of a new learning approach. It would also seem logical to wonder if the old saying has some truth to it—if something isn't broken, why fix it? The traditional way of learning seems to have served hundreds of years of students quite well. New ways to approach the learning-teaching process may be interesting, but are they necessary? A PBL curriculum appears to require a lot more resources, so it must be more expensive than the usual didactic approach. When we looked at the literature, there were many studies reported that attempted to answer the question that we asked, so we selected a few to focus on some of the more measurable or generalizable outcomes.

From the earliest outcome studies of the McMaster medical program, there were strong indicators that the "product", the new McMaster medical graduates, seemed to be at a similar level as their colleagues from other schools when measured by the national certification examination. In addition, the attrition rate was very low (under 1%) and graduates rated their own satisfaction as high. Reports from the clinical, research, and academic arenas were positive from the time of the first graduating class (Pallie & Carr, 1987; Neufeld, 1983; Neufeld, Woodward & MacLeod, 1989). However, two meta-analyses completed about a decade ago indicated that there were definite advantages to learning in a problem-based manner, but that students from these programs tended not to perform as well as their colleagues from traditional schools in the results of certification examinations designed to test basic science knowledge (Albanese & Mitchell, 1993; Vernon & Blake, 1993).

Viet Vu, van der Vleuten and Lacombe (1998) reported on a longitudinal study they completed, focusing on exploring the thinking processes undertaken by medical students from both traditional and reform curriculum schools. As we already have discussed briefly, students from traditional environments tend to adopt a "surface" approach to learning, relying heavily upon cramming information in readiness for being tested in examinations. Conversely, students from PBL environments tend to pursue more of a "depth"-related learning style, resulting in a higher likelihood of meaningful learning taking place and in their understanding of the material being studied. This observation was first highlighted by Newbie and Clarke (1986) in their study exploring elements of the learning style of medical students in Australia. To explore students' learning behaviours in a more profound manner, Viet Vu and colleagues utilized a self-report measure called the Inventory of Learning Process (ILP) (Viet Vu, van der Vleuten, & Lacombe, 1998, p. S25). The ILP has been evaluated for reliability and validity and is relatively easy to administer and score. This study resulted in some interesting findings that supported the initial assumptions about differences in learning approaches. The overall results showed no difference between the two student groups when addressing the learning behaviours adopted by 2nd-year medical students. However, with students who had experienced 1 traditional study year before entering the PBL stream, they were found to have adopted far fewer "deep" processing habits than their colleagues who had experienced their initial year of study in the PBL stream. In summary then, "these findings seem to imply that the emphasis in the PBL environment on the processes of analyzing, organizing, structuring, and integrating knowledge during the tutorials sessions do help students to maintain high levels of elaborative processing while increasing their ability in deep processing" (Viet Vu et al., 1998, p. S26).

Another group of researchers, Patel, Groen, and Norman (1991), explored the effects of problem-based and traditional medical school curricula on the students' problem-solving skills. A forward-thinking style of reasoning was uncovered predominantly in the PBL student group, which produced a more elaborate approach using biomedical information, which was not observed within the other student group from the more conventional school. However, this approach to reasoning tended to produce a higher incidence of errors during the reasoning process. Overall, it became clear that a PBL approach does result in students adopting a very systematic way of developing their knowledge base. Also, PBL integrates basic science and clinical information much more than the traditional way of thinking. Nevertheless, it would behoove educators to pay more attention to developing ways to minimize erroneous reasoning, particularly when considering the importance of self-directed learning principles within PBL curriculum models (Patel, Groen & Norman, 1991, p. 388).

Along similar lines to the previous articles discussed, Hmelo (1998) explored medical students' explanations of pathophysiological tasks during their first years in medical school with PBL, elective PBL, and traditional curricula. She found that the PBL students were able to generate explanations that tended to be more coherent, more accurate, and comprehensive than other students. Again, she found that these PBL students were more likely to use concepts from the basic sciences.

Cariaga-Lo, Richards, Hollingsworth, and Camp (1996) caution us to consider the importance of admission criteria to the ultimately successful outcome of students completing a PBL program. Duek, Wilkerson, and Adolfini (1996) explore concerns voiced by students who led their own tutorial groups, without the presence of faculty in the tutor role. One of the central findings revolved around the need for guidance and a sense of structure to assist in driving learning forward in a positive manner. The initial purpose for attempting the tutor-less format was to ease the pressure on faculty educational contributions; however, this does not seem to be a preferable strategy when there is little structure within the overall curriculum. Having a tutor there as a guide and support seems to be a very important element for the learners.

PBL is cited frequently as an invaluable way to prepare health care practitioners. Consequently, it is important to find out from clinical preceptors how they view these students within the clinical environment. Blake and Parkison looked particularly at a group of medical students who had just completed a newly formed pre-clinical curriculum with PBL as a main component (1998). The physician evaluators were asked to complete a five-point Likert scale assessing eight areas of knowledge and performance. In comparison with other 3rd-year students, the PBL student cohort was rated as "better" by a majority of assessors on clinical reasoning, problem-solving, knowledge of pathophysiology and disease processes, and on completing a patient history (Blake & Parkison, 1998, p. 69). Distlehorst and Robbs (1998) also reported that students in their 3-year retrospective study were not disadvantaged in any way by being engaged in a PBL curriculum. Outcome measures considered included the United States Medical Licensing Examination (USMLE), their clinical clerkships, and clinical practice examinations. At another university, the medical school underwent a process of curriculum redesign. Across several graduating classes, Blake, Hosokawa, and Riley (2000) compared the last 2 years of the traditional curriculum with the first 4 years of the new. The authors considered data from GPA and MCAT scores for all six classes, together with data from all U.S. and Canadian USMLE first-time takers. The major curriculum revisions did not disadvantage the students at all; in fact, they may have assisted in the achievement of higher scores.

With most health care practitioners, there is a growing expectation for practicing within a self-regulated professional climate. Many disciplines adopt the completion of a national certification examination as a main gate-keeping strategy. Therefore, it is critical that educators remain closely aware of the demands of these national examinations and ensure that their graduates are well prepared to take the exam.

In a recent article addressing thorny issues of evaluation and research in education, Norman and Schmidt (2000) respond to work by Colliver that presents a particularly pessimistic view of PBL and its outcomes for students. Colliver (2000) asserted that there is no convincing evidence to suggest that PBL is any better than any other way of designing learning opportunities. In fact, we could perhaps reply that there was never any intention to prove it was any better, but simply that it is yet another way to approach learning that has some energizing elements for learners. Norman and Schmidt agree that, in many cases, proponents of PBL have tended to be overly evangelical at times, but that the problem does not lie in the bases of cognitive psychology. In fact, they state clearly that they concur with Colliver that the underlying premises of PBL do require close study, and that this closer study is in itself most challenging to design and undertake. They offer a suggestion that further study should be focused upon a broad research agenda, encompassing theory building and testing in rigorous ways from laboratory environments to the real world. Rothman challenges everyone involved in PBL to recognize what he terms the "serious limitations" inherent within the PBL approach and to move towards other educational innovations (Rothman, 2000). In response to this gauntlet, Kaufman (2000) suggests strongly that a new approach to evaluation within PBL curricula is called for. Perhaps, he suggests, the focus should be upon the "product" of a PBL environment. "I would assert that PBL has been one of the most successful innovations in...education and has established its credibility. It is time to step back to reflect on 'accepted' practice and to develop the next generation of PBL" (Kaufman, 2000, p. 511).

KEY POINTS

- PBL is a viable and valuable way to learn.
- Performance on certification examinations may not be as high as students who have experienced a more traditional learning environment.
- PBL students tend to process their learning at a deeper level.
- A PBL approach tends to result in a more systematic way of developing a knowledge base.
- There is sound integration of basic science and clinical information.
- Faculty should be attuned to the need to minimize instances of erroneous reasoning.
- Learning the PBL way is often more enjoyable.
- It is time to move on, accepting PBL as a mainstream option for teaching and learning.

REFERENCES

Albanese, M. A., & Mitchell, S. (1993). Problem-based learning: A review of literature on its outcomes and implementation issues. *Acad Med, 68*, 61-82.

Blake, R. L., & Parkison, L. (1998). Faculty evaluation of the clinical performances of students in a problem-based learning curriculum. *Teach Learn Med, 10*(2), 69-73.

Blake, R. L., Hosokawa, M. C., & Riley, S. L. (2000). Student performances on step 1 and step 2 of the United States medical licensing examination following implementation of a problem-based learning curriculum. *Acad Med, 75*(1), 66-70.

Cariaga-Lo, L. D., Richards, B. F., Hollingsworth, M. A., & Camp, D. L. (1996). Non-cognitive characteristics of medical students: Entry to problem-based and lecture-based curricula. *Med Educ, 30*, 179-186.

Colliver, J. (2000). Effectiveness of problem-based learning curricula. *Acad Med, 75*,259-266.

Distlehorst, L. H., & Robbs, R. S. (1998). A comparison of problem-based learning and standard curriculum students: Three years of retrospective data. *Teach Learn Med, 10*(3), 131-137.

Duek, J., Wilkerson, L., & Asinolfi, T. (1996). Learning issues identified by students in tutorless problem-based tutorials. *Adv Health Sci Educ, 1*, 29-40.

Federman, D. D. (1999). Little-heralded advantages of problem-based learning. *Acad Med, 74*, 93-94.

Hmelo, C. E. (1998). Cognitive consequences of problem-based learning for the early development of medical expertise. *Teach Learn Med, 10*(2), 92-100.

Kaufman, D. M. (2000). Problem-based learning—time to step back? *Med Educ, 34*, 509-511.

Neufeld, V. R. (1983). Adventures of an adolescent: Curriculum changes at McMaster University. In C. Friedman & E. S. Purcell, (Eds.). *New Biology and Medical Education* (pp. 256-270). New York: Josiah Macy Jr. Foundation.

Neufeld, V. R., Woodward, C. A., & MacLeod, S. M. (1989) The McMaster M.D. program: A case study of renewal in medical education. *Acad Med, 64*, 423-447.

Newbie, D. I., & Clarke, R. M. (1986). The approaches to learning of students in a traditional and in an innovative problem-based medical school. *Med Educ, 20*, 267-271.

Norman, G. R., & Schmidt, H. G. (2000). Effectiveness of problem-based learning curricula: theory, practice and paper darts. *Med Educ, 34*, 721-728.

Pallie, W., & Carr, D. H. (1987). The McMaster medical education philosophy in theory, practice and historical perspective. *Medical Teacher, 9*(1), 59-71.

Patel, V. L., Groen, G. J., & Norman, G. R. (1991). Effects of conventional and problem-based medical curricula on problem solving. *Acad Med, 7*, 380-389.

Rothman, A. I. (2000). Problem-based learning—Time to move forward? *Med Educ, 34*, 509-511.

Vernon, D. T. A., & Blake, R. L. (1993). Does problem-based learning work? A meta-analysis of evaluative research. *Acad Med, 68*, 550-563.

Viet Vu, N., van der Vleuten, C. P. M., & Lacombe, G. (1998). Thinking about student thinking. *Acad Med, 73*(10), S25-S27.

LEARNING RESOURCES

Bligh, J., Lloyd-Jones, G., & Smith, G. (2000). Early effects of a new problem-based clinically oriented curriculum on students' perceptions of teaching. *Med Educ, 34*, 487-489.

Friedman, C. P., de Bliek, R., Greer, D. S., Mennin, S. P., Norman, G. R., Sheps, C. G., Swanson, D. B., & Woodward, C. A. (1990). Charting the winds of change: Evaluating innovative medical curricula. *Acad Med*, 8-14.

Woodward, C. A. (1986). McMaster University graduates. *J Med Educ, 61*, 859-860.

Woodward, C. A. (1996). Problem-based learning in medical education: Developing a research agenda. *Adv Health Sci Educ, 1*, 83-94.

Tutorial 7

Feedback on Problem-Based Learning

OBJECTIVES

- What do students and faculty think about their experiences with a reform curriculum model such as problem-based learning?
- What are the strengths and weaknesses?
- Are there any stories that could provide some good illustrations of what it means to learn and teach in a PBL context?

This time, while we were completing our research, we took a very different approach to finding the information we needed. While there were some articles that related experiences of mounting a PBL curriculum and others that described different models of PBL, it began to feel rather like the same thing over and over. So, we decided to include a number of personal narratives, as well as results from an unpublished faculty survey and the results of our own email survey. The resources that were accessed included faculty members from McMaster University, as well as visiting faculty from other centres. Similarly, information from students was sought and obtained at a personal level.

FORMAL LITERATURE

We found several articles that addressed feedback from students regarding their PBL experiences. When reading these articles, focused on student opinion and feedback, there are a few prevailing themes. These include understanding of PBL expectations, how efficient it is as a way of learning, the importance of the tutor in small groups, and satisfaction with the process. Caplow, Donaldson, Kardash, and Hosokawa (1997) examined students' ideas about PBL within a medical curriculum using weekly student journals, focus groups, interviews with tutors, and open-ended student questionnaires. The students strongly support the need for a period of acclimatizing them to the new ideas before being launched into learning seriously in the new way.

In addition, the importance of a strong linkage between the PBL objectives and the way in which learning is completed and evaluated was paramount to this student group. Also, the tutor role was seen to be both a help and hindrance to the learners. This reinforces the importance of a common training experience for all tutors that can set the underlying ground rules for assuming that critical role, and minimize the tendency for wide diversity in interpreting the role expectations. The authors concluded that "more research is needed on process dimensions, such as the socialization of students to an educational innovation and mediating variables such as tutor behaviours and roles" (Caplow et al., 1997, p. 447). It would seem that much of the criticism levied in this article could

be mitigated by well-articulated orientation and training expectations for faculty and students alike. This does not appear to be a
staunch discrediting of PBL as such.

However, on an even less supportive note, Biley and Smith
(1999) write about their ethnographic study of 17 nursing students
following their PBL experiences in undergraduate preparation. In
general, these students were dissatisfied with their educational
experience because they felt they would have preferred to be taught
by experts rather than have to find their own way through the
morass of information. Again, much of this discomfort seemed to
emerge from their concerns about the style of their tutor, a sense of
being isolated and not receiving support in their learning attempts,
and the perceived weak link between PBL evaluation and the need
to successfully complete an external credentialing examination.
Upon consideration, it would seem that many of these concerns
could be eradicated by more attention being paid to the development of a firm structure around these students, with a formal tutor
preparation program and evaluation of faculty performance.

In their report of a multiyear qualitative evaluation of an occupational therapy curriculum, Hammel, Royeen, Bagatell, Chandler,
Jensen, Loveland, and Stone (1999) describe feedback received
from three class cohorts of newly formed PBL curriculum for the
preparation of occupational therapists. PBL was the educational
approach and philosophy of choice because it addresses the national outcome competencies as stated by the Pew Health Professions
Commission (Hammel et al., 1999, p. 199). Respondents included
both students and faculty who had been involved in the PBL curriculum. Student respondents described PBL as a process, and one
that prepares learners for a future professional role that demands a
lifelong learning approach to continuing competence. These students saw faculty as facilitators of learning and understanding, making clear that the small or large student groups are responsible for
what they do ultimately to address their learning needs. "When
asked how their learning compared with their feelings about prior
educational experiences, students reported that PBL involved less
memorization of facts coupled with increasing application, synthesis, reinforcement, and monitoring of comprehension" (Hammel et
al., 1999, p. 202). Students were very aware that there is so much

to know that it is not possible to gain all the information they would like to in the present context; therefore, it is critical that they learn how to learn, and take those skills with them into their future professional lives. Perhaps one of the most difficult things for them was to become accustomed to the idea that there is never just one right answer, and that all answers should be grounded in clinical reasoning and an evidence-based approach to interpreting the data. Weaknesses in PBL that were offered reflect some previously heard comments, such as not knowing when they know enough, feeling the need for more specific feedback from faculty, yet balancing that against being given too much structure. The ability to become comfortable with ambiguity was cited as an important skill to gain. Another balance that is critical is the balance between group work and working alone. While the students appreciated and valued the importance of learning together, there was a firm commitment to the parallel need for individual study time. Information gleaned from the faculty focus groups provided an interesting picture of faculty members' dilemmas of making the transition from a traditional academic role to one of facilitation and guidance.

An even more positive outcome was reported by Kaufman and Holmes (1996) in their article addressing outcomes of a PBL curricular transition from the perspectives of both faculty and students. This institution had taken a great deal of time and effort to develop a transition strategy incorporating in-service training, mechanisms for providing feedback, support for developing and obtaining resources, and participation in decisions being made. Many of the faculty members have become converts to this student-centred approach to education, and most students have developed a strong commitment to the curriculum. They have become true advocates of the PBL process and provide thoughtful and useful feedback to the faculty concerning possible areas and strategies for improvement.

Lieberman, Stroup-Benham, Peel, and Camp (1997) completed a prospective comparison of traditional and problem-based curricula, finding that their results tended to mirror those of previous studies with a similar focus. Their overall findings showed a favorable impact of the PBL curriculum on the students' perceptions of the academic environment, including the positive nature of the intellectual climate, teaching roles and support, administrative systems, and the more intangible aspects of student life such as satisfaction, stress, and comfort.

Nursing students reported an overall satisfaction with learning by a problem-based approach in a study conducted by White, Amos, and Kouzekanani (1999). Students felt that the PBL approach was more based in reality and thus had a direct relevance and application to professional practice. Qualitative and quantitative analysis indicated positive results in the direction of problem-based curricula, specifically highlighting the power of PBL to "inject a renewed enthusiasm into the classroom environment" (White et al., 1999, p. 36).

> *Given all your reading and thinking about this literature, where has that left you in regards to your views about PBL as students?*
>
> Well, I guess that there are still some areas of concern for us. For example, the whole question of feeling as if faculty should tell us what we need to know since they are the experts, that is still lingering in my mind. Also, the importance of being able to schedule our own time and to stick to meeting objectives with less structure imposed—that is actually rather a frightening proposition!
>
> *I think that these are good questions, and perhaps they will be answered best once you get involved in the learning process. Also, perhaps reminding you of the basic principles one more time would be helpful. Remember that you can expect that faculty will respect your attempts to learn and understand. They will not play games with you about wanting you to go and find out information that can more easily be discussed and imparted in the context of your tutorial group. When faculty won't simply give an answer, it is because you can benefit from working it out for yourselves. That is what learning is about. Teaching and providing knowledge becomes valuable and pertinent when key questions are posed by learners to faculty; that is when lectures or large group sessions are invaluable.*

*We can discuss these issues again, and you can also dis-
cuss them more broadly with other faculty as you establish
other relationships that are helpful to you. So, where to from
here?*

We should move on to looking at the results from the
unpublished faculty survey that was conducted by
Salvatori (2000) in order to gain some understanding of
the opinions of occupational therapy faculty members
about teaching within a PBL environment.

UNPUBLISHED FACULTY SURVEY

Questions in the survey were focused on the number of years of
teaching experience, roles fulfilled, any PBL training received, perceived strengths of PBL for students and faculty, the most positive
element of PBL, weaknesses and disadvantages of PBL for faculty
and students, the most unlikable aspect of PBL, anything they
would change, and an overall rating on a 10-point scale.

Starting from the last question, nine faculty members responded
with overall ratings from 8 to 10 on the 10-point scale, with a totally even distribution of three responses in each numeric.

Strengths/Benefits for Students

PBL was seen to promote self-direction, independence, problem
solving, and focused learning on client-based issues using adult
learning principles. Students learn "how" to learn rather than
"what" to learn, which results in developing professional and personal confidence, preparing them well for various practice roles.
PBL promotes integration of ideas rather than simply memorization, while facilitating students in the development of their own
learning path. PBL builds on experiences and prior learning, within a context of collaboration, thus creating a climate of tolerance of
differences. A PBL approach helps students to hypothesize, collect
targeted data, think critically, and enhance clinical reasoning.

Strengths/Benefits for Faculty

When involved in a PBL context, there are increased opportunities for the use of creativity in teaching; it is more interactive and interesting, and closely linked with the practice community, thus enriching the faculty member's opportunities to stay current in the area of real world practice. Faculty can gain a better understanding of how students integrate knowledge and develop clinical reasoning together with critical thinking. The whole environment is conducive to the development of close relationships with learners and the learning process. New skill sets are developed, including facilitation skills in both small and large groups, as well as providing a creative opportunity to reshape lecturing styles and teaching in other locations such as skill labs and special resource sessions.

Weaknesses/Disadvantages for Students

Students are often very stressed and anxious when they begin the process, thus creating feelings of hostility and anger due to the sense of instability and the unknown. The use of learning scenarios can often suggest to the students that there is a right answer when, in fact, the scenario should be seen as a springboard for defining learning issues based on defined objectives. Students need to be motivated, confident about their learning needs, and organized. They also need to be aware of others' needs as well as their own, and in that way, the small group tutorial can become a very rich ground for gaining knowledge and understanding. When a group is not functioning well, then time must be taken to resolve the issues at hand, time that could be better spent in investing in other group members' learning. This can also engender frustration and annoyance at the errant peer.

Weaknesses/Disadvantages for Faculty

Faculty reported that they feel often that they could do more, despite knowing that this educational approach feels more labor-intensive. Recruiting tutors becomes problematic, particularly given the changing nature of workplaces and the sense of less support for involvement in educational activities. There are particular challenges in facilitating difficult groups, and this requires very different skills from those with which faculty were prepared when they began their academic careers. Sometimes it is easy to slip into the familiar role of teaching rather than facilitating. By definition, faculty members know they are vulnerable and cannot fake knowledge or confidence. This is hard to internalize and deal with at times.

Dislikes About Problem-Based Learning

In creating a student-centred learning culture, students can become egocentric and feel omnipotent. This would need to be addressed through reminding everyone of the basic principles of PBL, particularly those of mutual respect and trust. Faculty members felt that they got to know a few students very well, but did not know a larger number at all. This was not a preferred way to relate to a student class. This method is very challenging, more so than the traditional teaching roles, particularly when students are not open to feedback, do not embrace the PBL culture, and continue to go against the grain, thinking they can do what they wish.

Likes About Problem-Based Learning

Faculty members particularly appreciate the chance to observe learning and growth in students. Contextual learning appears to be far superior to the more traditional ways, and increases motivation to learn and understand. Working in small groups provides opportunities to truly encourage the development of lifelong learning skills. The flexibility of the approach is very appealing, and suits the preparation of health professional practitioners very well indeed. Students who internalize this approach well develop in personal and professional ways, learning to work well with others, tolerating a range of difference, and being eager consumers of information while challenging what they read. It is truly a dynamic and challenging process, the best way to teach.

Any Suggested Changes

Faculty members felt that attention should be paid to several key factors:
- There should be efforts made not to exceed six learners in each tutorial group; this could be difficult, given the need for more tutors with this model.

- Tutorial evaluation should be reexamined in order to ensure that the tutorial is a "growth" context, with always something new to learn.
- Reemphasize the tutor training opportunities in order to create a culture of lifelong learning among academic and clinical faculty.

This information is very interesting, and it is particularly so given that these faculty members are providing some excellent suggestions for the future; by doing so, they suggest to me that they are definitely aware of how much work has to go into developing a clear and visible structure in order to support learners in a problem-based culture.

We found it of special interest that the outcomes of the more formal research studies are reflected even within this small survey sample from one faculty only. It will be very interesting to see whether our own simple little survey will provide similar feedback. Also, we are intrigued to see what stories the respondents to our survey will provide to us. It is often so much easier to learn from graphic representations of ideas.

Well then, let's take a close look at the responses you received. By the way, what were the questions you posed?

OUR SURVEY

We asked everyone (faculty and students alike) three things:
1. What did they see were the strengths of PBL?
2. What were the weaknesses?
3. Are there any special stories or illustrations they could provide to help us understand more about the whole experience of teaching and learning in a PBL world?

We included faculty members who were involved from the practice setting as well as full-time teaching faculty. Perhaps the best way for us to gain an overview of their comments is to use some direct quotes—several respondents were very articulate.

About Being a Student in a Problem-Based Learning System

Feedback concerning student roles and expectations was helpful and informative. It was good to know that faculty appreciate the difficulties in learning a process like PBL after years of didactic teaching.

"As a learner (it can be) initially frustrating; precarious at times, in the early development of the group because you have to trust weak as well as strong, motivated students to bring what they had agreed to support learning objectives. It works best when work isn't done in a piecemeal fashion...If learners understand the scope of the problem and then are 'experts' in a couple of the sub-topics, the valuable discussion can occur, and...lead to understanding and retention far beyond rote learning of detail" (Faculty Member #1).

When I was looking through the feedback, I was reminded of a story from one of my faculty colleagues concerning a time when we were both facilitating a workshop in another province. As she states:

"*I will never forget when you and I facilitated a PBL workshop and we trained students to be a tutorial group using PBL. They had never done it before but caught on instantly, and were great as simulated students doing PBL. They were amazed at how quickly they got it and so delighted at being able to direct their learning. I thought we would have a lot more difficulty preparing the students. This emphasized for me the logic behind PBL with adult learners who are interested in learning and have the ability to be self-directed*" (Faculty Member #2).

I also remember that incident very vividly, and recall the enthusiasm and excitement of the students, and also the amazement and respect of the faculty members for their students' achievements—the same faculty members who had struggled with the same process, using the same problem scenario the day before.

This is a great example of a positive outcome with students who were invested in the process. Another respondent to the survey stated her concerns related to students "who don't really get it—the unconsciously incompetent" (Faculty Member #3). She is referring to learners who don't realize what they don't know or what they do know, and therefore have great difficulty in making their learning useful and applicable to the skill set they are gaining. From this individual's viewpoint, this happens rarely, and it is the role of the tutor to remediate the problem; however, it certainly is important. I would imagine that this could negatively impact on a tutorial group's learning as a whole, and also could result in students completing programs of study but with little ability to generalize their learning into their professional lives. However, again to put this in perspective, it happens rarely.

Also, not only is it the role of the tutor to first identify the issue and then build a program to resolve the concerns, but it is also the role of the rest of the tutorial group to pull together and confront the matter, supporting the student with the problems and moving things along. I have found that, in fact, peer group members are very good at making sure that learning happens.

One of the students providing feedback to us was quite concerned that the power of interpersonal communication can make or break a group experience.

"Sometimes tutorial groups fall prey to interpersonal communication and other group process problems. This sometimes detracts from the information sharing and learning experience. Also, it is hard to find middle ground for certain evaluation criteria; for example, when to be critical of information presented and when to stay quieter" (Student #1).

This student further elaborates and shares his concerns about ensuring that evaluation and assessment is handled in a balanced manner. Of particular note should be the importance of tutor and student group members together manifesting the principles of PBL through openness and honesty, and a true commitment to collaborative learning. The acceptance of feedback is never easy, and sometimes it is necessary to support individual group members as they attempt to accept what they are told and act constructively upon it.

Another student cited a very compelling situation within the last PBL group of the year. "We explored problems that addressed issues of sexuality and palliative care, two areas which are poorly addressed in most curricula" (Student #2). And: "My comfort level with addressing such issues and knowledge of what sexuality encompassed was minimal. After tackling this particular problem, we realized just how important addressing such issues is in terms of client self-esteem and additionally, how sexuality is such an intricate part of meaningful occupation".

Perhaps the real key here is that approaching learning through a problem-based approach allows students to explore questions and directions that are not explicit within educational curricula, but are critical within the real life context of professional practice.

About Problem-Based Learning as a Whole

A clinical preceptor, an occupational therapist of many years' experience, reflected on many of the positive elements of PBL for students' learning, and summed it up as follows: "With PBL, everyone wins; the students learn to learn, the University becomes known as an innovator, the city is vibrant with the learning process, and all this trickled down to a better health care milieu for our clients" (Clinical Preceptor #1).

About Curriculum Design and Learning Resources

Not always on the more positive side, another respondent pondered the use of the word "problem", "because it does suggest that there is an answer and I think we see that confusion among our students all the time—about their difficulty in dealing with ambiguities in the scenarios we provide" (Faculty Member #4).

Undoubtedly, this comment is central to much of the criticism received about PBL. The term "problem" does indicate that there is an answer to be sought; whereas, if the term "scenario" were used instead, it would allow more freedom of thinking. Students would feel less pressured to seek the answer that the problem designer intended, rather than using it as an opportunity to expand thinking and gain learning.

About Tutorials

"Toshiko, a visiting faculty member from Japan, participated in our Term 1 tutorials last year. She quietly observed and reflected on the tutorial dynamics around her. Finally, at a point near the end of term, she stated that the tutor is 'like a sherpa climbing a mountain'. That analogy has stayed with me since. I visualize the sherpa as a guide who helps climbers attain their dreams by scaling quite magnificent heights…I am quite happy to think of myself as a sherpa with a wise face, wrinkled and sun-drenched from years of exposure to the elements, guiding a group of people who have yet to realize what the sherpa can do…quietly and persistently" (Faculty Member #5).

This is a wonderfully graphic image of a tutor at his or her best, and we would hope that this is the kind of tutor with whom we have the chance to work! Also, there was a visiting professor from Israel who was a tutor for two of our groups last year. She has gone back to Israel now and is in the process of putting a PBL curriculum into place at her university there. She sent us a very thoughtful email, so I have copied it for the group as a resource document.

A MESSAGE FROM ISRAEL

Tal Jarus, Incoming Chair
Occupational Therapy Program
University of Tel Aviv, Israel, 2002

My adventure with PBL started in March 2000, when I arrived at McMaster University, School of Rehabilitation Science for my sabbatical. My first question was, What, no lectures? No physiology class? And what about psychology? And how can they know the "stuff" without anatomy? Needless to say, I was lost. Now, 2 years later I am trying to revise part of our curriculum into a PBL format and am frustrated with the resistance I am receiving from colleagues about sticking to the physiology, psychology, and anatomy. But, I am putting the cart before the horse.

After spending a year and a half in the occupational therapy department at McMaster University, I was "converted" to the PBL way. During that period, I spent time learning about their curriculum, practicing being a tutor of a small group, and being mentored in planning possible changes to the curriculum at my university back home. While we were sitting in the faculty lounge, we sketched a plan for making the revisions. We knew it would need time, so we developed a 3-year process. Sitting there, in such peaceful surroundings, it seemed both logical and feasible. In the first year, we will introduce the proposed revisions to faculty and work on changing one course, Introduction to Occupational Therapy. The second year, we will revise the whole first year curriculum into a PBL system, and in the third planning year, we will revise the second curriculum year.

Landing back in this hot and hectic country, suddenly it looked like mission impossible. The first concern was convincing the faculty to make the switch, remembering how skeptical I was just 1 year ago. Everyone is very strong-minded here, knowing exactly how and what to teach; some of us have been around the faculty for many years, and see no need (and have no will) to change. On the other hand, most of the faculty members have heard about PBL, know it is an interesting trend in educational systems in North America and beyond, and were quite eager to hear about it. In addition, there is no strong economic basis for the change, since funding for the program continues at a standard rate regardless of educational methodology or philosophy. Cultural issues were a great concern also; the pressure of reading and learning from literature written in a foreign language being the most critical.

Almost a year has passed since I arrived home, and my goals are partially achieved. After holding a workshop about PBL and a few focused staff meetings, the change of one course in the first year was accepted quite well and is now underway. The outline of the course is ready. It is divided into two units, using inquiry seminar and small group tutorial formats. Issues are emerging regarding the recruitment of the number of tutors required, and budget will always be a concern. However, we are moving ahead. Presently though, the concept of changing the whole curriculum to PBL seems far out of reach.

From a more personal, emotional perspective, it has been an interesting journey with its ups and downs. There are many times where I have had bad moments, especially when I cannot find enough tutors or when there are externally-imposed threats about no resources for small group teaching. But there are a lot of thrills when building something new, something you believe in. Once I realized that the team with whom I am working was as enthusiastic as I was, I became excited. When I imagine how the students will work in the small groups, hearing their brains work and sharing their struggle, I become energized. But come back another year from now, once the program is on its way and ask me how I feel—this will be the true test and another story.

KEY POINTS

- Stimulates creativity and ingenuity in finding resources relevant to the learning issues.
- Allows us as learners to take ownership of the problem, and thereby take charge of our own learning.
- Giving and receiving feedback is crucial to competent professional practice over time and the best training ground for this is in small group tutorial learning.
- Takes away anxiety about the "right" answer and encourages lifelong learning.
- Makes learning more interesting and even fun.
- Can be anxiety-provoking at the beginning until we learn how much is enough and how deep is deep enough.
- Essential that there is a link between learning this way and being prepared for more formal certification examinations.
- Keeps the focus on clients, patients, and families and not on the disease or body part.
- Promotes close relationships between faculty and learners in making learning a collaborative process.
- Teaching within a range of small to large groups is both energizing and challenging, never boring.

REFERENCES

Biley, F. C., & Smith, K. L. (1999). Making sense of problem-based learning: The perceptions and experiences of undergraduate nursing students. *J Adv Nurs, 30*(5), 1205-1212.

Caplow, J. A. H., Donaldson, J. F., Kardash, C. A., & Hosokawa, M. (1997). Learning in a problem-based medical curriculum: Students' conceptions. *Med Educ, 31,* 440-447.

Hammel, J., Royeen, C. B., Bagatell, N., Chandler, B., Jensen, G., Loveland, J., & Stone, G. (1999). Student perspectives on problem-based learning in an occupational therapy curriculum: A multiyear qualitative evaluation. *Am J Occup Ther, 53*(2), 199-206.

Kaufman, D. M., & Holmes, D. B. (1996). Tutoring in problem-based learning: Perceptions of teachers and students. *Med Educ, 30,* 371-377.

Lieberman, S. E., Stroup-Benham, C. A., Peel, J. L., & Camp, M. G. (1997). Medical student perception of the academic environment: A prospective comparison of traditional and problem-based curricula. *Acad Med, 72*(10), S13-S15.

Salvatori, P. (2000). Implementing a problem-based learning curriculum in occupational therapy: A conceptual model. *Australian Occupational Therapy Journal, 47,* 119-133.

White, M. J., Amos, E., & Kouzekanani, K. (1999). Problem-based learning: An outcomes study. *Nurse Educator, 24*(2), 33-36.

LEARNING RESOURCES

Foley, R. P., Polson, A. L., & Vance, J. M. (1997). Review of the literature in the clinical setting. *Teach Learn Med, 9*(1), 4-9.

Moore, G. T. (1991). The effect of compulsory participation of medical students in problem-based learning. *Med Educ, 25,* 140-143.

Tutorial 8

Wrap-Up

Well, it certainly does not seem as if we have been togeth-er that long as a learning group, and yet I have the sense that you have managed to cover a great deal of ground. I wonder if we could spend just a few moments reflecting on the past 8 weeks, and identifying the scope of what you have gained from the experience as well as any further learning issues that may remain.

We have enjoyed the opportunity to learn about PBL in a PBL way. We have extrapolated many of the key issues from all the literature we have uncovered, but we know there are so many other things that remain either unclear or unfinished for us. Perhaps the best approach to this is to ensure that we have the full collection of resources that we have used throughout the whole process of learning made into a resource document. That way we can ensure that we do not lose these resources and that we also have them close at hand.

I think that is an excellent notion. As your tutor, I must say this has been such an excellent group to facilitate, but I wonder if you have given any thought to what it might have been like to learn in a group where things did not go so smoothly?

We certainly have! Sometimes, when we were follow-ing up on resources and trying to understand what we were reading, we had some disagreements—some of them quite colorful! It was most fortunate that we did not have any conflicts when we were in our group ses-sions, because we could maximize our time together. However, perhaps having some problems might have given us the chance to identify some strategies for deal-ing with such situations. It is a shame that we don't have any more time together to discuss these potential prob-lems and to research resources that would help us deal with similar concerns when they arise in other groups in the future.

Well, never say that I don't give you anything! I have pulled together some documents that I have used in the past, as well as a few references. Consider these a good-bye pres-ent.

KEY POINTS

- How well did we do today in accomplishing our task?
- How well did we do today in working as a group?
- What were the major strengths of this session?
- What were the main areas in which we could improve?
- What do you think we should do next?
- What things did you not say that you might have?
- What do you see yourself doing differently?

That is great, thanks!

From an evaluative perspective, do you feel that the group has worked well together?

There is no doubt that small group learning is very different from working alone or responding to a lecture setting. However, we were surprised at how well we adapted to it, and how quickly we felt a close association develop among all group members. Perhaps we should keep most of this conversation until we get to the patio at the restaurant and then we can enjoy a very meaningful time for evaluation!

Excellent...let's go.

In the real world, the evaluation of a small group tutorial experience would be in-depth, involving each member reflecting on strengths and areas for improvement, and then receiving feedback in these areas from both colleagues and tutor. Also, the final evaluation is often seen as an occasion warranting celebration; consequently, the session may be held at the tutor's home, in a restaurant, or other social setting. This is not essential; however, such an approach to the final opportunity to work together is indeed congruent with the guiding principles of PBL. In the interest of our own learning then, we will assume that our "group" will now retire to the warmth and comfort of some pleasant place in order to explore in more detail this tutorial experience.

LEARNING RESOURCES

Cole, M. B. (1998). *Group dynamics in occupational therapy*. Thorofare, NJ: SLACK Incorporated.

Sampson, E., & Marthas, M. (1991). *Group process for the health professions* (3rd ed). New York and Toronto: John Wiley & Sons.

Yalom, I. (1995). *Theory and practice of group psychotherapy* (4th ed). New York: Basic Books.

Appendix A

Learning Resource Compendium

Abdul-Ghaffar, T. A., Lukowiak, K., & Nayar, N. (1999). Challenges of teaching physiology in a PBL school. *Adv Physiol Educ, 22*(1), S140-S147.

Albanese, M. A., & Mitchell, S. (1993). Problem-based learning: A review of literature on its outcomes and implementation issues. *Acad Med, 68*, 61-82.

Alpert, I. L., & Perkins, R. M. (1999). Should students in problem-based learning sessions be graded by the facilitator? Research brief. *Acad Med, 74*(11), 1256.

Barrows, H. S. (2000). *Practice-based learning applied to medical education.* Springfield, IL: Southern Illinois University School of Medicine.

Barrows, H., & Branda, L. (1975; revised Branda, L. 1991; 1994). *Role of the tutor/facilitator on problem-based, small group and self-directed learning.* Internal document, copyright 1993, McMaster University: Hamilton.

Barrows, H. S., & Tamblyn, R. M. (1980). *Problem-based learning: An approach to medical education.* (Series: Medical Education, Vol. 1.) New York: Springer Publishing Company.

Biley, F. C., & Smith, K. L. (1998). "The buck stops here": Accepting responsibility for learning and actions after graduation from a problem-based learning nursing education curriculum. *J Adv Nurs, 27*, 1021-1029.

Blake, R. L., & Parkison, L. (1998). Faculty evaluation of the clinical performances of students in a problem-based learning curriculum. *Teach Learn Med, 10*(2), 69-73.

Blake, R. L., Hosokawa, M. C., & Riley, S. L. (2000). Student performances on step 1 and step 2 of the United States medical licensing examination following implementation of a problem-based learning curriculum. *Acad Med, 75*(1), 66-70.

Bligh, J., Lloyd-Jones, G., & Smith, G. (2000). Early effects of a new problem-based clinically oriented curriculum on students' perceptions of teaching. *Med Educ, 34*, 487-489.

Branda, L. (1990). Implementing problem-based learning. *J Dent Educ, 54*(9), 548-549.

Branda, L. A., & Sciarra, A. F. (1995). Faculty development for problem-based learning. *Annals of Community-Oriented Education*, 8, 195-208.

Caplow, J. A. H., Donaldson, J. F., Kardash, C. A., & Hosokawa, M. (1997). Learning in a problem-based medical curriculum: Students' conceptions. *Med Educ, 31*, 440-447.

Cariaga-Lo, L. D., Richards, B. F., Hollingsworth, M. A., & Camp, D. L. (1996). Non-cognitive characteristics of medical students: Entry to problem-based and lecture-based curricula. *Med Educ, 30*, 179-186.

Colliver, J. (2000). Effectiveness of problem-based learning curricula. *Acad Med, 75*, 259-266.

De Grave, W. S., Dolmans, D. H. J. M., & van der Vleuten, C. P. M. (1999). Profiles of effective tutors in problem-based learning: Scaffolding student learning. *Med Educ, 33*, 901-906.

Distlehorst, L. H., & Robbs, R. S. (1998). A comparison of problem-based learning and standard curriculum students: Three years of retrospective data. *Teach Learn Med, 10*(3), 131-137.

Duek, J. E., Wilkerson, L., & Asinolfi, T. (1996). Learning issues identified by students in tutorless problem-based tutorials. *Adv Health Sci Educ, 1*, 29-40.

Faculty of Health Sciences (1994). *Guide to professional behaviours in tutorial meetings.* Hamilton, Ontario: McMaster University.

Federman, D.D. (1999) Little-heralded advantages of problem-based learning. *Acad Med, 74*, 93-94.

Ferrier, B. (1990). Problem-based learning: Does it make a difference? *J Dent Educ, 54*, 9, 550-551.

Finucane, P., Allery, L. A., & Hayes, T. M. (1995). Comparison of teachers at a "traditional" and an "innovative" medical school. *Med Educ, 29*, 104-109.

Foldevi, M., Sommansson, G., & Trell, E. (1994). Problem-based medical education in general practice: Experience from Linkoping, Sweden. *Br J Gen Pract, 44*, 473-476.

Foley, R. P., Polson, A. L., & Vance, J. M. (1997). Review of the literature in the clinical setting. *Teach Learn Med, 9*(1), 4-9.

Friedman, C. P., de Bliek, R., Greer, D. S., Mennin, S. P., Norman, G. R., Sheps, C. G., Swanson, D. B., & Woodward, C. A. (1990). Charting the winds of change: Evaluating innovative medical curricula. *Acad Med*, 8-14.

Goodall, H. L. Jr.(1990). *Small group communication in organizations* (2nd ed.) Dubuque, IA: Wm. C. Brown Publishers.

Greening, T. (1998). Scaffolding for success in problem-based learning. Retrieved April 23, 2003 from *Medical Education Online*: www.med-ed-online.org/f0000012.htm

Hak, T., & MaGuire, P. (2000). Group process: The black box of studies on problem-based learning. *Acad Med, 75*(7), 769-772.

Hammel, J., Royeen, C. B., Bagatell, N., Chandler, B., Jensen, G., Loveland, J., & Stone, G. (1999). Student perspectives on problem-based learning in an occupational therapy curriculum: A multiyear qualitative evaluation. *Am J Occup Ther, 53*(2), 199-206.

Hitchcock, M. A., & Mylona, Z-H. E. (2000). Teaching faculty to conduct problem-based learning. *Teach Learn Med, 12*(1), 52-57.

Hmelo, C. E. (1998). Cognitive consequences of problem-based learning for the early development of medical expertise. *Teach Learn Med, 10*(2), 92-100.

Jacobs, T. (1997). Integrating assessment in problem-focused curricula. *British Journal of Occupational Therapy, 60*(4), 174-178.

Jacobs, T. (1997). Developing integrated education programmes for occupational therapy: The problem of subject streams in a problem-based course. *British Journal of Occupational Therapy, 60*(3), 134-138.

Jensen, A. D., & Chilberg, J. C. (1991) *Small group communication: Theory and application.* Belmont, CA: Wadsworth Publishing Company.

Johnson, S. M., Finucane, P. M., & Prideaux, D. J. (1999) Problem-based learning: Process and practice. *Aust N Z J Med, 29,* 350-354.

Kanter, S. L. (1998). Fundamental concepts of problem-based learning for the new facilitator. *Bulletin of the American Library Association, 86*(3), 391-395.

Kaufman, D. M. (2000). Problem-based learning–time to step back? *Med Educ, 34,* 509-511.

Kaufman, D. M., & Holmes, D. B. (1996). Tutoring in problem-based learning: Perceptions of teachers and students. *Med Educ, 30,* 371-377.

Kaufman, D. M., Mensink, D., & Day, V. (1998). Stressors in medical school: Relation to curriculum format and year of study. *Teach Learn Med, 10*(3), 138-144.

Lieberman, S. E., Stroup-Benham, C. A., Peel, J. L., & Camp, M. G. (1997). Medical student perception of the academic environment: A prospective comparison of traditional and problem-based curricula. *Acad Med, 72*(10), S13-S15.

Lloyd-Jones, G., Margetson, D., & Bligh, J. G. (1998). Problem-based learning: A coat of many colours. *Med Educ, 32,* 492-494.

Marchese, T. (1998). The new conversations about learning. Retrieved April 23, 2003 from http://www.aahe.org/pubs/TM-essay.htm

Margetson, D. (1996). Beginning with essentials: Why problem-based learning begins with problems. *Education for Health, 1,* 61-69.

Maudsley, G. (1999a). Do we all mean the same thing by "problem-based learning"? A review of the concepts and a formulation of the ground rules. *Acad Med, 74*(2), 178-185.

Maudsley, G. (1999b). Roles and responsibilities of the problem based learning tutor in the undergraduate medical curriculum. *British Medical Journal, 318,* 657-661.

Mayo, W. P., Donnelly, M. B., & Schwartz, R. W. (1995). Characteristics of the ideal problem-based learning tutor in clinical medicine. *Evaluation & the Health Professions, 18*(2), 124-136.

MCHS(OT) Programme (2002). *Term 1 Handbook.* Hamilton, Ontario, McMaster University.

Mennin, S. P., & Kalishman, S. (1998). Student assessment. *Acad Med, 73*(9), S46-S54.

Moore, G. T. (1991). The effect of compulsory participation of medical students in problem-based learning. *Med Educ, 25,* 140-143.

Nendaz, M. R., & Tekian, A. (1999). Assessment in problem-based learning medical schools: A literature review. *Teach Learn Med, 11*(4), 232-243.

Neufeld, V. R. (1983). Adventures of an adolescent: Curriculum changes at McMaster University. In C. Friedman & E. S. Purcell (Eds.). *New Biology and Medical Education* (pp. 256-270). New York: Josiah Macy Jr. Foundation.

Neufeld, V. R., Woodward, C. A., & MacLeod, S. M. (1989). The McMaster M.D. program: A case study of renewal in medical education. *Acad Med, 64,* 423-447.

Neville, A. (1995). Student evaluation in problem-based learning. *Pedagogue, 5,* 2-7.

Neville, A., Cunnington, J., & Norman, G. R. (1996). Development of clinical reasoning exercises in a problem-based curriculum. *Acad Med, 71,* S105-S107.

Newbie, D. I., & Clarke, R. M. (1986). The approaches to learning of students in a traditional and in an innovative problem-based medical school. *Med Educ, 20,* 267-271.

Norman, G. (1988). Problem-solving skills, solving problems and problem-based learning. *Med Educ, 22,* 279-286.

Norman, G. (1993). Where is the learning in problem-based learning? *Pedagogue, 4,* 2.

Norman, G. R., & Schmidt, H. G. (2000). Effectiveness of problem-based learning curricula: Theory, practice and paper darts. *Med Educ, 34,* 721-728.

Norman, G. R. (1991). What should be assessed? In D. Boud & G. Feletti (Eds.). *The challenge of problem-based learning* (pp. 254-259). New York: St. Martin's Press.

Pallie, W., & Brain, E. (1978). "Modules" in morphology for self study: A system for learning in an undergraduate medical programme. *Med Educ, 12,* 107-113.

Pallie, W., & Carr, D. H. (1987). The McMaster medical education philosophy in theory, practice and historical perspective. *Medical Teacher, 9*(1), 59-71.

Pallie, W., & Miller, D. (1982). Communicating morphological concepts in health sciences. *J Biocommun, 9*(3), 26-32.

Patel, V. L., Groen, G. J., & Norman, G. R. (1991). Effects of conventional and problem-based medical curricula on problem solving. *Acad Med, 7,* 380-389.

Peterson, M. (1997). Skills to enhance problem-based learning. Retrieved April 23, 2003 from *Medical Education Online*: www.med-ed-online.org/f0000009.htm

Professional Behaviours Working Group. (1994). *Guide to professional behaviours in tutorial meetings*. Hamilton, Ontario: Faculty of Health Sciences, McMaster University.

Programme for Educational Development. (1987). *Evaluation methods: A research manual*. Hamilton, Ontario: McMaster University.

Programme for Faculty Development. (1997). *Problem-based learning*. Internal document. Hamilton, Ontario: Faculty of Health Sciences, McMaster University.

Programme for Faculty Development. (2001). *Problem based learning in small groups workshop evaluation resource package*. Hamilton, Ontario: Faculty of Health Sciences, McMaster University.

Rendas, A. B., Pinto, P. R., & Gamboa, T. (1998). Problem-based learning in pathophysiology: Report of a project and its outcome. *Teach Learn Med, 10*, 34-39.

Rothman, A. I. (2000). Problem-based learning–time to move forward? *Med Educ, 34*, 509-511.

Saarinen, H., & Salvatori, P. (1994). Dialogue: Educating occupational and physiotherapists for the year 2000: What, no anatomy courses? *Physiotherapy Canada, 46*(2), 81-86.

Salvatori, P. (2000). Implementing a problem-based learning curriculum in occupational therapy: A conceptual model. *Australian Occupational Therapy Journal, 47*, 119-133.

Schmidt, H. (1983). Problem-based learning: Rationale and description. *Med Educ, 17*: 11-16.

Schmidt, H. G., & Bouhuijs, P. A. J. (1980). *Onderwijs in taakgerichte groepen* (Task-oriented small group learning). Het Spectrum, Utrecht.

Smith, R. M. (1993). The triple-jump examination as an assessment tool in the problem-based medical curriculum at the University of Hawaii. *Acad Med, 68*, 366-372.

Solomon, P., & Crowe, J. (2001). Perceptions of student peer tutors in a problem-based learning programme. *Medical Teacher, 23*(2), 181-186.

Solomon, P., & Finch, E. (1998). A qualitative study identifying stressors associated with adapting to problem-based learning. *Teach Learn Med, 10*(2), 58-64.

Sullivan, M. E., Hitchcock, M. A., & Dunnington, G. L. (1999). Peer and self assessment during problem-based tutorials. *Am J Surg, 177*, 266-269.

Swanson, D. B., Case, S. M., & van der Vleuten, C. P. M. (1991). Strategies for student assessment. In D. Boud & D. B. G. Feletti (Eds.). *The challenge of problem-based learning* (pp. 260-273). New York: St. Martin's Press.

Thomas, R. E. (1997). Problem-based learning: Measurable outcomes. *Med Educ, 31,* 320-329.

Tipping, J., Freeman, R. F., & Rachlis, A. R. (1995). Using faculty and student perceptions of group dynamics to develop recommendations for PBL training. *Acad Med, 70*(11), 1050-1052.

Valle, R., Petra, I., Martinez-Gonzales, A., Rojas-Ranirez, J. A., Morales-Lopes, S., & Pina-Garza, B. (1999). Assessment of student performance in problem-based learning tutorial sessions. *Med Educ, 33,* 818-822.

Vernon, D. T. A., Blake, R. L. (1993). Does problem-based learning work? A meta-analysis of evaluative research. *Acad Med, 68,* 550-563.

Viet Vu, N., van der Vleuten, C. P. M., & Lacombe, G. (1998). Thinking about student thinking. *Acad Med, 73*(10), S25-S27.

Wakefield, J. G., Keane, D., Hay, J. (1992). Evaluation of tutor effectiveness: A review of the literature. *Annals of Community-Oriented Medicine, 5,* 209-213.

Walton, H. J. & Matthews, M. B. (Eds). (1987). *Medical education booklet No.23: Essentials of problem-based learning.* Dundee, UK: World Federation for Medical Education, Ninewells Hospital and Medical School.

Walton, H. J. & Matthews, M. B. (1989). Essentials of problem-based learning. *Med Educ, 23,* 542-558.

White, M. J., Amos, E., & Kouzekanani, K. (1999). Problem-based learning: An outcomes study. *Nurse Educator, 24*(2), 33-36.

Woodward, C. A. (1986). McMaster university graduates. *Journal of Medical Education, 61,* 859-860.

Woodward, C. A. (1996). Problem-based learning in medical education: Developing a research agenda. *Adv Health Sci Educ, 1,* 83-94.

Student Resource Package

This section provides a complete set of the resources provided throughout the text. It has been put together as an easy reference guide.

Tutorial 1

A model for a problem-based small group process (after introductions are made and ground rules are set or reinforced):

- Choose the learning scenario.
- Define any unfamiliar language or concepts within the written scenario (and determine whether any of these are learning issues in themselves).
- Brainstorm around any issues that come to mind when reading the scenario.
- Distill from this open exploration key areas for potential learning.
- Organize these key areas within a logical conceptual framework.
- Select priority issues for focused exploration.
- Develop a learning plan with specific questions.
- Define potential learning resources.
- Clarify that all group members understand and subscribe to the learning plan.
- Evaluate the learning experience of that particular session.

KEY POINTS

- All members introduce themselves.
- Group ground rules are set.
- The learning scenario for exploration is chosen.
- Brainstorming is undertaken.
- Potential learning issues are distilled from the brainstorming and organized within a logical framework.
- A learning plan is developed from these learning issues taking into consideration the timeframe available (i.e., reasonable learning objectives that can be accomplished in the time between group sessions).
- Potentially useful resources are identified.
- Evaluation of the session is undertaken, focusing on how well the group functioned as a whole and how each individual functioned, based on self, peer, and tutor feedback.

LEARNING RESOURCE

Federman, D.D. (1999) Little-heralded advantages of problem-based learning. *Acad Med, 74,* 93-94.

Tutorial 2

KEY POINTS

- It began back in the late 1960s in Canada.
- Since that time, it has become an important alternative to more traditional learning and educational models.
- It began as a reform movement to educate medical students.
- Now it is used across many other professional groups.
- Underlying values and assumptions include: partnership, honesty, openness, respect and trust.
- Common characteristics include: a learner-centred focus, faculty acting more as guides, scenarios being a springboard to learning and new knowledge being acquired through self-directed learning.

LEARNING RESOURCES

Barrows, H. S., & Tamblyn, R. M. (1980). *Problem-based learning: An approach to medical education.* (Series: Medical Education, Vol. 1.) New York: Springer Publishing Company.

Goodall, H. L. Jr. (1990). *Small group communication in organizations* (2nd ed.). Dubuque, IA: Wm. C. Brown Publishers.

Hak, T., & MaGuire, P. (2000). Group process: The black box of studies on problem-based learning. *Acad Med, 75*(7), 769-772.

Jensen, A. D., & Chilberg, J. C. (1991). *Small group communication: Theory and application.* Belmont, CA: Wadsworth Publishing Company.

Johnson, S. M., Finucane, P. M., & Prideaux, D. J. (1999). Problem-based learning: Process and practice. *Austr N Z J Med, 29,* 350-354.

Kanter, S. L. (1998). Fundamental concepts of problem-based learning for the new facilitator. *Bulletin of the American Library Association, 86*(3), 391-395.

Lloyd-Jones, G., Margetson, D., & Bligh, J. G. (1998). Problem-based learning: A coat of many colours. *Med Educ, 32,* 492-494.

Marchese, T. (1998). The new conversations about learning. Retrieved April 23, 2003 from http://www.aahe.org/pubs/TM-essay.htm

Margetson, D. (1996). Beginning with essentials: Why problem-based learning begins with problems. *Education for Health, 1,* 61-69.

Maudsley, G. (1994). Do we all mean the same thing by "problem-based learning"? A review of the concepts and a formulation of the ground rules. *Acad Med, 74*(2), 178-185.

Neufeld, V. R. (1983). Adventures of an adolescent: Curriculum changes at McMaster University. In C. Friedman & E. S. Purcell, (Eds.). *New Biology and Medical Education* (pp. 256-270). New York: Josiah Macy Jr. Foundation.

Neufeld, V. R., Woodward, C. A., & MacLeod, S. M. (1989). The McMaster M.D. program: A case study of renewal in medical education. *Acad Med, 64,* 423-447.

Norman, G. (1988). Problem-solving skills, solving problems and problem-based learning. *Med Educ, 22,* 279-286.

Norman, G. (1993). Where is the learning in problem-based learning? *Pedagogue, 4,* 2.

Programme for Faculty Development. (1997). *Problem-based learning.* Internal document, Programme for Faculty Development, Faculty of Health Sciences, Hamilton, Ontario: McMaster University.

Saarinen, H., & Salvatori, P. (1994). Dialogue: Educating occupational and physiotherapists for the year 2000: What, no anatomy courses? *Physiotherapy Canada, 46*(2), 81-86.

Schmidt, H. (1983). Problem-based learning: Rationale and description. *Med Educ, 17,* 11-16.

Schmidt, H. G., & Bouhuijs, P. A. J. (1980). Onderwijs in taakgerichte groepen (Task-oriented small group learning). Het Spectrum, Utrecht.

Thomas, R. E. (1997). Problem-based learning: Measurable outcomes. *Med Educ*, *31*, 320-329.

Walton, H. J., & Matthews, M. B. (Eds.) (1987). *Medical education booklet No.23: Essentials of problem-based learning*. Dundee, UK: World Federation for Medical Education, Ninewells Hospital and Medical School.

Walton, H. J., & Matthews, M. B. (1989). Essentials of problem-based learning. *Med Educ*, *23*, 542-558.

Tutorial 3

KEY POINTS

- The physical environment is flexible, easily changeable.
- The curriculum looks different, with integrated courses incorporating content from many distinct bodies of knowledge, using problems/learning scenarios as a springboard for learning; large group sessions as a means to refine questions and acquire knowledge in response to the questions.
- Faculty and students relate as partners, exemplifying behaviors that reflect honesty, openness, respect, and trust.
- Institutional systems, including policies, procedures and guidelines, illustrate the application of PBL principles.

LEARNING RESOURCES

Abdul-Ghaffar, T. A., Lukowiak, K., & Nayar, N. (1999). Challenges of teaching physiology in a PBL school. *Adv Physiol Educ*, *22*(1), S140-S147.

Branda, L. (1990). Implementing problem-based learning. *J Dent Educ*, *54*(9), 548-549.

Federman, D. D. (1999) Little-heralded advantages of problem-based learning. *Acad Med*, *74*, 93-94.

Ferrier, B. (1990). Problem-based learning: Does it make a difference? *J Dent Educ*, *54*(9), 550-551.

Greening, T. (1998). Scaffolding for success in problem-based learning. Retrieved April 23, 2003 from *Medical Education Online*: www.med-ed-online.org/f0000012.htm

Jacobs, T. (1997). Developing integrated education programmes for occupational therapy: The problem of subject streams in a problem-based course. *British Journal of Occupational Therapy, 60*(3), 134-138.

Pallie, W., & Brain, E. (1978). "Modules" in morphology for self-study: A system for learning in an undergraduate medical programme. *Med Educ, 12,* 107-113.

Pallie, W., & Miller, D. (1982). Communicating morphological concepts in health sciences. *J Biocommun, 9*(3), 26-32.

Professional Behaviours Working Group. (1994). *Guide to Professional Behaviours in Tutorial Meetings.* Hamilton, Ontario: McMaster University.

Salvatori, P. (2000). Implementing a problem-based learning curriculum in occupational therapy: A conceptual model. *Australian Occupational Therapy Journal, 47,* 119-133.

Tutorial 4

KEY POINTS

- The process is student-led and not teacher-centred, therefore students are expected to be active participants in identifying their own learning needs and pursuing the needed knowledge and skills.
- Faculty are facilitators and guides, relinquishing control of teaching and joining in partnership with the learners in a mutual journey of exploration.

LEARNING RESOURCES

Barrows, H., & Branda, L. (1975; revised Branda, L., 1991; 1994). *Role of the tutor/facilitator on problem-based, small group and self-directed learning.* Internal document, copyright 1993. Hamilton, Ontario: McMaster University.

Branda, L. A., & Sciarra, A. F. (1995). Faculty development for problem-based learning. *Annals of Community-Oriented Education, 8,* 195-208.

De Grave, W. S., Dolmans, D. H. J. M., & van der Vleuten, C. P. M. (1999). Profiles of effective tutors in problem-based learning: Scaffolding student learning. *Med Educ*, *33*, 901-906.

Federman, D.D. (1999) Little-heralded advantages of problem-based learning. *Acad Med*, *74*, 93-94.

Finucane, P., Allery, L. A., & Hayes, T. M. (1995). Comparison of teachers at a 'traditional' and an 'innovative' medical school. *Med Educ*, *29*, 104-109.

Hitchcock, M. A., & Mylona, Z-H. E. (2000). Teaching faculty to conduct problem-based learning. *Teach Learn Med*, *12*(1), 52-57.

Maudsley, G. (1999). Roles and responsibilities of the problem based learning tutor in the undergraduate medical curriculum. *British Medical Journal*, *318*, 657-661. www.bmj.com

Mayo, W. P., Donnelly, M. B., & Schwartz, R. W. (1995). Characteristics of the ideal problem-based learning tutor in clinical medicine. *Evaluation & the Health Professions*, *18*(2), 124-136.

MCHS(OT) Programme (2002). *Term 1 Handbook*. Hamilton, Ontario: McMaster University.

Wakefield, J. G., Keane, D., & Hay, J. (1992). Evaluation of tutor effectiveness: A review of the literature. *Annals of Community-Oriented Medicine*, *5*, 209-213.

On Learner Roles:

Biley, F. C., & Smith, K. L. (1998). 'The buck stops here': Accepting responsibility for learning and actions after graduation from a problem-based learning nursing education curriculum. *J Adv Nurs*, *27*, 1021-1029.

Kaufman, D. M., Mensink, D., & Day, V. (1998). Stressors in medical school: Relation to curriculum format and year of study. *Teach Learn Med*, *10*(3), 138-144.

Peterson, M. (1997). Skills to enhance problem-based learning. Retrieved April 23, 2003 from *Medical Education Online*: www.med-ed-online.org/f0000009.htm

Solomon, P., & Crowe, J. (2001). Perceptions of student peer tutors in a problem-based learning programme. *Medical Teacher*, *23*(2), 181-186.

Solomon, P., & Finch, E. (1998). A qualitative study identifying stressors associated with adapting to problem-based learning. *Teach Learn Med*, *10*(2), 58-64.

Tipping, J., Freeman, R. F., & Rachlis, A. R. (1995). Using faculty and student perceptions of group dynamics to develop recommendations for PBL training. *Acad Med*, *70*(11), 1050-1052.

Tutorial 5

KEY POINTS

- Evaluation methods must be congruent with PBL principles, therefore methods used in traditional systems should not be utilized.
- Examples of process-oriented methods include the triple-jump exercise, self-assessment, peer assessment, and self-reflection.
- Examples of outcome-oriented methods include academic papers, modified essay questions, practica experiences, and the clinical reasoning exercise.

LEARNING RESOURCES

Alpert, I. L., & Perkins, R. M. (1999). Should students in problem-based learning sessions be graded by the facilitator? Research brief. *Acad Med, 74*(11), 1256.

Barrows, H. S. (2000). *Practice-based learning applied to medical education.* Springfield, IL: Southern Illinois University School of Medicine.

De Grave, W. S., Dolmans, D. H. J. M., & van der Vleuten, C. P. M. (1998). Tutor intervention profile: Reliability and validity. *Med Educ, 32*, 262-268.

Faculty of Health Sciences (1994). *Guide to professional behaviours in tutorial meetings.* Hamilton, Ontario: McMaster University.

Foldevi, M., Sommansson, G., Trell, E. (1994). Problem-based medical education in general practice: Experience from Linkoping, Sweden. *Br J Gen Pract, 44*, 473-476.

Jacobs, T. (1997). Integrating assessment in problem-focused curricula. *British Journal of Occupational Therapy, 60*(4), 174-178.

Kanter, S. L. (1998). Fundamental concepts of problem-based learning for the new facilitator. *Bulletin of the American Library Association, 86*(3), 391-395.

Mennin, S. P., & Kalishman, S. (1998). Student assessment. *Acad Med, 73*(9), S46-S54.

Nendaz, M. R., & Tekian, A. (1999). Assessment in problem-based learning medical schools: A literature review. *Teach Learn Med, 11*(4), 232-243.

Neville, A. (1995). Student evaluation in problem-based learning. *Pedagogue, 5*, 2-7.

Neville, A., Cunnington, J., & Norman, G. R. (1996). Development of clinical reasoning exercises in a problem-based curriculum. *Acad Med, 71,* S105-S107.

Norman, G. R. (1991). What should be assessed? In D. Boud & G. Feletti (Eds.). *The challenge of problem-based learning* (pp. 254-259). New York: St. Martin's Press.

Programme for Educational Development. (1987). *Evaluation methods:A research manual.* Hamilton, Ontario: McMaster University.

Programme for Faculty Development. (2001). *Problem-based learning in small groups workshop evaluation resource package.* Hamilton, Ontario: Faculty of Health Sciences, McMaster University.

Rendas, A. B., Pinto, P. R., & Gamboa, T. (1998). Problem-based learning in pathophysiology: Report of a project and its outcome. *Teach Learn Med, 10,* 34-39.

Smith, R. M. (1993). The triple-jump examination as an assessment tool in the problem-based medical curriculum at the University of Hawaii. *Acad Med, 68,* 366-372.

Sullivan, M. E., Hitchcock, M. A., & Dunnington, G. L. (1999). Peer and self-assessment during problem-based tutorials. *Am J Surg, 177,* 266-269.

Swanson, D. B., Case, S. M., & van der Vleuten, C. P. M. (1991). Strategies for student assessment. In D. Boud & D. B. G. Feletti (Eds.). *The challenge of problem-based learning* (pp. 260-273). New York: St. Martin's Press.

Valle, R., Petra, I., Martinez-Gonzales, A., Rojas-Ranirez, J. A., Morales-Lopes, S., & Pina-Garza, B. (1999). Assessment of student performance in problem-based learning tutorial sessions. *Med Educ, 33,* 818-822.

Tutorial 6

KEY POINTS

- PBL is a viable and valuable way to learn.
- Performance on certification examinations may not be as high as students who have experienced a more traditional learning environment.
- PBL students tend to process their learning at a deeper level.
- A PBL approach tends to result in a more systematic way of developing a knowledge base.

(continued)

- There is sound integration of basic science and clinical information.
- Faculty should be attuned to the need to minimize instances of erroneous reasoning.
- Learning the PBL way is often more enjoyable.
- It is time to move on, accepting PBL as a mainstream option for teaching and learning.

LEARNING RESOURCES

Albanese, M. A., & Mitchell, S. (1993). Problem-based learning: A review of literature on its outcomes and implementation issues. *Acad Med, 68*, 61-82.

Blake, R. L., & Parkison, L. (1998). Faculty evaluation of the clinical performances of students in a problem-based learning curriculum. *Teach Learn Med, 10*(2), 69-73.

Blake, R. L., Hosokawa, M. C., & Riley, S. L. (2000). Student performances on step 1 and step 2 of the United States medical licensing examination following implementation of a problem-based learning curriculum. *Acad Med, 75*(1), 66-70.

Bligh, J., Lloyd-Jones, G., & Smith, G. (2000). Early effects of a new problem-based clinically oriented curriculum on students' perceptions of teaching. *Med Educ, 34*, 487-489.

Cariaga-Lo, L. D., Richards, B. F., Hollingsworth, M. A., & Camp, D. L. (1996). Non-cognitive characteristics of medical students: Entry to problem-based and lecture-based curricula. *Med Educ, 30*, 179-186.

Colliver, J. (2000). Effectiveness of problem-based learning curricula. *Acad Med, 75*,259-266.

Distlehorst, L. H., & Robbs, R. S. (1998). A comparison of problem-based learning and standard curriculum students: Three years of retrospective data. *Teach Learn Med, 10*(3), 131-137.

Duek, J. E., Wilkerson, L., & Asinolfi, T. (1996). Learning issues identified by students in tutorless problem-based tutorials. *Adv Health Sci Educ, 1*, 29-40.

Federman, D.D. (1999) Little-heralded advantages of problem-based learning. *Acad Med, 74*, 93-94.

Friedman, C. P., de Bliek, R., Greer, D. S., Mennin, S. P., Norman, G. R., Sheps, C. G., Swanson, D. B., & Woodward, C. A. (1990). Charting the winds of change: Evaluating innovative medical curricula. *Acad Med*, 8-14.

Hmelo, C. E. (1998). Cognitive consequences of problem-based learning for the early development of medical expertise. *Teach Learn Med, 10*(2), 92-100.

Kaufman, D. M. (2000). Problem-based learning—Time to step back? *Med Educ, 34*, 509-511.

Newbie, D. I., & Clarke, R. M. (1986). The approaches to learning of students in a traditional and in an innovative problem-based medical school. *Med Educ, 20*, 267-271.

Norman, G. R., & Schmidt, H. G. (2000). Effectiveness of problem-based learning curricula: Theory, practice and paper darts. *Med Educ, 34*, 721-728.

Pallie, W., & Carr, D. H. (1987). The McMaster medical education philosophy in theory, practice and historical perspective. *Medical Teacher, 9*(1), 59-71.

Patel, V. L., Groen, G. J., & Norman, G. R. (1991). Effects of conventional and problem-based medical curricula on problem solving. *Acad Med, 7*, 380-389.

Rothman, A. I. (2000). Problem-based learning—Time to move forward? *Med Educ, 34*, 509-511.

Vernon, D. T. A., & Blake, R. L. (1993). Does problem-based learning work? A meta-analysis of evaluative research. *Acad Med, 68*, 550-563.

Viet Vu, N., van der Vleuten, C. P. M., & Lacombe, G. (1998). Thinking about student thinking. *Acad Med, 73*(10), S25-S27.

Woodward, C. A. (1986). McMaster University graduates. *Journal of Medical Education, 61*, 859-860.

Woodward, C. A. (1996). Problem-based learning in medical education: Developing a research agenda. *Adv Health Sci Educ, 1*, 83-94.

Tutorial 7

KEY POINTS

- Stimulates creativity and ingenuity in finding resources relevant to the learning issues.
- Allows us as learners to take ownership of the problem, and thereby take charge of our own learning.
- Giving and receiving feedback is crucial to competent professional practice over time, and the best training ground for this is in small group tutorial learning.

(continued)

- Takes away anxiety about the "right" answer and encourages lifelong learning.
- Makes learning more interesting and even fun.
- Can be anxiety-provoking at the beginning until we learn how much is enough and how deep is deep enough.
- Essential that there is a link between learning this way and being prepared for more formal certification examinations.
- Keeps the focus on clients, patients and families and not on the disease or body part.
- Promotes close relationships between faculty and learners in making learning a collaborative process.
- Teaching within a range of small to large groups is both energizing and challenging, never boring.

LEARNING RESOURCES

Biley, F. C., & Smith, K. L. (1999). Making sense of problem-based learning: The perceptions and experiences of undergraduate nursing students. *J Adv Nurs, 30*(5), 1205-1212.

Caplow, J. A. H., Donaldson, J. F., Kardash, C. A., & Hosokawa, M. (1997). Learning in a problem-based medical curriculum: Students' conceptions. *Med Educ, 31*, 440-447.

Foley, R. P., Polson, A. L., & Vance, J. M. (1997). Review of the literature in the clinical setting. *Teach Learn Med, 9*(1), 4-9.

Hammel, J., Royeen, C. B., Bagatell, N., Chandler, B., Jensen, G., Loveland, J., & Stone, G. (1999). Student perspectives on problem-based learning in an occupational therapy curriculum: A multiyear qualitative evaluation. *Amer J Occup Ther, 53*(2), 199-206.

Kaufman, D. M., Holmes, D. B. (1996). Tutoring in problem-based learning: Perceptions of teachers and students. *Med Educ, 30*, 371-377.

Lieberman, S. E., Stroup-Benham, C. A., Peel, J. L., & Camp, M. G. (1997). Medical student perception of the academic environment: A prospective comparison of traditional and problem-based curricula. *Acad Med, 72*(10), S13-S15.

Moore, G. T. (1991). The effect of compulsory participation of medical students in problem-based learning. *Med Educ, 25*, 140-143.

White, M. J., Amos, E., & Kouzekanani, K. (1999). Problem-based learning: An outcomes study. *Nurse Educator, 24*(2), 33-36.

Tutorial 8

KEY POINTS

- How well did we do today in accomplishing our task?
- How well did we do today in working as a group?
- What were the major strengths of this session?
- What were the main areas in which we could improve?
- What do you think we should do next?
- What things did you not say that you might have?
- What do you see yourself doing differently?

LEARNING RESOURCES

Cole, M. B. (1998). *Group dynamics in occupational therapy.* Thorofare, NJ: SLACK Incorporated.

Sampson, E., & Marthas, M. (1991). *Group process for the health professions* (3rd ed). New York and Toronto: John Wiley & Sons.

Yalom, I. (1995). *Theory and practice of group psychotherapy* (4th ed.). New York: Basic Books.

INDEX

Build Your Library

Along with this title, we publish numerous products on a variety of topics. We are sure that you will find the below titles to be an essential addition to your library. Order your copies today or contact us for a copy of our latest catalog for additional product information.

MEASURING OCCUPATIONAL PERFORMANCE: SUPPORTING BEST PRACTICE IN OCCUPATIONAL THERAPY

Mary Law, PhD, OT Reg. (Ont.); Carolyn M. Baum, PhD, OTR/L, FAOTA; and Winnie Dunn, PhD, OTR, FAOTA

320 pp., Soft Cover, 2000, ISBN 1-55642-298-9, Order #32989, $37.95

Authored by three preeminent leaders in the field of occupational therapy, *Measuring Occupational Performance* provides a comprehensive approach to assessing occupational performance. The key actions that occupational therapists must implement when conducting assessments, reporting findings, and interpreting measurement information for intervention planning are outlined.

USING ENVIRONMENTS TO ENABLE OCCUPATIONAL PERFORMANCE

Lori Letts, PhD, OT Reg. (Ont.); Patty Rigby, MHSc, OT Reg. (Ont.); and Debra Stewart, MSc, OT Reg. (Ont.)

336 pp., Hard Cover, 2003, ISBN 1-55642-578-3, Order #35783, $35.95

Using Environments to Enable Occupational Performance is a unique new text that specifically focuses on how environments (physical, social, cultural, institutional) can be used by occupational therapists to enable occupational performance with all types of clients. Each chapter contains "real world" scenarios from occupational therapists about how the environment can be used to optimize occupational performance. This feature is a beneficial element for both students and clinicians as they are learning these applications.

Contact Us

SLACK Incorporated, Professional Book Division
6900 Grove Road, Thorofare, NJ 08086
1-800-257-8290/1-856-848-1000, Fax: 1-856-853-5991
orders@slackinc.com or www.slackbooks.com

ORDER FORM

QUANTITY	TITLE	ORDER #	PRI
	Measuring Occupational Performance	32989	$38.9
	Using Environments to Enable Occupational Performance	35783	$35.
	Subtotal		$
	Applicable state and local tax will be added to your purchase		$
	Handling		$4.5(
	Total		$

Name: _____

Address: _____

City: _____ State: _____ Zip: _____

Phone: _____ Fax: _____

Email: _____

- Check Enclosed (Payable to SLACK Incorporated)_____
- Charge my: ___ [card images] ___ VISA ___ MasterCard

Account #: _____

Exp. date: _____ Signature:_____

NOTE: Prices are subject to change without not
Shipping charges will apply.
Shipping and handling charges are non-refundable.

CODE: 3